High-Dimensional Partial Differential Equations in Science and Engineering

Volume 41

CRM PROCEEDINGS & LECTURE NOTES

Centre de Recherches Mathématiques
Université de Montréal

High-Dimensional Partial Differential Equations in Science and Engineering

André Bandrauk
Michel C. Delfour
Claude Le Bris
Editors

The Centre de Recherches Mathématiques (CRM) of the Université de Montréal was created in 1968 to promote research in pure and applied mathematics and related disciplines. Among its activities are special theme years, summer schools, workshops, postdoctoral programs, and publishing. The CRM is supported by the Université de Montréal, the Province of Québec (FCAR), and the Natural Sciences and Engineering Research Council of Canada. It is affiliated with the Institut des Sciences Mathématiques (ISM) of Montréal, whose constituent members are Concordia University, McGill University, the Université de Montréal, the Université du Québec à Montréal, and the École Polytechnique. The CRM may be reached on the Web at www.crm.umontreal.ca.

American Mathematical Society
Providence, Rhode Island USA

The production of this volume was supported in part by the Fonds pour la Formation de Chercheurs et l'Aide à la Recherche (Fonds FCAR) and the Natural Sciences and Engineering Research Council of Canada (NSERC).

2000 *Mathematics Subject Classification.* Primary 65Mxx, 35Gxx;
Secondary 76–XX.

Library of Congress Cataloging-in-Publication Data

High-dimensional partial differential equations in science and engineering / André Bandrauk, Michel C. Delfour, Claude Le Bris, editors.
p. cm. — (CRM proceedings & lecture notes, ISSN 1065-8580 ; v. 41)
Includes bibliographical references.
ISBN-13: 978-0-8218-3853-2 (alk. paper)
1. Differential equations, Partial—Numerical Solutions—Congresses. 2. Science—Mathematics—Congresses. 3. Engineering—Mathematics—Congresses. I. Bandrauk, André D. II. Delfour, Michel C., 1943– III. Le Bris, Claude. IV. Title. V. Series.

QA377.H54 2007
518′.64—dc22 2007060763

∞ The paper used in this book is acid-free and falls within the guidelines
established to ensure permanence and durability.
This volume was submitted to the American Mathematical Society
in camera ready form by the Centre de Recherches Mathématiques.
Visit the AMS home page at http://www.ams.org/

10 9 8 7 6 5 4 3 2 1 12 11 10 09 08 07

Contents

Preface

High-dimensional spatio-temporal partial differential equations are a major challenge to scientific computing of the future. Up to now deemed prohibitive, they have recently become manageable by combining recent developments in numerical techniques, appropriate computer implementations, and the use of computers with parallel and even massively parallel architectures. This opens new perspectives in many fields of applications. Kinetic plasma physics equations, many body Schrödinger equation, Dirac and Maxwell equations for molecular electronic structure and nuclear dynamic computations, options pricing equations in mathematical finance, and Fokker-Planck and fluid dynamics equations for complex fluids, are examples of equations that can now be handled. The objective of the workshop was to bring together experts of international stature in that broad spectrum of areas to confront their approaches and possibly bring out common problem formulations and research directions in the numerical solutions of high dimensional pde's in various fields of science and engineering with special emphasis on chemistry and physics.

The meeting took place at the Centre de recherches mathématiques of the Université de Montréal from August 7 to 12, 2005. This ambitious interdisciplinary, interuniversity, and international event was the product of a fruitful and sustained collaboration between André Bandrauk (Canada Research Chair, Université de Sherbrooke), Michel C. Delfour (CRM and DMS, Université de Montréal), and Claude Le Bris (CERMICS, École Nationale des Ponts et Chaussées, France) who earlier co-organized the very successful workshop "Quantum Control: Mathematical and Numerical Challenges" in 2002. It was co-sponsored by the three institutions and the French *Ministère de la recherche et des nouvelles technologies* (ACI Nouvelles interfaces des mathématiques), *IBM Canada Higher Education and Research Sector*, the *Regroupement québécois sur les matériaux de pointe* (RQMP), and the *vice rectorat à la recherche* (Université de Montréal).

The conference involved 63 registered speakers, experts, and students in the general areas of high-dimensional partial differential equations with the multidisciplinary participation of chemists, engineers, mathematicians, physicists, and specialists of finance. It emphasized the participation of graduate students and postdoctoral fellows in applied mathematics, theoretical chemistry and physics by providing 19 travel and lodging grants and reserving time in the program to present the results of their research.

It was a privileged encounter between American, Canadian and European teams in a very convivial atmosphere as expressed by the invited speaker Andreas Savin (Laboratoire de Chimie Théorique, CNRS et Université Pierre et Marie Curie, France)

> *That was not only a perfectly organized conference, but also a pleasant surprise, produced by the effort of mathematicians to be clear to a chemist ...,*[1]

July 4th, 2006

André Bandrauk
Michel C. Delfour
Claude Le Bris

[1]Translated from the French: *Ça n'a pas seulement été une conférence parfaitement organisée, mais aussi une surprise agréable, produite par l'effort des mathématiciens d'être clairs pour un chimiste ...,*

List of Participants

Bandrauk, André D.
Université de Sherbrooke
andre.bandrauk@usherbrooke.ca

Barmaki, Samira
Université de Sherbrooke
samira.barmaki@usherbrooke.ca

Béland, Michel
RQCHP
michel.beland@rqchp.qc.ca

Benmoussa, Khalid
Université Laval
kbenmous@giref.ulaval.ca

Bourgeois, Étienne
Université de Montréal
e.bourgeois@UMontreal.CA

Brabec, Thomas
University of Ottawa
brabec@uottawa.ca

Canaff, Charles
RQCHP
charles.canaff@UMontreal.CA

Cancès, Éric
CERMICS
cances@cermics.enpc.fr

Chelkowski, Szczepan
Université de Sherbrooke
s.chelkowski@usherbrooke.ca

Chen, Goong
Texas A&M University
gchen@math.tamu.edu

Côté, Michel
Université de Montréal
Michel.Cote@UMontreal.CA

Daum, Fred
Raytheon
frederick_e_daum@raytheon.com

Degond, Pierre
INSA
degond@mip.ups-tlse.fr

Dehaes, Mathieu
École Polytechnique de Montréal
mathieu.dehaes@polymtl.ca

Dehghanian, Effat
Université de Sherbrooke
e.dehghanian@usherbrooke.ca

Delfour, Michel C.
Université de Montréal
delfour@CRM.UMontreal.CA

Dion, François
Université Laval
fdion2@chm.ulaval.ca

Ernzerhof, Matthias
Université de Montréal
Matthias.Ernzerhof@UMontreal.CA

Esteban, Maria J.
CEREMADE
esteban@ceremade.dauphine.fr

Fortin, André
Université Laval
andre.fortin@giref.ulaval.ca

Gélinas, Guillaume
Université de Montréal
guillaumegelinas@hotmail.com

Gidofalvi, Gergely
The University of Chicago
ggidofal@uchicago.edu

Giroux, Gaston
Université de Sherbrooke
gastongiroux2000@yahoo.ca

Griebel, Michael
Institut für Numerische Simulation
griebel@ins.uni-bonn.de

Habashi, Wagdi G.
McGill University
wagdi.habashi@mcgill.ca

Hurd, Tom R.
McMaster University
hurdt@mcmaster.ca

Kane, Abdoulaye Sabou
Université Laval
akane@mat.ulaval.ca

Kapral, Raymond
University of Toronto
rkapral@gatto.chem.utoronto.ca

Lagmago Kamta, Gérard
Université de Sherbrooke
gerard.lagmago.kamta@usherbrooke.ca

Lapeyre, Bernard
CERMICS
Bernard.Lapeyre@cermics.enpc.fr

Le Bris, Claude
École Nationale des Ponts et Chaussées
lebris@cermics.enpc.fr

Lorin de la Grandmaison, Emmanuel
Université de Montréal
lorin@CRM.UMontreal.CA

Maday, Yvon
Université Pierre & Marie Curie
maday@ann.jussieu.fr

Mazziotti, David A.
University of Chicago
damazz@uchicago.edu

Mechkour, Houari
ESIEE-MOSIM
mechkouh@esiee.fr

Metatla, Noureddine
Université de Sherbrooke
Noureddine.Metatla@usherbrooke.ca

Mohlenkamp, Martin
Ohio University
mjm@math.ohiou.edu

Morales, Hector
Rensselaer Polytechnic Institute
moralj@rpi.edu

Nguyen, Nam
Université de Sherbrooke
n.nguyen@usherbrooke.ca

Owens, Robert G.
Université de Montréal
owens@dms.UMontreal.CA

Paraschivoiu, Marius
Concordia University
paraschi@me.concordia.ca

Patera, Anthony T.
MIT
patera@mit.edu

Pau, George Shu Heng
MIT
gpau@mit.edu

Peslherbe, Gilles H.
Concordia University
ghp@alcor.concordia.ca

Pommier, David
INRIA
david.pommier@inria.fr

Quanyan, Zhu
McGill University
qzhu4@ece.mcgill.ca

Richer, Jacques
RQCHP
jacques.richer@UMontreal.CA

Roy Mahapatra, Debiprosad
Wilfrid Laurier University
droymahapatra@wlu.ca

Salomon, Julien
Université Pierre & Marie Curie
salomon@ann.jussieu.fr

Savin, Andreas
Université Pierre & Marie Curie
andreas.savin@lct.jussieu.fr

Scheffe, Molly
Wavefront Associate
sscheffe@bwh.harvard.edu

Seco, Luis A.
University of Toronto
seco@math.utoronto.ca

Sedik, Elwallid
Université de Sherbrooke
elwallid.sedik@usherbrooke.ca

Shtinkova, Ralitsa
Bulgarian Academy of Science
sr.shtinkova@gmail.com

Sinelnikov, Evgueni
Université Laval
esinelnikov@chm.ulaval.ca

Stubbs, Daniel
RQCHP
daniel.stubbs@UMontreal.CA

Tchitchekova, Deyana
Chimie Computationnelle et
Photonique
chichekova@hotmail.com

Trabelsi, Saber
Université Paris Diderot
strabelsi@ann.jussieu.fr

Urquiza, Jos Manuel
Université de Montréal
urquiza@CRM.UMontreal.CA

Wyatt, Robert E.
University of Texas
swyattre@mail.utexas.edu

Xiao, Ping
Memorial University of Newfoundland
ping@cs.mun.ca

Zhou, Hsiao-Ling
Georgia State University
zhou@phy-astr.gsu.edu

Zhuang, Min
Université de Montréal
Min.Zhuang@UMontreal.CA

List of Speakers

Bandrauk, André
Université de Sherbrooke
Andre.Bandrauk@USherbrooke.ca

Brabec, Thomas
University of Ottawa
brabec@uottawa.ca

Cancès, Éric
CERMICS
cances@cermics.enpc.fr

Chen, Goong
Texas A&M University
gchen@math.tamu.edu

Côté, Michel
Université de Montréal
Michel.Cote@UMontreal.CA

Degond, Pierre
Paul Sabatier
degond@mip.ups-tlse.fr

Delfour, Michel C.
Université de Montréal
delfour@CRM.UMontreal.CA

Ernzerhof, Matthias
Université de Montréal
Matthias.Ernzerhof@UMontreal.CA

Esteban, Maria J.
CEREMADE
esteban@ceremade.dauphine.fr

Fortin, André
Université Laval
andre.fortin@giref.ulaval.ca

Griebel, Michael
IAM
griebel@iam.uni-bonn.de

Habachi, Wagdi
University McGill
wagdi.habashi@staff.mcgill.ca

Hurd, Tom R.
University McMaster
hurdt@mcmaster.ca

Kapral, Raymond E.
University of Toronto
rkapral@chem.utoronto.ca

Lapeyre, Bernard
CERMICS
Bernard.Lapeyre@cermics.enpc.fr

Maday, Yvon
Université Pierre & Marie Curie
maday@ann.jussieu.fr

Mazziotti, David A.
University of Chicago
damazz@uchicago.edu

Owens, Robert G.
Université de Montréal
rg.owens@UMontreal.CA

Patera, Anthony T.
MIT
patera@mit.edu

Peslherbe, Gilles H.
University of Concordia
ghp@alcor.concordia.ca

Savin, Andreas
Université Pierre & Marie Curie
andreas.savin@lct.jussieu.fr

Seco, Luis
University of Toronto
seco@math.utoronto.ca

Wyatt, Robert E.
University of Texas at Austin
wyattre@mail.utexas.ed

Centre de Recherches Mathématiques
CRM Proceedings and Lecture Notes
Volume **41**, 2007

Singularity-Free Methods for the Time-Dependent Schrödinger Equation for Nonlinear Molecules in Intense Laser Fields — A Non-Perturbative Approach

André D. Bandrauk and HuiZhong Lu

Abstract. We present a new method for the non-perturbative numerical solution of the Time-Dependent Schrödinger Equation, TDSE, for molecules in intense laser fields, using cylindrical/polar coordinates systems. For cylindrical coordinates systems, after use of a split-operator method which separates the z direction propagation and the (x, y) plane propagation, we approximate the wave function in each (x, y) section by a Fourier serie $(\Sigma c_m(\rho)e^{im\phi})$ which offers an exponential convergence in the ϕ direction and is naturally applicable for polar coordinates system. The coefficients $\big(c_m(\rho)\big)$ are then calculated by a Finite Difference Method (FDM) in the ρ direction. We adopt the Crank–Nicholson method for the temporal propagation. The final linear system consists of a set of independent one-dimensional (ρ) linear systems and the matrix for every one-dimensional linear systems is *sparse*, so the whole linear system may be very efficiently solved. The most important advantage of the new method is that it is *unitary* at all steps and is analytically convergent every where even near the $\rho = 0$ origin, in contrast to most other numerical method based on cylindrical/polar coordinates systems which are not unitary and introduce a numerical singularity at $\rho = 0$. We also illustrate its generalization to the case of dimension $D = 4$, to treat two electrons in the H_2 molecule.

1. Introduction

The current advent of ultrashort intense laser pulses [1] necessitates the development of non-perturbative numerical methods to solve problems of interaction of lasers with molecules [2]. Because of efficiency, in many cases one uses cylindrical coordinates for the numerical solution of the 3-D Time Dependent Schrödinger Equation, TDSE, for linear molecules exposed parallel to an intense laser pulse: H_2^+ [3–6], H_2 [7]. Generally, we use the split-operator method or the ADI (Alternating Direction Implicit Method [8]) method which separate the propagation in the different directions. This process reduces a lot of computing time and is very precise

2000 *Mathematics Subject Classification.* Primary: 65Z05, Secondary: 65N99.

The first author thanks the Canada Research Chair in Computational Chemistry and Photonics.

We thank NSERC and CFI for supporting this research both in personnel and supercomputer facilities.

This is the final form of the paper.

[8]. After splitting, one must propagate the wave function in each section plane in the polar coordinate. Polar coordinates have also been used in 2-D simulations of molecular alignment [5]. But in the polar coordinate system, $\rho = \sqrt{x^2+y^2}$, z, one encounters often the unitarity,precision problem in the ρ direction when one uses Finite Difference Methods (FDM) and the difficulty of very expensive algorithms in CPU-time when one uses a development based on special functions such as Bessel function [3].

Because of its simplicity,the FDM is the most used in the ρ direction [3–6]. In order to avoid the boundary condition problem — there is no natural physical boundary condition at $\rho = 0$, one usually adopts the following nonlinear (stretching) transformation [8]:

$$\rho = T(\tilde{\rho}), \tag{1.1}$$

$$\widetilde{\psi}(\tilde{\rho}) = \sqrt{T'T}\psi(\rho), \tag{1.2}$$

$$i\frac{\partial\widetilde{\psi}}{\partial t} = -\frac{1}{2}\left[\frac{1}{T'^2}\frac{\partial^2}{\partial\tilde{\rho}^2} - 2\frac{T'''}{T'^3}\frac{\partial}{\partial\tilde{\rho}}\right]\widetilde{\psi} + \left[\frac{5}{4}\frac{T''^2}{T'^4} - \frac{1}{2}\frac{T'''}{T'^3} + \frac{1}{4T^2} + V\right]\widetilde{\psi}, \tag{1.3}$$

$$\widetilde{\psi}(0) = 0, \quad \widetilde{\psi}(\tilde{\rho}_{\max}) = 0, \tag{1.4}$$

The equation is restricted in the ρ direction after the use of the split-operator or ADI method in the polar coordinate (ρ, ϕ). Two problems (unitarity,convergence) occur as:

(1) when we discretize the differential operators in (1.3) with standard central difference schemes we find immediately the problem of unitarity: if $\rho/\tilde{\rho}$ is not constant, the resulting propagation matrix will not be symmetric, so the final discrete propagator is not unitary. This is just what occurs with transformations like $\rho = \tilde{\rho}^{3/2}$ [6]: $1/T' = 2/[3\sqrt{\tilde{\rho}}]$ shows that the coefficients of the second order derivative (kinetic energy) change very quickly near the origin so the propagation matrix is far from being symmetrical near $\tilde{\rho} = 0$.

(2) there is a problem of convergence which is very little discussed. It is well known that the convergence of the FDM is based on the relation:

$$\widetilde{\psi}(\tilde{\rho}+h) = \sum_{i=0}^{n}\frac{h^i}{n!}\frac{d^i\widetilde{\psi}}{d\tilde{\rho}^i}(\tilde{\rho}) + \int_{\tilde{\rho}}^{\tilde{\rho}+h}\frac{(\tilde{\rho}+h-x)^n}{n!}\frac{d^{n+1}\widetilde{\psi}(x)}{d\tilde{\rho}^{n+1}}\,dx, \tag{1.5}$$

where the coefficient of the remaining term (integration) must be limited to obtain convergence. Now, after the transformation above we have $\rho = \tilde{\rho} \Longrightarrow \widetilde{\psi}(\tilde{\rho}) = \sqrt{\tilde{\rho}}\psi(\tilde{\rho})$, $\rho = \tilde{\rho}^{3/2} \Longrightarrow \widetilde{\psi}(\tilde{\rho}) = \sqrt{3/2}\tilde{\rho}\psi(\tilde{\rho}^{3/2})$. When $\psi(\rho)$ is an analytical function, we can make the development:

$$\begin{aligned}\frac{\partial^{i+1}\widetilde{\psi}}{\partial\tilde{\rho}^{i+1}} &= \frac{\partial^{i+1}}{\partial\tilde{\rho}^{i+1}}[\sqrt{\tilde{\rho}}\psi(\tilde{\rho})] \\ &= \frac{\partial^{i+1}}{\partial\tilde{\rho}^{i+1}}\sum_{k=0}^{\infty}c_k\tilde{\rho}^{k+1/2} \\ &= \sum_{k=0}^{\infty}d_k\tilde{\rho}^{(2k-2i-1)/2} \quad for \quad \rho = \tilde{\rho},\end{aligned} \tag{1.6}$$

or

$$\begin{aligned}
\frac{\partial^{i+1}\tilde{\psi}}{\partial\tilde{\rho}^{i+1}} &= \frac{\partial^{i+1}}{\partial\tilde{\rho}^{i+1}}[\sqrt{3/2}\tilde{\rho}\psi(\tilde{\rho}^{3/2})] \\
&= \frac{\partial^{i+1}}{\partial\tilde{\rho}^{i+1}}\sum_{k=0}^{\infty} c_k\tilde{\rho}^{1+3k/2} \\
&= \sum_{k=0}^{\infty} e_k\tilde{\rho}^{(3k-2i)/2} \quad for \quad \rho = \tilde{\rho}^{3/2}.
\end{aligned} \tag{1.7}$$

The two preceding equations show directly that with transformed wave function the integration term in (1.5) will not be finite (bounded) near $\tilde{\rho} = 0$ for general cases. We also conclude that if $c_k = 0$ for all k odd, equation (1.6) gives an analytical function so the equality (1.5) upon which the FDM's convergence is based is satisfied. This means that when $\psi(\rho)$ is an even function (i.e. $\psi(\rho) = \psi(-\rho)$) the transformation $\rho = \tilde{\rho}^{3/2}$ will offer a convergent numerical scheme which is the special case of 2-D (ρ, ϕ) section non-passing by protons for 3-D H_2^+ exposed to a linearly polarize laser pulse which is parallel to the axis of the molecule [6] .

It is worth to note that when the wave function is independent of ϕ (e.g. — molecule and laser parallel), some previous numerical methods [4] have continued the function to the $\rho < 0$ area by symmetry thus avoiding the boundary condition problem at $\rho = 0$. Even without the preceding singular transformation, the original differential operator $\partial^2/\partial\rho^2 + 1/\rho\ \partial/\partial\rho$ is not in self-adjoint form, so the discrete propagator is still non-unitary. Another weakness of this method of propagation is that it is not available for wave functions which are dependent on ϕ as is the case for the one-electron H_2^+ in a perpendicular laser field or for the two-electrons H_2 exposed to a general laser pulse.

The consequence of the above two problems is that the non-unitarity may make the numerical method unstable and the non-analytic behaviour of the transformed wave-function near the origin makes the method very poor in precision. For the continuation to $\rho < 0$ method [4], we have found that: for H_2^+ parallel to a linear laser pulse, after preparation of the initial condition by an imaginary time step method [3], and coming back to real time with zero laser intensity, the population diminishes to 95% because of the non-unitarity of the discretized operator.

To overcome the preceding difficulties of FDM in the ρ direction, our idea is to first write the operator $\partial^2/\partial\rho^2 + 1/\rho\ \partial/\partial\rho$ in self-adjoint form — $1/\rho\ \partial/\partial\rho\ \rho\ \partial/\partial\rho$ which help us to obtain a unitary discrete operator (for detail, see next section). To avoid the problem of boundary conditions at $\rho = 0$, we propagate the wave function in the whole plane without separating ρ and ϕ so we do not use the above wave function transformations which cause problems of convergence.

We must emphasise that the separation $(1 + c\ \partial^2/\partial\rho^2 + c\ 1/\rho\ \partial/\partial\rho)(1 + c\ 1/\rho^2\ \partial^2/\partial\phi^2) = 1 + c(\partial^2/\partial x^2 + \partial^2/\partial y^2) + c^2(\partial^2/\partial\rho^2 + c\partial/\partial\rho/\rho)(\partial^2/\partial\phi^2/\rho^2)$ results that the error term (3rd term of this equality) is very difficult to be limited in numerical simulation and this is just the error term appearing in the split-operator method which separates ρ and ϕ direction propagation. We will demonstrate the advantage of our choice to propagate the wave function in the whole plane without separating ρ and ϕ.

2. The Numerical Method

As explained in the preceding section, after using the splitting method in 3-D which separates the z-coordinate, the main numerical difficulty comes from the free propagation of the wave function in the remaining 2-D plane in polar coordinates (ρ, ϕ).

For details for the splitting method we refer to [9], and we concentrate here on studying the approximation of the free propagation equation in 2-D:

$$i\frac{\partial F}{\partial t} = -\alpha\left[\frac{\partial^2}{\partial \rho^2} + \frac{1}{\rho}\frac{\partial}{\partial \rho} + \frac{1}{\rho^2}\frac{\partial^2}{\partial \phi^2}\right]F. \tag{2.1}$$

We adopt the Crank–Nicholson method for the temporal approximation of (2.1):

$$\begin{aligned}\left\{1 - i\frac{\delta t}{2}\alpha\left[\frac{1}{\rho}\frac{\partial}{\partial \rho}\rho\frac{\partial}{\partial \rho} + \frac{1}{\rho^2}\frac{\partial^2}{\partial \phi^2}\right]\right\} F(\rho, \phi, t_{l+1}) \\ = \left\{1 + i\frac{\delta t}{2}\alpha\left[\frac{1}{\rho}\frac{\partial}{\partial \rho}\rho\frac{\partial}{\partial \rho} + \frac{1}{\rho^2}\frac{\partial^2}{\partial \phi^2}\right]\right\} F(\rho, \phi, t_l).\end{aligned} \tag{2.2}$$

For the spatial approximation of (2.2), we use the truncated Fourier series,

$$\begin{aligned}F(\rho, \phi, t_l) &= \frac{1}{N}\sum_{n=0}^{N-1} c_n^l(\rho)e^{in\phi} \\ &= \frac{1}{N}\left[\sum_{n=0}^{N/2-1} c_n^l(\rho)e^{in\phi} + \sum_{n=N/2}^{N-1} c_n^l(\rho)e^{i(n-N)\phi}\right],\end{aligned} \tag{2.3}$$

where $\{c_n(\rho)\}$ is the unknown set of functions on ρ. The preceding formula shows that for fixed ρ, $\{c_n^l(\rho)\}_{n=0}^{N-1}$ is just the discrete Fourier coefficients of $F(\rho, \phi, t_l)$ as a function of ϕ. So we have:

$$\begin{aligned}c_n^l(\rho) &= \sum_{m=0}^{N-1} F(\rho, \frac{m}{N}2\pi, t_l)e^{-in2\pi m/N} \qquad \text{for} \quad n = 0,\ N-1, \\ &\equiv \sum_{m=0}^{N-1} F\left(\rho, \frac{m}{N}2\pi, t_l\right)e^{-i(n-N)2\pi m/N} \ \text{for} \quad n = N/2,\ N-1.\end{aligned} \tag{2.4}$$

For a regular function F, we can demonstrate that $c_{n\neq 0}(\rho = 0)$ should be zero by developping the function in the series:

$$F(\rho, \phi) = \sum_{i,j} a_{i,j}x^i y^j. \tag{2.5}$$

At $\rho = 0$, only the $a_{0,0}$ term of the series contributes to the integration in (2.4), but for $n \neq 0$ we have $\int a_{0,0}e^{in\phi}\,d\phi = 0$. It should be noted that the $c_{n\neq 0}(\rho = 0) = 0$ condition insures also that the function in the series (2.3) is well defined at the origin ($\rho = 0$). To study the coefficient $c_0(\rho)$, we first rewrite it from (2.4) as:

$$\begin{aligned}c_0(\rho) &= \sum_{m=0}^{N-1} F\left(\rho, \frac{m}{N}2\pi\right) = \frac{N}{2\pi}\int_0^{2\pi} F(\rho, \phi)\,d\phi \\ &= \frac{N}{2\pi}\Sigma a_{j,k}\int_0^{2\pi} x^j y^k\,d\phi.\end{aligned} \tag{2.6}$$

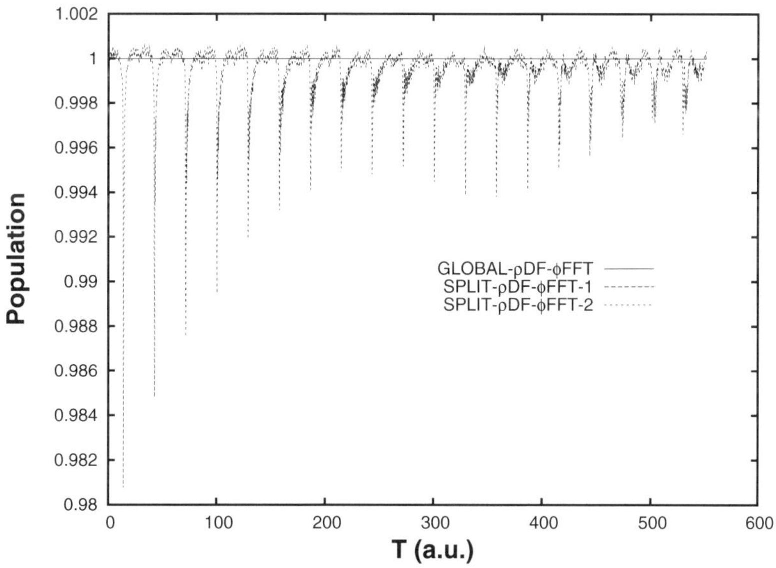

FIGURE 1. Evolution of the total electron population for methods (2–4) in polar coordinates for 2-D H_3^{++}, I=$10^{14}W/\text{cm}^2$, λ = 800nm.

3.1. Unitarity of the 3 methods in polar coordinate. In order to test the unitarity, we have initialized the wave function with $\psi(k*h,\phi) = (k*h)*(-1)^k*(1+e^{i\phi}+e^{i\phi*N/2})$ which is very singular but every unitary discrete propagator should conserve it's norm. The discrete propagators are obtained with the three methods in polar coordinates for one electron H_3^{++} in 2-D and a laser pulse of intensity $I = 10^{14}W/\text{cm}^2$ and $\lambda = 800\text{nm}$. In Fig. 1, we compare the evolution of the total population without absorber near boundaries for all three methods: SPLIT–ρDF–ϕFFT–1, SPLIT–ρDF–ϕFFT–2, GLOBAL–ρDF–ϕFFT. We can't distinguish the curves for SPLIT–ρDF–ϕFFT–1 and GLOBAL–ρDF–ϕFFT, but we see a clear fluctuation for the result of SPLIT–ρDF–ϕFFT–2 which demonstrates the non-unitarity of this method of transformation.

3.2. Problem of numerical singularity. To illustrate the precision of the three different methods in polar coordinates, we have solved numerically the 2-D Schrödinger equation for H_3^{++} exposed to a linearly polarized (in x axis direction: $\alpha = 90^o$) laser field [10]: intensity $I = 10^{14}W/\text{cm}^2$, wavelength $\lambda = 800\text{nm}$. The 3 protons' position are shown in Fig. 2 with $R = 6$, $\theta = 60^o$ and $\alpha = 90^o$. It is well known that the convergence on the phase of the wave function is much more difficult to obtain than it's amplitude [9].

Here, the converged solution has been obtained by SPLIT–xFFT–yFFT and we show the numerical results for the 3 methods in polar coordinates for the real (R) and imaginary (Im) parts in Fig. 2 for $t = 1$ and 5 cycles. We see that the two methods of nonlinear transformation ($\rho = \tilde{\rho}$, Fig. 4(c) and $\rho = \tilde{\rho}^{3/2}$, Fig. 4(d)) have introduced numerical singularities at the origin. In contrast, the present method does not introduce any numerical singularity at $\rho = 0$ and gives very good convergence. It's to be noted that all three methods in polar coordinates have used

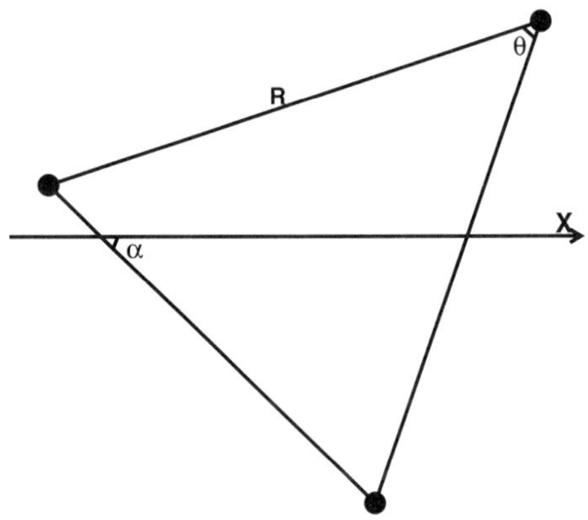

FIGURE 2. Proton positions (•) and coordinates for H_3^{++}.

the same number of discretization points in each direction (ρ, ϕ). As illustration, we show also the population contour for $t = 5$ cycles in Fig. 3 for all four methods which confirm again that the SPLIT–ρDF–ϕFFT–1 method (Fig. 4(c)) has not yet converged to the exact solution. The SPLIT–ρDF–ϕFFT–2 (Fig. 4(d)) method performs well away from the singularity. The difference between the SPLIT–ρDF–ϕFFT–1&2 methods shown in Fig. 4 come from the fact: $\tilde{\rho}$ is more singular than $\tilde{\rho}^{3/2}$.

3.3. New numerical results with circularly polarized laser field. Finally, we show some numerical results for the one-electron H_3^{++} with different configurations exposed to a circularly polarized laser field: intensity $I = 2 \times 10^{14} W/\text{cm}^2$, wavelength $\lambda = 800\text{nm}$. We have calculated the ionization for different combination of R and θ: $(R, \theta) = (4 + 0.25 * k, j * 7.5^o)$ for $0 \leq k \leq 24$ et $1 \leq j \leq 24$. In Fig. 4, we show the contour of the ionized population as function of (R, θ) at $t = 10$ cycles. This figure shows maxima in ionization rates at $\theta \sim 40^o$ and $\sim 100^o$ of the H–H–H angle. This is manifestation of a non-perturbative quantum phenomena of charge resonance enhanced ionization, CREI, by quantum tunnelling [10]. In Fig. 5, we show the population contour for some special configuration at $t = 10$ cycles $(27 \times 10^{-15} s = 27$ femtosecond) where the ionization is maximum. This is an example of electron "spinning" induced by the laser field.

4. The Two Electron Problem in H_2

We have developed a new numerical method for the numerical solution of the Time-Dependent Schrödinger Equation in polar (2-D)/cylindrical (3-D) coordinates for one-electron system. Our above numerical results have demonstrated the advantage of the present method compared to other methods in these coordinates: unitarity of the discrete propagator, and no numerical singularity. We next extend this method to solve the problem of two electrons interaction in H_2 with a laser field in 2-D. We first map the processors on a 2-D Cartesian grid, then the data will be distributed through the 2-D grid along ρ_1, ρ_2 directions. With this distribution

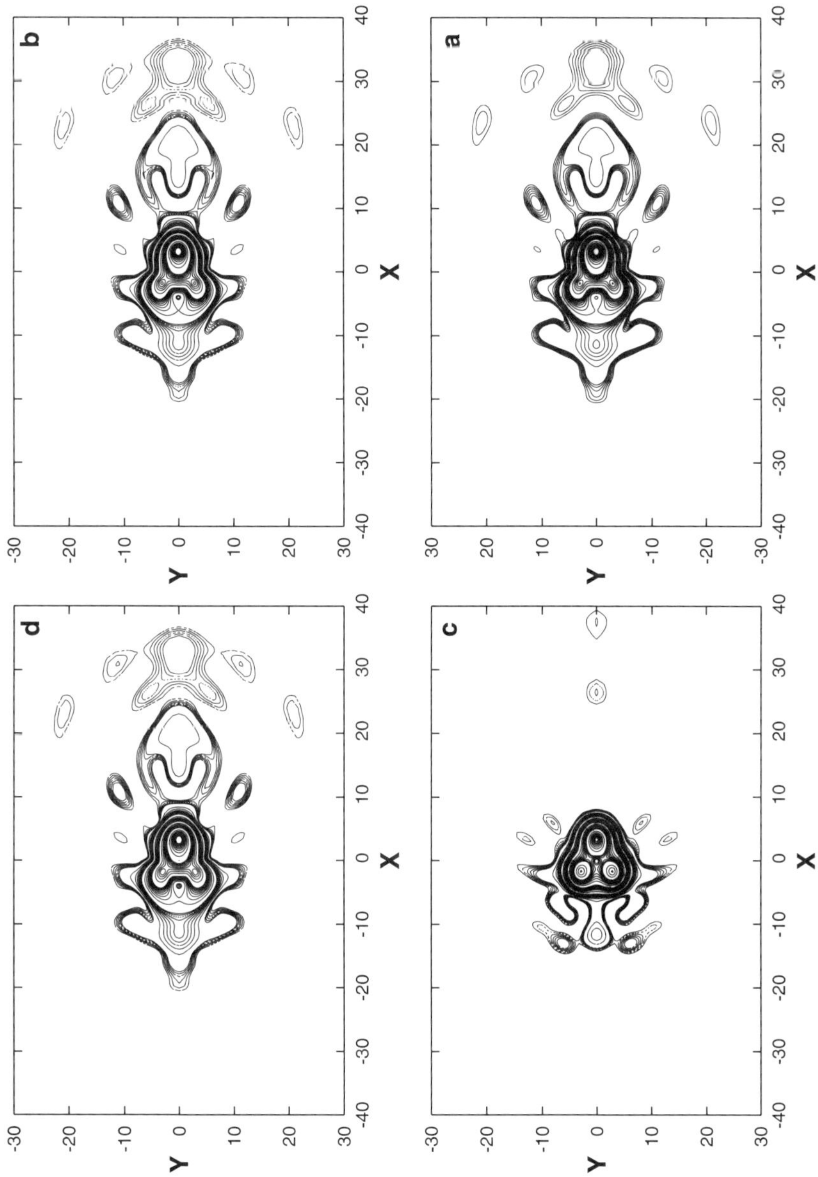

FIGURE 3. Population iso-line contour at $t = 5$ cycles for H_3^{++} exposed to linearly polarized laser field with $I = 10^{14} W/\text{cm}^2$, $\lambda =$ 800nm: (a) SPLIT–xFFT–yFFT; (b) GLOBAL–ρDF–ϕFFT; (c) SPLIT–ρDF–ϕFFT–1; (d) SPLIT–ρDF–ϕFFT–2.

FFT transformations will be local in each processor and the solution of the 1-D discrete linear system obtained with splitting methods in (ρ_1, ρ_2) or the 2-D discrete system without splitting will be worked out in parallel. We illustrate the method presently for the linear Born–Oppenheimer (this corresponds to static nuclei, i.e. R fixed) 2-electron H_2 molecule. We have previously calculated the exact nonlinear

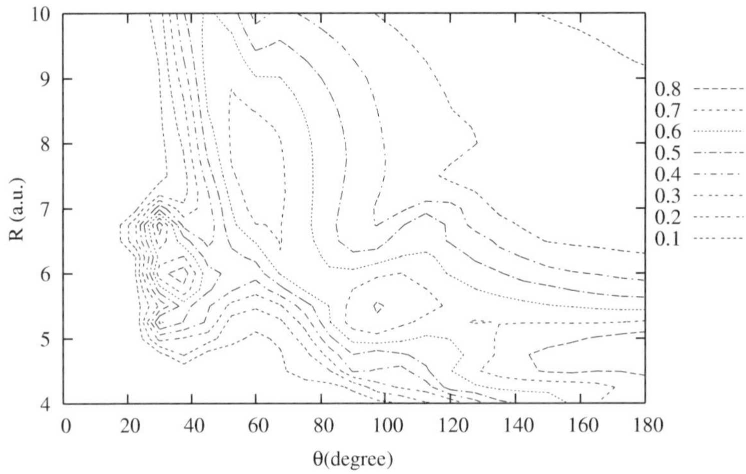

FIGURE 4. Ionization of H_3^{++} exposed to circularly polarized laser filed in different (R, θ) configuration for $t = 10$ cycles, $I = 2 \times 10^{14} W/\mathrm{cm}^2$, $\lambda = 800\mathrm{nm}$. Enhanced ionization occurs near $(R, \theta) = (5.9, 38^o)$, $(5.5, 98^o)$ and $(6.75, 32^o)$.

non-perturbative optical response of H_2 in 1-D, which corresponds to a 2-D partial differential equation for electrons [11, 12].

The H_2 problem becomes a 4-D partial differential equation with (ρ_1, θ_1) and (ρ_2, θ_2) coordinates for each electron as defined for the one-electron case above. The TDSE is appropriately written as

$$i \frac{\partial \psi}{\partial t}(\rho_1, \theta_1, \rho_2, \theta_2) = \left[-\frac{1}{2}(\triangle_1 + \triangle_2) + V_C + V_E \right] \psi(\rho_1, \theta_1, \rho_2, \theta_2), \tag{4.1}$$

where $\triangle$ is the free propagation 2-D Laplacian on the right hand side of equation (2.1), V_C is the electron-electron repulsion and electron-proton attractive Coulomb potentials,

$$V_C = \frac{1}{\sqrt{\rho_1^2 + \rho_2^2 - 2\rho_1\rho_2 \cos\theta_1 - \theta_2 + c}} - \sum_{j=1}^{2} \frac{1}{\sqrt{\rho_j^2 \pm R\rho_j + R^2/4 + d}} \tag{4.2}$$

R is the internuclear distance between protons, c and d are regularization parameters to remove Coulomb singularities (this corresponds to averaging over the third dimension perpendicular to the plane of the molecule, i.e., the Z direction). The laser-electron radiative coupling is described by the time-dependent potential $V_E(\rho_1, \theta_1, \rho_2, \theta_2, E(t))$,

$$V_E = -(\rho_1 \cos\theta_1 + \rho_2 \cos\theta_2) E(t). \tag{4.3}$$

We solve the 4-D TDSE, equation by a third order Split-Operator method [9],

$$\begin{aligned} \psi(t + \delta t) = \exp\left[-i\frac{\delta t}{2}(V_C + V_E)\right] \exp\left[-i\frac{\delta t}{2}\triangle_1\right] \exp\left[-i\frac{\delta t}{2}\triangle_2\right] \\ \exp\left[-i\frac{\delta t}{2}\triangle_2\right] \exp\left[-i\frac{\delta t}{2}\triangle_1\right] \exp\left[-i\frac{\delta t}{2}(V_C + V_E)\right] + \mathcal{O}(\delta t^3). \end{aligned} \tag{4.4}$$

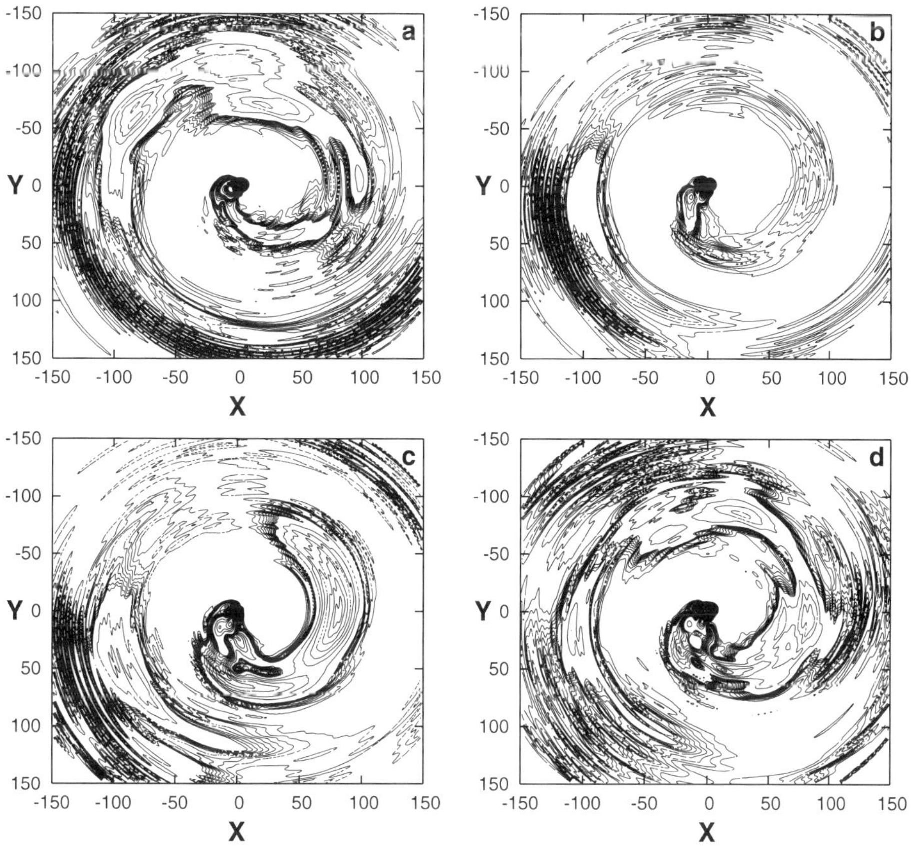

FIGURE 5. Population iso-line contours of H_3^{++} exposed to a circularly polarized laser field, at $t = 10$ cycles, $I = 2 \times 10^{14} W/\text{cm}^2$ for the configurations: (a) $R = 5.5$ a.u., $\theta = 97.5^o$; (b) $R = 6$ a.u., $\theta = 37.5^o$; (c) $R = 6.75$ a.u., $\theta = 30^o$; (d) $R = 7.75$ a.u., $\theta = 60^o$.

Since the $\triangle_j$'s in (4.4) contain singularities at $\rho = 0$ (see eq. (2.1)), we use the same methodology as described in section 2 to calculate the 2-D free propagation equation for one-electron thus obtaining the 4-D free propagator for the two electrons in H_2.

We have solved the 4-D TDSE, equation (4.1) from (4.4) at equilibrium $R = 1.675$a.u., for intensities $I = 1.5 \times 10^{15} W/\text{cm}^2$, wavelength $\lambda = 800$nm, (Ti-Sa laser) and for a total time of $t = 6$ cycles $= 16.5 \times 10^{-15}$sec (16.5 femtoseconds (fs)). The regularization parameters were chosen to be $c = 1$, $d = 0.35$ allowing to reproduce the ground state potential energy of the H_2 molecules [11, 12]. The time steps δt are taken to be $\delta t = 0.03$a.u. $= 73 \times 10^{-18}$sec and the spatial discretization is $\delta\rho = 0.25 a.u. = 0.0125$nm for a grid size $0 \leq \rho \leq 180$a.u. $= 18$nm.

Figure 6 illustrates the radial two electron distribution,

$$|\psi(\rho_1, \rho_2)|^2 = \int d\theta_1 \int d\theta_2 |\psi(\rho_1, \rho_2, \theta_1, \theta_2)|^2. \tag{4.5}$$

after $t = 4$ cycles $= 11$fs. One sees clearly that as $\rho_1(\rho_2)$ increases to large values, corresponding to one-electron ionizations $\rho_2(\rho_1)$ remains small and but is expected

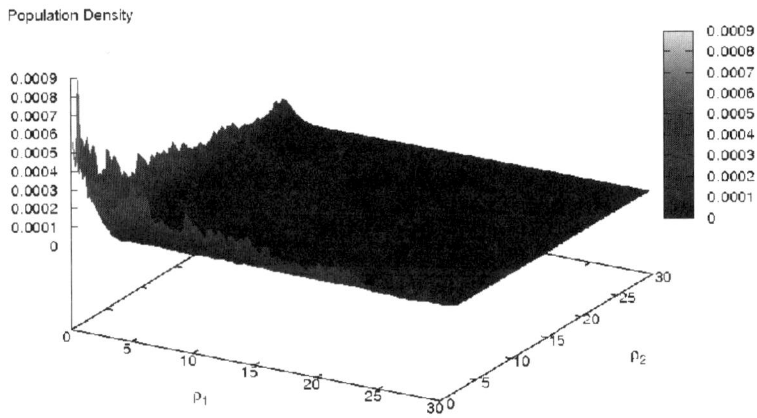

FIGURE 6. Radial two electron distribution $|\psi(\rho_1, \rho_2)|^2$, eq. (4.5), for the H_2 molecule at $t = 4$ cycles, $I = 1.5 \times 10^{15} W/\text{cm}^2$, 800nm, laser pulse.

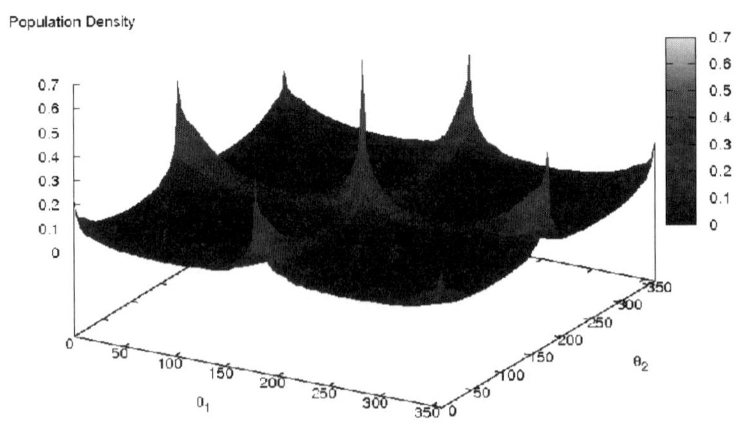

FIGURE 7. Angular two electron distribution $|\psi(\rho_1, \rho_2)|^2$, eq. (4.6), for same parameters as Fig. 6.

to expand due to "shakeup" excitation [13]. The two electron angular distribution

$$|\psi(\theta_1, \theta_2)|^2 = \int \rho_1 d\rho_1 \int \rho_2 d\rho_2 |\psi(\rho_1, \theta_1, \rho_2, \theta_2)|^2 \tag{4.6}$$

manifests angular correlation. The main peaks in Fig. 7 occurs at $\theta_1 = \theta_2 = \pi$ corresponding to correlated forward double displacement by the laser. Another prominent peak occurs at $\theta_1 = \pi$, $\theta_2 = 2\pi$ which corresponds to forward-backward electron induced separation in the strong ($I = 1.5 \times 10^{15} W/\mathrm{cm}^2$) low frequency ($\omega = 0.057$au, $\lambda = 800$nm) six cycle laser pulse. The shakeup excitation and angular correlations are expected to occur on "attosecond" (10^{-18}sec) time scales, the natural time scale for electron reorganization in atoms and molecules [14]. The present 4-D simulations are a first step in describing these ultrarapid electron phenomena by large dimensional numerical simulations. Current evolving computer technology should allow for solving the two electron TDSE, eq. (4.1) in 3-D with moving nuclei (i.e. beyond Born–Oppenheimer). This was previously solved for the one-electron system H_2^+ and corresponded to a 4-D problem (3-D for the electron and 1-D for the protons parallel to the E field, [15]). Inclusion of the proton motion in 1-D and 3-D electron motion in H_2 gives rise to a 7-D partial differential equation, necessitating at least a peta byte (10^{15}B) of computer memory. Clearly quantum electron-nuclear dynamics corresponds to partial differential equations with dimension > 3. This is a new computational and mathematical challenge.

References

1. T. Brabec and F. Krausz, *Intense few-cycles laser fields*, Rev. Modern Phys. **72** (2000), 545–591.
2. A.D. Bandrauk (ed.), *Molecules in laser fields*, M. Dekker Pub., N.Y., 1994.
3. S. Chelkowski, T. Zuo, and A.D. Bandrauk, *Ionization rates of H_2^+ in an intense laser field by numerical integration of the time-dependent Schrödinger equation*, Phys. Rev. A **46** (1992), 5342–5345.
4. T. Zuo, A.D. Bandrauk, M. Ivanov, and P.B. Corkum, *Control of high-order harmonic generation in strong laser fields*, Phys. Rev. A **51** (1995), 3991–3998.
5. C. Dion, *Simulation numérique de l'isomérisation de* HCN *en champ laser intense*, M.Sc., 1995.
6. H. Kono, A. Kita, Y. Ohtsuki, and Y. Fujimura, *An efficient quantum m echanical method for the electronic dynamics of the three-dimensional hydrogen atom interacting with a linearly polarized strong laser pulse*, J. Comput. Phys. **130** (1997), 148–159.
7. K. Harumiya, H. Kono, Y. Fujimura, I. Kawata, and A.D. Bandrauk, *Intense laser-field ionization of H_2 enhanced by two-electron dynamics*, Phys. Rev. A **66** (2002), 043403-1–043403-14.
8. A.D. Bandrauk and H.Z. Lu, *Handbbok of numerical analysis — Computational Chemistry*, (P.G. Ciarlet and C. LeBris, eds.), Elsevier Science, vol. 10, B.V. Amsterdam, 2003, pp. 803–830.
9. A.D. Bandrauk and H. Shen, *Exponential split operator methods for solving coupled time-dependent Schrödinger equations*, J. Chem. Phys. **99** (1993), 1185–1193.
10. A.D. Bandrauk and J. Ruel, *Charge resonance Enhance ionization — geometric and orientational Effects*, Phys. Rev. A **59** (1999), 2153–2156.
11. A.D. Bandrauk and H.Z Lu, *Electron correlation and double ionization of a 1-D H_2 in an intense laser field*, J. Modern Opt. **53** (2006), 35–53.
12. A.D. Bandrauk and H.Z Lu, *Laser induced recollision in H_2 and electron correlation*, Phys. Rev. A **72** (2005), 023408-1–023408-8.
13. I.V. Litvinyuk, F. Légaré, P.W. Dooley, D.M. Villeneuve, P.B. Corkum, J. Zanghellini, A. Pegarkov, C. Fabian, and T. Brabec, *Shakeup excitation during optical tunnel ionization*, Phys. Rev. Lett. **94** (2005), 033003.
14. J. Breidbach and L.S. Cederbaum, *Universal attosecond response to the removal of an electron*, Phys. Rev. Lett. **94** (2005), 033901–033902.
15. S. Chelkowski, A. Conjusteau, T. Zuo, and A.D. Bandrauk, *Dissociative ionization of H_2^+ in an intense laser field: charge-resonance-enhanced ionization, Coulomb explosion, and harmonic generation at 600*nm, Phys. Rev. A **54** (1996), 3235–3244.

Département de Chimie, Université de Sherbrooke, Sherbrooke, Québec, J1K 2R1, Canada.
E-mail address: Andre.Bandrauk@USherbrooke.ca

Centre de Calcul Scientifique, Université de Sherbrooke, Sherbrooke, Québec, J1K 2R1, Canada.
E-mail address: HuiZhong.Lu@USherbrooke.ca

Centre de Recherches Mathématiques
CRM Proceedings and Lecture Notes
Volume **41**, 2007

Feasibility and Competitiveness of a Reduced Basis Approach for Rapid Electronic Structure Calculations in Quantum Chemistry

Éric Cancès, Claude Le Bris, Yvon Maday, Ngoc Cuong Nguyen, Anthony T. Patera, and George Shu Heng Pau

ABSTRACT. We apply the reduced basis methodology to electronic structure calculations with a view to significantly speeding up this computation when it must be performed many times — as in each time step of an ab initio MD simulation or inside a geometry optimization procedure. The feasibility and the efficiency of the approach are demonstrated on several test cases. Promising directions for further research and application to real-scale computations are indicated and discussed.

1. Introduction

As opposed to other more generic problems of the engineering sciences, computational quantum chemistry problems are known to be sufficiently specific to justify a dedicated numerical approach. Indeed, the methods developed over the past sixty years on such problems are very special in nature. This is in particular the case because (i) the problems to be addressed are incredibly complex, even when compared with infamous engineering problems of outstanding difficulty; and (ii) the methods have been developed by the community of chemists and physicists without almost any real involvement of experts in numerical analysis and scientific computing. Although it is often the case that efficiency is reached at the price of non-genericity, in quantum chemistry the situation is extreme. As a result, there exist many approaches that have proved successful in several engineering domains but which have not been at all tested or adapted in the context of computational chemistry. Reduced basis methods are one instance of such approaches. It is the purpose of the present article to report on some exploratory applications of this general methodology to the specific quantum chemistry context. As will be seen, and although definite conclusions on the validity of the approach on real cases are

---

2000 *Mathematics Subject Classification.* 35J60, 35Q40.

The authors would like to thank François Jollet for the many helpful discussions. N.C. Nguyen, A.T. Patera, and G.S.H. Pau acknowledge the support by DARPA and AFOSR under Grant FA9550-05-1-0114 and by the Singapore-MIT Alliance.

This is the final form of the paper.

yet to be obtained, the results are definitely promising and certainly sufficient to motivate further efforts.

Let us briefly give some insight into the mathematical nature of the calculations that are at the heart of a standard quantum chemistry computation. Basically, all quantum chemistry calculations rely on a first central computation: electronic ground state of the physical system under study, a molecular system of finite size or a condensed phase. Excited states predictions, linear response theory calculations, conformations determinations, ab initio molecular dynamics simulations, etc., all rely on this basic foundation. This central computation is a challenge on its own: in principle, being given a set of nuclei (which are typically treated as classical point particles) located at (possibly momentarily) fixed positions in space, the task is to determine the wavefunction corresponding to the electronic ground state; this solves the quantum N-body problem, parametrized by the positions of nuclei. Mathematically, the latter problem is the minimization of a quadratic form on functions varying in $L^2(\mathbb{R}^{3N})$, subject to the constraint that these functions are antisymmetric and appropriately normalized. For almost all physically relevant cases $N \geq 3$, the N-body problem is untractable and thus the dominant strategy in computational chemistry consists in deriving models approximating this N-body problem. These approximated models are in turn discretized and solved numerically.

Hartree–Fock (HF) type models [6, 34] and Density Functional Theory (DFT) type models [6, 8, 25] are the main two categories of approximation models. They both consist of a non-quadratic constrained minimization problem. At the discretized level, it is solved in practice by considering a *nonlinear eigenvalue problem*, which is the specific form of the Euler–Lagrange equation associated with the constrained minimization problem. A solution to this problem is searched for iteratively so as to (hopefully) minimize the energy; the procedure is called the Self-Consistent-Field iteration. The existence of this nonlinear eigenvalue problem at the core of quantum chemistry calculations is the origin of the tremendous overall computational cost of the problem. To realize this, it suffices to recall that in a geometry optimization calculation (i.e., the search for the global minimizer — configuration of nuclei and ground state electronic wavefunction — that minimizes the global energy), or in ab initio molecular dynamics calculations (i.e., the integration in time of the Newton equations of motion for the nuclei equipped with their electronic cloud assumed in its parametrized ground-state) [10, 19], each iteration of the search (or of the numerical integrator, as appropriate) involves as the inner loop the solution of this nonlinear eigenvalue problem. There is thus much interest in accelerating this inner loop calculation, and it is a long standing problem of computational chemistry.

What often makes the problem tractable in practice is that the chemists have developed specific case-dependent basis sets which are remarkably efficient. Owing to the very nature of the basis set functions, the size of the matrices is kept sufficiently small so that (at least for systems of up to a hundred of electrons) a solution may be calculated for an acceptable price at excellent accuracy. This is fortunate, because such calculations are very demanding in terms of accuracy.

However, the efficient basis sets developed are often dependent on the positions of the nuclei. They typically correspond to solutions of hydrogenoic-like problems (Slater-type orbitals, ...), or approximations to thereof, developed for the purpose of computational efficiency (contracted Gaussian basis sets,...). In a situation where the positions of the nuclei vary (and this is the case both in the context of geometry optimization or time-dependent simulations), such basis sets are often delicate and slow to manipulate. They are sometimes replaced by more standard basis sets, independent of the positions of the nuclei, that are less efficient for a fixed configuration of nuclei but more efficient when the latter vary; a typical example is a plane wave basis set. Such basis sets are also typically used for solid phase calculations, where the problem is posed on the unit cell of a lattice and subject to periodic boundary conditions. The efficiency of these more general basis sets comes at a price: the size of the basis set required to reach a good accuracy is then much larger than was the case for nuclei-dependent basis sets.

All of these considerations indicate that, in the context of varying nuclei positions, there is an opportunity for better basis sets to be developed. This motivates the application of the reduced basis methodology. The reduced basis method was first introduced in the late 1970s in the context of nonlinear structural analysis [1,24] and subsequently abstracted, analyzed, and extended to a much larger class of parametrized partial differential equations [9,14,16,27,28]. The foundation of the reduced basis method is built upon three important realizations, which we phrase here in the particular context of quantum chemistry.

The first realization is that in the molecular dynamics simulation of a physical system the quantity of primary importance is typically *not* the field variable u^{e} — such as the wavefunctions or the electron density, but rather certain selected *output of interest* s^{e} — such as the ground state energies of the system and the forces exerted on the nuclei. Both the field variable u^{e} and output s^{e} depend on the parameters, or *inputs*, μ, which serve to identify a particular configuration of the system. Typical inputs include nuclei positions, dimensions of the simulation cells, and the dielectric constant. The relevant system behavior is thus described by an implicit input-output relationship, $s^{\mathrm{e}}(\mu)$, evaluation of which demands solution of the underlying partial differential equation governing electronic structure. In practice, the exact solution $u^{\mathrm{e}}(\mu)$ (respectively, output $s^{\mathrm{e}}(\mu)$) is not available and must thus be replaced by a "truth" approximation $u(\mu)$ (respectively, $s(\mu)$) which resides in a finite dimensional approximation space of dimension $\mathcal{N}$. Nevertheless, as mentioned earlier, calculation of the truth approximation $u(\mu)$ and output $s(\mu)$ can be expensive for certain classes of quantum chemistry problems.

The second realization is related to the choice of basis sets alluded to earlier. The critical observation is that $u(\mu)$ in fact resides on a very low-dimensional manifold $\mathcal{M} \equiv \{u(\mu) \mid \mu \in \mathcal{D}\}$ induced by the parametric dependence; here $\mathcal{D} \in \mathbb{R}^P$ is the parameter space in which our input μ — a P-tuple of parameters — varies. Furthermore, the field variable $u(\mu)$ will often be quite regular in μ — the parametrically induced manifold $\mathcal{M}$ is smooth — even when the field variable enjoys only limited regularity with respect to the spatial coordinate.[1] The reduced basis method explicitly recognizes and exploits dimension reduction afforded by the low-dimensional — and smooth — parametrically induced solution manifold. More precisely, rather than general basis sets such as the plane wave basis set, the basis

[1]The smoothness is deduced from the equation for the sensitivity derivatives; the stability and continuity properties of the partial differential operator are crucial. Clearly the latter are delicate matters in the context of eigenvalue problems.

set consists of solutions of the partial differential equation at N^u selected parameter points μ_i, $1 \leq i \leq N^u$. Then, the set of all solutions $u(\mu)$ as μ varies can be approximated very well by its projection on a finite and low-dimensional vector space spanned by the $u(\mu_i)$: for sufficiently well chosen μ_i, there exist coefficients c_i, $1 \leq i \leq N(\mu)$ such that the finite sum $\sum_{i=1}^{N^u} c_i u(\mu_i)$ is very close to $u(\mu)$ for any μ.

The third realization lies in what we term the "computational opportunities". Here, we rely on two observations. The first observation derives from the mathematical formulation: very often, the partial differential operator can be expressed as the sum of Q products of (known, easily evaluated) parameter-*dependent* functions and parameter-*independent* operators; we shall denote this structure as "affine" parameter dependence. The second observation derives from our context: molecular dynamic simulation requires rapid and repetitive evaluations of the "input-output" relations and thus places a predominant emphasis on very low *marginal* cost — the additional effort associated with each new evaluation. These two observations present a clear investment-amortization opportunity: we can exploit the underlying affine parametric structure to design effective *offline-online* computational procedures which willingly accept greatly increased initial preprocessing — offline — expense in exchange for greatly reduced marginal — online — "in service" cost.

In the more recent past the reduced basis approach and in particular associated *a posteriori* error estimation procedures have been successfully developed for (i) linear elliptic and parabolic PDEs that are affine in the parameter [13, 20, 21, 29, 37] ; and (ii) elliptic PDEs that are at most quadratically nonlinear in the first argument [23,35,36]. In these cases a very efficient offline-online computational strategy can be developed. The operation count for the online stage — in which, given a new parameter value, we calculate the reduced basis output and associated error bound — depends only on N^u (typically small) and Q, but it is *independent* of $\mathcal{N}$, the dimension of the underlying "truth" approximation.

Unfortunately, in quantum chemistry computations based on Hartree–Fock type models or Density Functional Theory type models, the underlying PDEs do not have the same nice structure that allows us to apply the methodology developed in earlier works. There are three main issues: (i) the equations can contain both non-affine terms and also very nasty nonlinear terms, for example associated with an exchange-correlation term; (ii) the solution sought is not scalar — for each μ, we look for a set of eigensolutions; and (iii) the parametrization of the PDEs can be complex, for example due to a set of moving nuclei and periodic boundary conditions. In fact, the above three issues are the main difficulties faced by any numerical approximation of the PDEs obtained in computational chemistry and are the deciding factors when determining the appropriate numerical approach to employ. From this particular mathematical standpoint, Hartree–Fock type models and Density Functional Theory type models are, with all due respect, similar.

In this paper, we will address these issues albeit in an idealized "model" context. In Section 2, we review the recently developed "empirical interpolation procedure" [2, 12] through a simple model elliptic problem. Then in Section 3 and Section 4, we apply the approach within the quantum chemistry context: in Section 3 to a (non-affine) HF hydrogen molecular system; and in Section 4 to a (nonlinear) DFT one-dimensional model. Section 3 also addresses, or at least introduces, issues of geometric and parametric complexity. And Section 4 proposes a "solution" to the many-electron difficulty. We will not handle issues related to *a posteriori* error estimators in this paper, but they are essential to our overall reduced basis methodology and will be addressed in our future work.

Note that our reduced basis approximation is built (offline) upon the "truth" approximation; in subsequent sections, the "truth" approximation is based on the finite element approximation. The choice is purely for convenience; certainly there are other and better choices — and hence any discussions of offline effort are, at best, notional. We could in fact use other basis sets; in [26], the planewave method is the choice of "truth" approximation.

2. Methodology

2.1. Abstract formulation. To simplify our presentation, we will describe our methodology based on a nonlinear elliptic problem. Extensions to nonlinear eigenvalue problems of interest in quantum chemistry will be discussed in subsequent sections. The abstract statement of the nonlinear elliptic problem is as follows: given any $\mu \in \mathcal{D} \subset \mathbb{R}^P$, we evaluate the output $s^{\mathrm{e}}(\mu)$ as

$$s^{\mathrm{e}}(\mu) = \ell\big(u^{\mathrm{e}}(\mu)\big) \tag{2.1}$$

where $u^{\mathrm{e}}(\mu) \in Y^{\mathrm{e}}$ is the solution of

$$\mu a_0(u^{\mathrm{e}}(\mu), v) + \int_\Omega g(u^{\mathrm{e}}(\mu); x; \mu) v = f(v), \quad \forall v \in Y^{\mathrm{e}}. \tag{2.2}$$

Here $\mathcal{D}$ is the parameter domain in which our P-tuple (input) parameter μ resides (here $P = 1$); $Y^{\mathrm{e}}(\Omega)$ is an appropriate Hilbert space with the associated inner product $(w, v)_{Y^{\mathrm{e}}} = \int_\Omega \nabla w \cdot \nabla v + wv$ and norm $\|\cdot\|_{Y^{\mathrm{e}}} = (\cdot,\cdot)_Y^{1/2}$; Ω is a bounded domain in $\mathbb{R}^d$ with Lipschitz continuous boundary $\partial\Omega$; $g(w; x; \mu)$ is a general non-affine nonlinear function of $w \in Y$, spatial coordinate x, and the parameter μ; $a_0(\cdot,\cdot)$ is a Y^{e}-continuous bilinear functional; and $f(\cdot)$ and $l(\cdot)$ are Y^{e}-continuous linear functionals. These functionals are not parameter-dependent. Our function space $Y^{\mathrm{e}}(\Omega)$ will satisfy $\big(H_0^1(\Omega)\big)^\nu \subset Y^{\mathrm{e}}(\Omega) \subset \big(H^1(\Omega)\big)^\nu \subset \big(L^2(\Omega)\big)^\nu$, where $\nu = 1$ for a scalar field variable and $\nu = d$ for a vector field variable. Here $H^1(\Omega)$ (respectively, $H_0^1(\Omega)$) is the usual Hilbert space of derivative square-integrable functions (respectively, derivative square-integrable functions that vanish on the domain boundary $\partial\Omega$) and $L^2(\Omega)$ is the Lebesgue space of square-integrable functions [30].

More often than not, the exact solution is not available and we thus replace $u^{\mathrm{e}}(\mu)$ with a "truth" approximation, $u(\mu)$, which resides in (say) a suitably fine piecewise-linear finite element approximation space $Y \subset Y^{\mathrm{e}}$ of *very* large dimension $\mathcal{N}$. The abstract statement for the resulting problem is thus: given any $\mu \in \mathcal{D}$, we evaluate $s(\mu) = \ell\big(u(\mu)\big)$, where $u(\mu) \in Y$ satisfies

$$\mu a_0(u(\mu), v) + \int_\Omega g(u(\mu); x; \mu) v = f(v), \quad \forall v \in Y. \tag{2.3}$$

We shall assume — hence the appellation "truth" — that the discretization is sufficiently rich such that $u(\mu)$ and $u^{\mathrm{e}}(\mu)$ and hence $s(\mu)$ and $s^{\mathrm{e}}(\mu)$ are indistinguishable at the accuracy level of interest. The reduced basis approximation shall be built upon this reference (or "truth") finite element approximation, and the error of this approximation will thus be evaluated with respect to $u(\mu) \in Y$. Our formulation must be stable and efficient as $\mathcal{N} \to \infty$. Note that Y inherits the inner product and norm from Y^{e}.

We shall make the following assumptions. First, we assume that the bilinear form $a_0(\cdot,\cdot)\colon Y \times Y \to \mathbb{R}$ is symmetric, $a_0(w,v) = a_0(v,w)$, $\forall w, v \in Y$. We shall also make two crucial hypotheses related to well-posedness. Our first hypothesis is that the bilinear form a_0 satisfies the following stability and continuity conditions:

$$0 < \alpha \equiv \inf_{v \in Y} \frac{\mu a_0(v,v)}{\|v\|_Y^2}, \quad \forall \mu \in \mathcal{D}, \tag{2.4}$$

$$\sup_{w \in Y} \sup_{v \in Y} \frac{\mu a_0(w,v)}{\|w\|_Y \|v\|_Y} \equiv \gamma < \infty, \quad \forall \mu \in \mathcal{D}, \tag{2.5}$$

and that $f \in L^2(\Omega)$. In the second hypothesis, we require $g\colon \mathbb{R} \times \Omega \times \mathcal{D} \to \mathbb{R}$ to be continuous in its arguments, increasing in its first argument, and $g(z;x;\mu) \geq 0$, $\forall z \in \mathbb{R}$, $\forall x \in \Omega$ and $\forall \mu \in \mathcal{D}$. With these assumptions, the problems (2.2) and (2.3) are then well-posed [12].

2.2. Reduced basis approximation.

2.2.1. *Empirical interpolation procedure.* We first introduce the nested samples, $S_N^u = \{\mu_1^u \in \mathcal{D}, \ldots, \mu_N^u \in \mathcal{D}\}$, $1 \leq N \leq N_{\max}$, and associated nested Lagrangian [28] reduced basis spaces $W_N^u = \operatorname{span}\{u(\mu_n^u), 1 \leq n \leq N\} = \operatorname{span}\{\zeta_n, 1 \leq n \leq N\}$, $1 \leq N \leq N_{\max}$, where $u(\mu_n^u)$ is the solution of (2.2) at $\mu = \mu_n^u$ and ζ_n, $1 \leq n \leq N$ are the orthonormalized bases of $u(\mu_n^u)$, $1 \leq n \leq N$ with respect to $(\cdot,\cdot)_Y$. The classical reduced basis approximation [13,20,21,29,37] is then obtained by a standard Galerkin projection: given $\mu \in \mathcal{D}$, we evaluate

$$s_N(\mu) = \ell(u_N(\mu)), \tag{2.6}$$

where $u_N(\mu) \in W_N^u$ satisfies

$$\mu a_0(u_N(\mu), v) + \int_\Omega g(u_N(\mu); x; \mu) v = f(v), \quad \forall v \in W_N^u. \tag{2.7}$$

Unfortunately, the presence of *strong* nonlinearity in g does not allow an efficient *offline-online* procedure outlined in [23, 36]. As a result, although the dimension of the system (2.7) is small, solving it is actually expensive: the evaluation of the nonlinearity term $\int_\Omega g(u_N(\mu); x; \mu) v$ will scale as some power of $\mathcal{N}$. Due to this $O(\mathcal{N})$ dependence, it is somewhat disingenuous to interpret (2.7) as a reduced-order model since the resulting computational advantage relative to classical approaches using advanced iterative techniques can be modest.

Our goal is to efficiently compute $s_N(\mu)$ such that the incurred computational cost is dependent on the dimension of reduced basis approximation spaces and the parametric complexity of the problems, but *independent* of $\mathcal{N}$. Towards this end, we develop a collateral reduced basis expansion for the nonlinear term by using the empirical interpolation procedure [2, 12] reviewed below.

In particular, we consider the approximation of the parameter-dependent nonlinear function $g(w;x;\mu)$ by a reduced basis expansion $g_M^w(x;\mu)$ based upon the empirical interpolation procedure outlined in [2, 12]. The procedure includes three steps. In the first step, we introduce nested samples $S_M^g = \{\mu_1^g \in \mathcal{D}, \ldots, \mu_M^g \in \mathcal{D}\}$, $1 \leq M \leq M_{\max}$ and associated nested approximation spaces $W_M^g = \operatorname{span}\{\xi_m \equiv g(u(x;\mu_m^g); x; \mu_m^g), 1 \leq m \leq M\}$. The construction of S_M^g and W_M^g is based on a greedy selection process. We first define the best approximation $g_M^*(x;\mu) \equiv \arg\min_{z \in W_M^g} \|g(u(x;\mu_m^g); \cdot; \mu) - z\|_{L^\infty(\Omega)}$ and the associated error $\varepsilon_M^*(\mu) \equiv \|g(u(x;\mu_m^g); \cdot; \mu) - g_M^*(\cdot;\mu)\|_{L^\infty(\Omega)}$. Then, we choose our first sample point to be

$\mu_1^g = \arg\max_{\mu\in\Xi^g} \|g(\cdot;\cdot;\mu)\|_{L^\infty(\Omega)}$, and define $S_1^g = \{\mu_1^g\}$, $\xi_1(x) \equiv g(u(x;\mu_1^g);x;\mu_1^g)$, and $W_1^g = \text{span}\{\xi_1\}$; here Ξ^g is a suitably large but finite-dimensional parameter set in $\mathcal{D}$. For $M \geq 2$, we determine $\mu_M^g = \arg\max_{\mu\in\Xi^g} \varepsilon_{M-1}^*(\mu)$, and define $S_M^g = S_{M-1}^g \cup \mu_M^g$, $\xi_M(x) = g(u(x;\mu_M^g);x;\mu_M^g)$, and $W_M^g = \text{span}\{\xi_m, 1 \leq m \leq M\}$. In essence, W_M^g comprises basis functions from the parametrically induced manifold $\mathcal{M}^g \equiv \{g(u(\cdot;\mu);\cdot;\mu) \mid \mu \in \mathcal{D}\}$.[2]

In the second step, we construct nested sets of interpolation points $T_M = \{t_1,\ldots,t_M\}$, $1 \leq M \leq M_{\max}$. We first set $t_1 = \arg\operatorname{ess\,sup}_{x\in\Omega} |\xi_1(x)|$, $q_1(x) = \xi_1(x)/\xi_1(t_1)$, $B_{11}^1 = 1$. Then for $M = 2,\ldots,M_{\max}$, we solve the linear system $\sum_{j=1}^{M-1} \sigma_j^{M-1} q_j(t_i) = \xi_M(t_i)$, $1 \leq i \leq M-1$, and set $r_M(x) = \xi_M(x) - \sum_{j=1}^{M-1} \sigma_j^{M-1} q_j(x)$, $x_M = \arg\operatorname{ess\,sup}_{x\in\Omega} |r_M(x)|$, $q_M(x) = r_M(x)/r_M(t_M)$, and $B_{ij}^M = q_j(t_i)$, $1 \leq i,j \leq M$.

Finally, for any given $w \in Y$ and M, we may approximate $g(w;x;\mu)$ by $g_M^w(x;\mu) = \sum_{m=1}^M \varphi_{Mm}(\mu) q_m(x)$, where $\sum_{j=1}^M B_{ij}^M \varphi_{Mj}(\mu) = g(w(t_i);t_i;\mu)$, $1 \leq i \leq M$ and $B_{ij}^M = q_j(t_i)$. Although this "composed" interpolant is defined for general $w \in Y$, we expect good approximation only for w (very) close to the manifold $\mathcal{M}^u \equiv \{u(\mu) \mid \mu \in \mathcal{D}\}$ on which W_M^g is constructed. Theoretical and numerical aspects of the empirical interpolation have been analyzed in great detail in [2, 12]. We summarize here the main results: (i) the process is stable — B^M is a well-conditioned lower triangular matrix with dominant unity diagonal; and (ii) $\varepsilon_M(\mu) \leq \varepsilon_M^*(\mu)\,(1+\Lambda_M)$, where $\varepsilon_M(\mu) = \|g(u(\cdot;\mu);\cdot;\mu) - g_M^{u(\mu)}(\cdot;\mu)\|_{L^\infty}$ [2,12] and the Lebesgue constant, Λ_M, is bounded above by $2^M - 1$.[3]

We may now replace $g(u_N(\mu);x;\mu)$ — as required in our reduced basis projection for $u_N(\mu)$ — with $g_M^{u_{N,M}}(x;\mu)$. Our reduced basis approximation is thus: given $\mu \in \mathcal{D}$, we evaluate

$$s_{N,M}(\mu) = \ell\big(u_{N,M}(\mu)\big), \tag{2.8}$$

where $u_{N,M}(\mu) \in W_N^u$ satisfies

$$\mu a_0(u_{N,M}(\mu), v) + \int_\Omega g_M^{u_{N,M}}(x;\mu) v = f(v), \quad \forall v \in W_N^u. \tag{2.9}$$

As shown in the next subsection, this reduced basis formulation enables us to develop a very efficient *offline-online* procedure for the computation of $s_{N,M}(\mu)$.

Finally, we address the issue related to the choice of our parameter samples S_N^u and thus the corresponding reduced basis spaces W_N^u. We could in fact generalize the greedy selection process outlined in Section 2.2.1 for the construction of any S_N^u [23, 29, 36]. We assume that we are given a sample S_N^u and hence reduced basis space W_N^u and associated reduced basis approximation (procedure to determine) $u_{N,M}(\mu)$ and $s_{N,M}(\mu)$, $\forall \mu \in \mathcal{D}$. Then, for a suitably fine grid Ξ_{Test} over the parameter space $\mathcal{D}$, we determine $\mu_N^* = \arg\max_{\mu\in\Xi_{\text{Test}}} \varepsilon_{N,M}^u(\mu)$ and append μ_N^* to S_N^u to form S_{N+1}^u and hence W_{N+1}^u. There is considerable flexibility in the

---

[2]Thanks to our truth approximation, the optimization for $g_{M-1}^*(x;\mu)$ and hence $\varepsilon_{M-1}^*(\mu)$ is a *standard linear program.* In actual practice, rather than the $L^\infty(\Omega)$-norm, we exploit the $L^2(\Omega)$-norm surrogate in our best approximation, the construction of S_M^g is considerably less expensive.

[3]The bound is very pessimistic and of little practical value. In applications, the actual asymptotic behavior of the Lebesgue constant is much lower than the upper bound $2^M - 1$, typically $O(M)$; however, the bound does provide a theoretical basis for some stability.

definition of $\varepsilon^u_{N,M}(\mu)$, the measure of the approximation error — for example, we may use

$$\text{error in } s_{N,M}(\mu)\colon\ |s_{N,M}(\mu) - s(\mu)|, \tag{2.10}$$

$$\text{error in } u_{N,M}(\mu)\colon\ \|u_{N,M}(\mu) - u(\mu)\|_Y, \tag{2.11}$$

$$\text{projection error:}\ \min_{\nu\in\mathbb{R}^N} \left\| \sum_{n=1}^N \nu_n \zeta_n - u(\mu) \right\|_Y . \tag{2.12}$$

The procedure is repeated until $\max_{\mu\in\Xi_{\text{Test}}} \varepsilon^u_{N,M}(\mu)$ is below a tolerance we desire. In essence, this strategy ensure "maximally independent" snapshots and hence a rapidly convergent reduced basis approximation. This, in conjunction with our orthogonalization procedure, also guarantees a well-conditioned reduced basis discrete system. This strategy is clearly not very effective in high-dimensional parameter spaces, as it requires solutions of (2.3) at all parameter points in Ξ_{Test}. Note however that if a rigorous *a posteriori* error estimator is available, a more efficient procedure is possible [23, 29].

2.2.2. *Offline-online procedure.* We now demonstrate how the incorporation of the empirical interpolation method into the reduced basis approximation leads to an efficient *online-offline* computational strategy — a strategy where the operation count in the online stage is independent of $\mathcal{N}$ and only dependent on N and M, both of which are much smaller than $\mathcal{N}$. To see this more clearly, we expand our reduced basis approximation and empirical interpolation approximation as

$$u_{N,M}(\mu) = \sum_{j=1}^N u_{N,Mj}(\mu)\zeta_j, \quad g_M^{u_{N,M}}(x;\mu) = \sum_{m=1}^M \varphi_{Mm}(\mu) q_m(x). \tag{2.13}$$

Inserting these representations into (2.9) yields

$$\mu \sum_{j=1}^N A^N_{ij} u_{N,Mj}(\mu) + \sum_{m=1}^M C^{N,M}_{im} \varphi_{Mm}(\mu) = F_{Ni}, \quad 1 \le i \le N; \tag{2.14}$$

where $A^N \in \mathbb{R}^{N\times N}$, $C^{N,M} \in \mathbb{R}^{N\times M}$, $F_N \in \mathbb{R}^N$ are given by $A^N_{ij} = a_0(\zeta_j, \zeta_i)$, $1 \le i$, $j \le N$, $C^{N,M}_{im} = \int_\Omega q_m \zeta_i$, $1 \le i \le N$, $1 \le m \le M$, and $F_{Ni} = f(\zeta_i)$, $1 \le i \le N$, respectively. Furthermore, $\varphi_M(\mu) \in \mathbb{R}^M$ is given by

$$\begin{aligned} \sum_{k=1}^M B^M_{mk} \varphi_{Mk}(\mu) &= g(u_{N,M}(t_m;\mu); t_m; \mu), \quad 1 \le m \le M \\ &= g\left(\sum_{n=1}^N u_{N,Mn}(\mu) \zeta_n(t_m); t_m; \mu \right), \quad 1 \le m \le M. \end{aligned} \tag{2.15}$$

We then substitute $\varphi_M(\mu)$ from (2.15) into (2.14) to obtain the following nonlinear algebraic system

$$\mu \sum_{j=1}^N A^N_{ij} u_{N,Mj}(\mu) + \sum_{m=1}^M D^{N,M}_{im} g\left(\sum_{n=1}^N \zeta_n(t_m) u_{N,Mn}(\mu); t_m; \mu \right) = F_{Ni}, \quad 1 \le i \le N, \tag{2.16}$$

where $D^{N,M} = C^{N,M}(B^M)^{-1} \in \mathbb{R}^{N\times M}$.

To solve (2.16) for $u_{N,Mj}(\mu)$, $1 \le j \le N$, we may apply a Newton iterative scheme: given a current iterate $\bar{u}_{N,Mj}(\mu)$, $1 \le j \le N$, we must find an increment $\delta u_{N,Mj}$, $1 \le j \le N$, such that

$$(2.17)\quad \sum_{j=1}^{N}(\mu A_{ij}^N + \overline{E}_{ij}^N)\delta u_{N,Mj}(\mu) = F_{Ni} - \mu\sum_{j=1}^{N} A_{ij}^N \bar{u}_{N,Mj}(\mu)$$
$$- \sum_{m=1}^{M} D_{im}^{N,M} g\left(\sum_{n=1}^{N}\zeta_n(t_m)\bar{u}_{N,Mn}(\mu); t_m; \mu\right), \quad 1 \le i \le N;$$

here $\overline{E}^N \in \mathbb{R}^{N\times N}$ must be calculated at every Newton iteration as

$$(2.18)\quad \overline{E}_{ij}^N = \sum_{m=1}^{M} D_{im}^{N,M} g'\left(\sum_{n=1}^{N}\zeta_n(t_m)\bar{u}_{N,Mn}(\mu); t_m; \mu\right)\zeta_j(t_m), \quad 1 \le i,\ j \le N,$$

where $g'(w;t;\mu)$ is the first derivative of g with respect to w. Finally, the output can be evaluated as

$$(2.19)\qquad s_{N,M}(\mu) = \sum_{j=1}^{N} u_{N,Mj}(\mu) L_{Nj},$$

where $L_N \in \mathbb{R}^N$ is the output vector with entries $L_{Nj} = \ell(\zeta_j)$, $1 \le j \le N$. We observe that we can now develop an efficient *offline-online* procedure for the rapid evaluation of $s_{N,M}(\mu)$ for each μ in $\mathcal{D}$.

In the offline stage — performed once — we generate nested reduced basis spaces $W_N^u = \{\zeta_1, \ldots, \zeta_N\}$, $1 \le N \le N_{\max}$, nested approximation spaces $W_M^g = \{q_1, \ldots, q_M\}$, $1 \le M \le M_{\max}$, and nested sets of interpolation points $T_M = \{t_1, \ldots, t_M\}$; we then form and store A^N, B^M, $D^{N,M}$, and F_N. In the online stage — performed many times for each new μ — we solve (2.17) for $u_{N,Mj}(\mu)$, $1 \le j \le N$.

The operation count of the online stage is essentially the predominant Newton update component: at each Newton iteration, we first assemble the right-hand side and compute $\overline{E}^N$ at cost $O(MN^2)$ — note we perform the sum in the parenthesis of (2.18) before performing the outer sum; we then form and invert the left-hand side (Jacobian) at cost $O(N^3)$. The online complexity depends only on N, M and number of Newton iterations; we thus recover online $\mathcal{N}$ independence.

2.3. A simple example. We consider a particular instantiation of our abstract statement in which

$$(2.20)\qquad a_0(w,v) = \int_\Omega \nabla w \cdot \nabla v, \quad f(v) = \int_\Omega v, \quad g(w) = |w|^{2/3} w,$$

and $\ell = f$; here $\Omega =]0,1[\in \mathbb{R}$, w and $v \in Y \equiv H_0^1(\Omega)$, and $\mu \in \mathcal{D} \equiv [0.001, 1]$. Our model problem is clearly well-posed. In addition, small μ results in boundary layer solutions as shown in Figure 1.

We now present numerical results for the reduced basis approximation. To begin, we introduce a parameter sample $\Xi_{\text{Test}} \subset \mathcal{D}$ of size 100 and define $\varepsilon_{N,M}^s = \max_{\mu\in\Xi_{\text{Test}}} |s(\mu) - s_{N,M}(\mu)|/|s(\mu)|$. We show in Figure 2 convergence of $\varepsilon_{N,M}^s$ with respect to N and M. We observe that the reduced basis approximation converges very rapidly. In addition, the quality of our reduced basis approximation depends on N and M in a strongly coupled manner: for a fixed value of M, the error

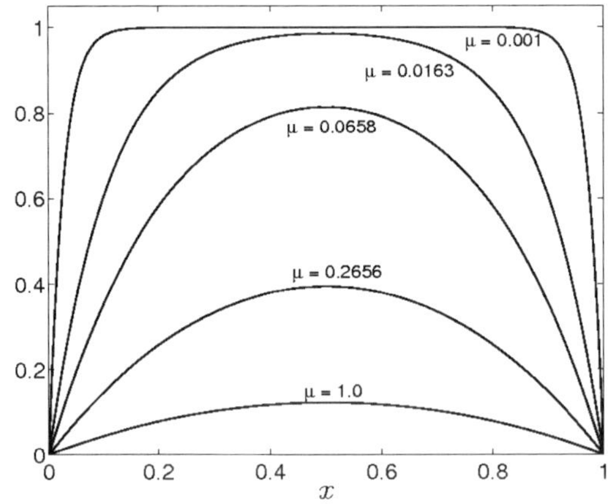

FIGURE 1. Solution $u(\mu)$ for several values of μ.

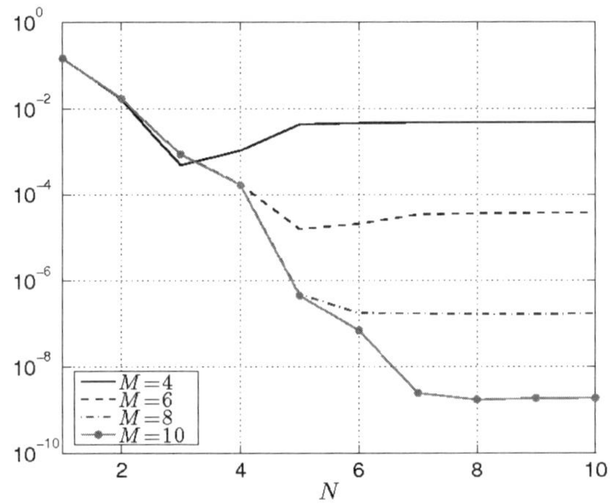

FIGURE 2. Convergence of the maximum relative error $\varepsilon^s_{N,M}$ as a function of N and M.

decreases monotonically with N for $N \leq N_M$, where N_M is such that there is no appreciable change in $\varepsilon^s_{N,M}$ for $N > N_M$. However, when M is increased, the achievable $\varepsilon^s_{N,M}$ decreases further and N_M increases; this strongly suggests that the reduced basis error may obviously be degraded by a poor initial choice of M; we can always choose M large enough that the incurred error does not affect our desired accuracy, for example through the use of *a posteriori* error estimator.

3. Hydrogen Molecule

In this section, we use the reduced basis method to rapidly determine the ground state energy of a hydrogen molecule. We then demonstrate the computational savings afforded by our reduced basis approximation in a molecular dynamics (MD) simulation of the hydrogen molecule. Although the hydrogen molecule is the simplest multi-electron molecule, numerical calculation of its properties was the first demonstration of the power of the Schrödinger equation in quantum chemistry.

3.1. Problem description.

3.1.1. *Exact statement.* We consider the interaction of an hydrogen molecular system consisting of a pair of electrons and two nuclei, each with effectively infinite mass and charge $Z = 1$. The positions of the nuclei in the Cartesian coordinates r are denoted by R_i, $i = 1, 2$, those of the electrons are denoted by r_i, $i = 1, 2$, and the internuclear separation is denoted by R. Then, the Hamiltonian of this molecular system reads as

$$\mathbf{H} = -\sum_{i=1}^{2}\left(\frac{\nabla_i^2}{2} + \sum_{j=1}^{2}\frac{Z}{|r_i - R_j|}\right) + \frac{1}{|r_1 - r_2|} + \frac{Z^2}{R}, \tag{3.1}$$

for which the ground state solution is described by a ground-state wavefunction ψ^{e}. By applying the Born–Oppenheimer approximation and the restricted Hartree–Fock (RHF) theory [6,15,17,18,34], which expresses the wavefunction as the determinant of two single-electron wavefunctions with similar spatial component, we obtain the RHF problem: the ground-state molecular orbital $\mathcal{U}^{\mathrm{e}}$ is given by

$$\mathcal{U}^{\mathrm{e}} = \arg \inf_{w \in H^1(\mathbb{R}^3)} \left\{ E^{\mathrm{e}}_{\mathrm{RHF}}(w), \int_{\mathbb{R}^3} w^2\, dr = 1 \right\}, \tag{3.2}$$

where the energy $E^{\mathrm{e}}_{\mathrm{RHF}}(w)$ is given by

$$\begin{aligned} E^{\mathrm{e}}_{\mathrm{RHF}}(w) = \frac{1}{2}\int_{\mathbb{R}^3} |\nabla w|^2\, dr - \int_{\mathbb{R}^3} \frac{w^2}{|r - R_1|}\, dr - \int_{\mathbb{R}^3} \frac{w^2}{|r - R_2|}\, dr \\ + \frac{1}{2}\int_{\mathbb{R}^3}\int_{\mathbb{R}^3} \frac{w^2(r) w^2(r')}{|r - r'|}\, dr'\, dr. \end{aligned} \tag{3.3}$$

Recall that $H^1(\mathbb{R}^3)$ is the Hilbert space in $\mathbb{R}^3$. For any $w \in H^1(\mathbb{R}^3)$, let

$$\Phi(w) = \int_{\mathbb{R}^3} \frac{w^2(r')}{|r - r'|}\, dr', \tag{3.4}$$

it thus follows that $\Phi(w)$ is the solution of the Poisson problem

$$-\nabla^2 \Phi(w) = 4\pi w^2 \quad \text{in} \quad \mathbb{R}^3. \tag{3.5}$$

We shall exploit this potential in what follows.

To further simplify (3.2), we exploit the symmetry of the solution $\mathcal{U}^{\mathrm{e}}$ and recast the problem into an infinite half-plane $\mathbb{R}^2_+$; we denote the axial and radial coordinates as y_1 and y_2, respectively. Let the origin of the coordinate system be the midpoint between the two nuclei; the positions of the two nuclei are thus $(-\mu/2, 0)$ and $(\mu/2, 0)$. Here, $\mu \equiv R$, the internuclear separation, is the only parameter of interest, and varies in the range $\mathcal{D} \equiv [0.5, 6]$. By introducing $u^{\mathrm{e}}_\infty(y) = \sqrt{2\pi}\mathcal{U}^{\mathrm{e}}(y)$, $\forall y = (y_1, y_2) \in \mathbb{R}^2_+$, we can rewrite the above RHF problem as

$$u^{\mathrm{e}}_\infty(y; \mu) = \arg \inf_{w \in H^1_0(\mathbb{R}^2_+)} \left\{ E^{\mathrm{e}}_\infty(w; \mu), \int_{\mathbb{R}^2_+} w^2 y_2 = 1 \right\}. \tag{3.6}$$

Here the energy functional $E^{\mathrm{e}}_\infty(w; \mu)$ is given by

$$E^{\mathrm{e}}_\infty(w; \mu) = \frac{1}{2}\int_{\mathbb{R}^2_+} |\nabla w|^2 y_2 - \int_{\mathbb{R}^2_+} g_{\mathrm{o}}(y; \mu) w^2 + \frac{1}{2}\int_{\mathbb{R}^2_+} \phi^{\mathrm{e}}_\infty w^2 y_2, \tag{3.7}$$

where

$$g_o(y;\mu) = \frac{y_2}{\sqrt{(y_1+\mu/2)^2+y_2^2}} + \frac{y_2}{\sqrt{(y_1-\mu/2)^2+y_2^2}}, \tag{3.8}$$

$$-\nabla^2\phi^{\mathrm{e}}_\infty = 2w^2 \quad \text{in} \quad \mathbb{R}^2_+. \tag{3.9}$$

The boundary conditions for both u^{e}_∞ and ϕ^{e}_∞ are homogeneous Neumman on $y_2 = 0$ and homogeneous Dirichlet at infinity.

We next truncate the infinite half-plane domain to a large parameter-dependent computational domain $\Omega_o(\mu) \equiv]-25-\mu/2, 25+\mu/2[\times]0,25[$ as shown in Figure 3(a); since the internuclear separation ($\mu \le 6$) is less than $\frac{1}{8}$ of the longitudinal span of the computational domain (> 50), Ω_o is sufficiently large that the truncation will not adversely affect our solution. As a result, the RHF problem for the hydrogen molecular system is being approximated by

$$u^{\mathrm{e}}_o(y;\mu) = \arg \inf_{w\in H^1_0(\Omega_o(\mu))} \left\{ E^{\mathrm{e}}_o(w;\mu), \int_{\Omega_o(\mu)} w^2 y_2 = 1 \right\}, \tag{3.10}$$

where

$$E^{\mathrm{e}}_o(w;\mu) = \frac{1}{2}\int_{\Omega_o(\mu)} |\nabla w|^2 y_2 - \int_{\Omega_o(\mu)} g_o(y;\mu) w^2 + \frac{1}{2}\int_{\Omega_o(\mu)} \phi^{\mathrm{e}}_o w^2 y_2, \tag{3.11}$$

$$-\nabla^2\phi^{\mathrm{e}}_o = 2w^2 \quad \text{in} \quad \Omega_o(\mu). \tag{3.12}$$

The outputs of interest are $s^{\mathrm{e}}_o(\mu) = E^{\mathrm{e}}_o(u^{\mathrm{e}}_o(\mu);\mu)$ and $\partial s^{\mathrm{e}}_o(\mu)/\partial\mu = \partial E^{\mathrm{e}}_o(u^{\mathrm{e}}_o(\mu);\mu)/\partial\mu$ from which we may calculate the ground state energy of the hydrogen molecular system $\mathcal{E}^{\mathrm{e}}_{\mathrm{oH}_2}(\mu) = 2s^{\mathrm{e}}_o(\mu) + 1/\mu$ and the force exerted on the nuclei, $F^{\mathrm{e}}_{\mathrm{oH}_2}(\mu) = 2\partial s^{\mathrm{e}}_o(\mu)/\partial\mu - 1/\mu^2$. The boundary conditions for both u^{e}_o and ϕ^{e}_o are homogeneous Neumman on $y_2 = 0$ and homogeneous Dirichlet on the rest of the boundary.

3.1.2. *Mapped formulation on reference domain.* We next introduce a reference domain $\Omega = \Omega_o(\mu = 3) =]-26.5, 26.5[\times]0,25[$ as shown in Figure 3(b). We decompose $\Omega_o(\mu)$ and Ω as

$$\overline{\Omega}_o(\mu) = \overline{\Omega}^1_o(\mu) \cup \overline{\Omega}^2_o(\mu) \cup \overline{\Omega}^3_o(\mu), \quad \overline{\Omega} = \overline{\Omega}^1 \cup \overline{\Omega}^2 \cup \overline{\Omega}^3, \tag{3.13}$$

where $\Omega^1_o(\mu) \equiv]-25-\mu/2, -\mu/2[\times]0,25[$, $\Omega^2_o(\mu) \equiv]-\mu/2, \mu/2[\times]0,25[$, $\Omega^3_o(\mu) \equiv]\mu/2, 25+\mu/2[\times]0,20[$, $\Omega^1 \equiv]-26.5,-1.5[\times]0,25[$, $\Omega^2 \equiv]-1.5,1.5[\times]0,25[$, $\Omega^3 \equiv]1.5, 26.5[\times]0,25[$. We now consider a piecewise affine mapping $\mathcal{F}$ from Ω to $\Omega_o(\mu)$: the mapping is $(y_1,y_2) = (x_1 + 1.5 - \mu/2, x_2)$ from Ω^1 to $\Omega^1_o(\mu)$; the mapping is $(y_1,y_2) = (\mu/3 x_1, x_2)$ from Ω^2 to $\Omega^2_o(\mu)$; the mapping is $(y_1,y_2) = (x_1 - 1.5 + \mu/2, x_2)$ from Ω^3 to $\Omega^3_o(\mu)$.

Our exact solution on the original domain, $u^{\mathrm{e}}_o(y;\mu)$, can then be expressed in terms of the solution on the mapped domain, $u^{\mathrm{e}}(x;\mu)$, as $u^{\mathrm{e}}_o(y;\mu) = u^{\mathrm{e}}(\mathcal{F}^{-1}(y);\mu)$. The solution on the mapped domain, $u^{\mathrm{e}}(x;\mu) \in Y^{\mathrm{e}} \equiv H^1_0(\Omega)$ satisfies

$$u^{\mathrm{e}}(x;\mu) = \arg \inf_{w\in Y^{\mathrm{e}}} \{E^{\mathrm{e}}(w;\mu), m(w,w;\mu) = 1\}, \tag{3.14}$$

where

$$E^{\mathrm{e}}(w;\mu) = \tfrac{1}{2}a_0(w,w;\mu) - a\big(w,w;g(x;\mu)\big) + \tfrac{1}{2}b(w,w,\phi^{\mathrm{e}}(w);\mu), \tag{3.15}$$

$$a_0(\phi^{\mathrm{e}}(w), v;\mu) = 2b(w,w,v;\mu), \quad \forall v \in Y^{\mathrm{e}}. \tag{3.16}$$

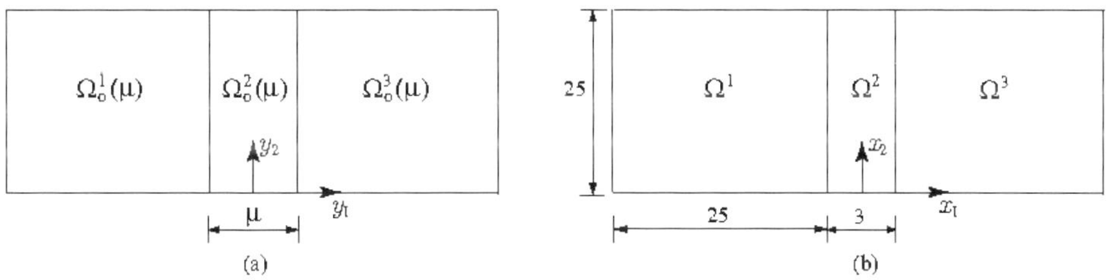

FIGURE 3. Truncated computational domains: (a) original domain, and (b) reference domain.

Here the forms are given by

$$a_0(w,v;\mu) = \int_{\Omega^1} \nabla w \nabla v x_2 + \frac{3}{\mu}\int_{\Omega^2} \frac{\partial w}{\partial x_1}\frac{\partial v}{\partial x_1} x_2 + \frac{\mu}{3}\int_{\Omega^2} \frac{\partial w}{\partial x_2}\frac{\partial v}{\partial x_2} x_2 \tag{3.17}$$

$$+ \int_{\Omega^3} \nabla w \nabla v x_2 \tag{3.18}$$

$$m(w,v;\mu) = \int_{\Omega^1} wvx_2 + \frac{\mu}{3}\int_{\Omega^2} wvx_2 + \int_{\Omega^3} wvx_2 \tag{3.19}$$

$$b(w,v,z;\mu) = \int_{\Omega^1} wvzx_2 + \frac{\mu}{3}\int_{\Omega^2} wvzx_2 + \int_{\Omega^3} wvzx_2 \tag{3.20}$$

$$a\big(w,v;g(x;\mu)\big) = a_1\big(w,v;g_1(x;\mu)\big) + a_2\big(w,v;g_2(x;\mu)\big) + a_3\big(w,v;g_3(x;\mu)\big) \tag{3.21}$$

where

$$a_1(w,v;g_1(x;\mu)) = \int_{\Omega^1} g_1(x;\mu)wv,$$
$$a_2(w,v;g_2(x;\mu)) = \int_{\Omega^2} g_2(x;\mu)wv,$$
$$a_3(w,v;g_3(x;\mu)) = \int_{\Omega^3} g_3(x;\mu)wv,$$

and

$$g(x;\mu) = \begin{cases} g_1(x;\mu) = \dfrac{x_2}{\sqrt{(x_1+1.5)^2+x_2^2}} + \dfrac{x_2}{\sqrt{(x_1+1.5-\mu)^2+x_2^2}}, & x\in\Omega_1, \\ g_2(x;\mu) = \dfrac{\mu}{3}\dfrac{x_2}{\sqrt{(\mu/3x_1+\mu/2)^2+x_2^2}} + \dfrac{\mu}{3}\dfrac{x_2}{\sqrt{(\mu/3x_1-\mu/2)^2+x_2^2}}, & x\in\Omega_2, \\ g_3(x;\mu) = \dfrac{x_2}{\sqrt{(x_1-1.5+\mu)^2+x_2^2}} + \dfrac{x_2}{\sqrt{(x_1-1.5)^2+x_2^2}}, & x\in\Omega_3. \end{cases}$$

Our outputs of interest are then evaluated as $s^e(\mu) = E^e(u^e(\mu);\mu)$ and $\partial s^e(\mu)/\partial\mu = \partial E^e(u^e(\mu);\mu)/\partial\mu$. In this reference domain, the ground state energy of the hydrogen molecular system is then given by $\mathcal{E}^e_{H_2}(\mu) = 2s^e(\mu) + 1/\mu$ and the force exerted on the nuclei is given by $F^e_{H_2}(\mu) = 2\partial s^e(\mu)/\partial\mu - 1/\mu^2$.

3.1.3. *Truth Approximation.* We now introduce the "truth" approximations $u(\mu) \in Y$ and $\phi(\mu) \in Y$, where $Y \subset Y^e$ is a truth linear finite element approximation space of dimension $\mathcal{N}$. By application of the Euler–Lagrange formulation, we obtain a generalized nonlinear eigenvalue system for $\big(u(\mu), \phi(\mu), \lambda(\mu)\big) \in Y \times Y \times \mathbb{R}$ given by

$$\tfrac{1}{2}a_0(u(\mu), v; \mu) - a\big(u(\mu), v; g(x;\mu)\big) + b(\phi(\mu), u(\mu), v; \mu) = \lambda(\mu) m(u(\mu), v; \mu), \quad \forall v \in Y, \tag{3.22}$$

$$\tfrac{1}{2}a_0(\phi(\mu), v; \mu) = b(u(\mu), u(\mu), v; \mu), \quad \forall v \in Y, \tag{3.23}$$

$$m(u(\mu), u(\mu); \mu) = 1. \tag{3.24}$$

Here $\lambda(\mu)$ is the Lagrange multiplier. It then follows that

$$s(\mu) = \lambda(\mu) - \tfrac{1}{2}b(\phi(\mu), u(\mu), u(\mu); \mu),$$

and after simple algebraic manipulations, we obtain

$$\frac{\partial s(\mu)}{\partial \mu} = \tfrac{1}{2}a_0'(u(\mu), u(\mu); \mu) - a'\big(u(\mu), u(\mu); g(x;\mu)\big) + b'(\phi(\mu), u(\mu), u(\mu); \mu) - \lambda(\mu) m'(u(\mu), u(\mu); \mu) - \tfrac{1}{4}a_0'(\phi(\mu), \phi(\mu); \mu); \tag{3.25}$$

where the prime denotes the partial derivative of the forms with respect to the parameter μ. The ground state energy of the hydrogen molecular system is then given by $\mathcal{E}_{\mathrm{H}_2}(\mu) = 2s(\mu) + 1/\mu$ and the force exerted on the nuclei is given by $F_{\mathrm{H}_2}(\mu) = 2\partial s(\mu)/\partial\mu - 1/\mu^2$.

The Newton's method is then used to solve the above system: given a current iterate $\big(\bar{u}(\mu), \bar{\phi}(\mu), \bar{\lambda}(\mu)\big) \in Y \times Y \times \mathbb{R}$, we find an increment $(\delta u(\mu), \delta\phi(\mu), \delta\lambda(\mu)) \in Y \times Y \times \mathbb{R}$ such that

$$\begin{aligned}
&\tfrac{1}{2}a_0(\delta u(\mu), v; \mu) - a\big(\delta u(\mu), v; g(x;\mu)\big) + b(\bar{\phi}(\mu), \delta u(\mu), v; \mu) - \bar{\lambda}(\mu) m(\delta u(\mu), v; \mu) \\
&\quad + b(\delta\phi(\mu), \bar{u}(\mu), v; \mu) - \delta\lambda(\mu) m(\bar{u}(\mu), v; \mu) \\
&\qquad = \bar{\lambda}(\mu) m(\bar{u}(\mu), v; \mu) - \tfrac{1}{2}a_0(\bar{u}(\mu), v; \mu) + a\big(\bar{u}(\mu), v; g(x;\mu)\big) \\
&\qquad\quad - b(\bar{\phi}(\mu), \bar{u}(\mu), v; \mu), \quad \forall v \in Y,
\end{aligned}$$

$$b(\delta u(\mu), \bar{u}(\mu), v; \mu) - \tfrac{1}{4}a_0(\delta\phi(\mu), v; \mu) = \tfrac{1}{4}a_0(\bar{\phi}(\mu), v; \mu) - \tfrac{1}{2}b(\bar{u}(\mu), \bar{u}(\mu), v; \mu), \ \forall v \in Y,$$

$$m(\delta u(\mu), \bar{u}(\mu); \mu) = \tfrac{1}{2}\big(1 - m(\bar{u}(\mu), \bar{u}(\mu); \mu)\big).$$

To accelerate the convergence, the already computed solution at a particular parameter $\mu = \mu_{\mathrm{old}}$ is taken as an initial guess when solving the system for $\mu = \mu_{\mathrm{new}}$.

3.2. Reduced basis formulation. We realize that the problem is nonlinear in $u(\mu)$ and $\phi(\mu)$, and non-affine in μ. Since the nonlinearity is only quadratic, the nonlinear terms can be efficiently treated by the usual reduced basis formulation [13, 23, 35]. To achieve an economical reduced basis approximation of $\mathcal{N}$*-independent* online complexity, however, the empirical interpolation method is needed to deal with the non-affine terms.

3.2.1. *Discrete equations.* To begin, we introduce two samples $S^u_{N^u} = \{\mu^u_1, \ldots, \mu^u_{N^u}\}$ and $S^\phi_{N^\phi} = \{\mu^\phi_1, \ldots, \mu^\phi_{N^\phi}\}$ and two associated reduced basis approximation spaces $W^u_{N^u} = \mathrm{span}\{\zeta_j \equiv u(\mu^u_j), 1 \le j \le N^u\}$ and $W^\phi_{N^\phi} = \mathrm{span}\{\varsigma_j \equiv \phi(\mu^\phi_j), 1 \le j \le N^\phi\}$. We next construct three samples $S^{g_1}_{M^{g_1}} = \{\mu^{g_1}_1, \ldots, \mu^{g_1}_{M^{g_1}}\}$, $S^{g_2}_{M^{g_2}} = \{\mu^{g_2}_1, \ldots, \mu^{g_2}_{M^{g_2}}\}$, and $S^{g_3}_{M^{g_3}} = \{\mu^{g_3}_1, \ldots, \mu^{g_3}_{M^{g_3}}\}$; associated approximation spaces $W^{g_1}_{M^{g_1}} = \mathrm{span}\{g_1(x;\mu^{g_1}_j), 1 \le j \le M^{g_1}\} = \mathrm{span}\{q^{g_1}_1, \ldots, q^{g_1}_{M^{g_1}}\}$, $W^{g_2}_{M^{g_2}} = \mathrm{span}\{g_2(x;\mu^{g_2}_j), 1 \le j \le M^{g_2}\} = \mathrm{span}\{q^{g_2}_1, \ldots, q^{g_2}_{M^{g_2}}\}$, and $W^{g_3}_{M^{g_3}} = \mathrm{span}\{g_3(x;\mu^{g_3}_j), 1 \le j \le M^{g_3}\} = \mathrm{span}\{q^{g_3}_1, \ldots, q^{g_3}_{M^{g_3}}\}$; and three sets of interpolations points

$T^{g_1}_{M^{g_1}} = \{t^{g_1}_1, \ldots, t^{g_1}_{M^{g_1}}\}$, $T^{g_2}_{M^{g_2}} = \{t^{g_2}_1, \ldots, t^{g_2}_{M^{g_2}}\}$, $T^{g_3}_{M^{g_3}} = \{t^{g_3}_1, \ldots, t^{g_3}_{M^{g_3}}\}$ following the empirical interpolation procedure outlined in Section 2.2.1. For simplicity of exposition, throughout this section we assume that $N^\phi = N^u \equiv N$ and $M^{g_1} = M^{g_2} = M^{g_3} \equiv M$.

By applying a standard Galerkin projection and replacing the non-affine functions $g_1(x;\mu)$, $g_2(x;\mu)$, and $g_3(x;\mu)$ with our coefficient function approximations $g_{1M}(x;\mu)$, $g_{2M}(x;\mu)$, and $g_{3M}(x;\mu)$, we obtain the reduced basis formulation: given $\mu \in \mathcal{D}$, we evaluate

$$s_{N,M}(\mu) = \lambda_{N,M}(\mu) - \tfrac{1}{2} b(\phi_{N,M}(\mu), u_{N,M}(\mu), u_{N,M}(\mu); \mu)\ , \tag{3.26}$$

$$\begin{aligned}\frac{\partial s_{N,M}(\mu)}{\partial \mu} = {} & \tfrac{1}{2} a_0'(u_{N,M}(\mu), u_{N,M}(\mu); \mu) \\ & - a_1'\big(u_{N,M}(\mu), u_{N,M}(\mu); g_{1M}(x;\mu)\big) \\ & - a_2'\big(u_{N,M}(\mu), u_{N,M}(\mu); g_{2M}(x;\mu)\big) \\ & - a_3'\big(u_{N,M}(\mu), u_{N,M}(\mu); g_{3M}(x;\mu)\big) \\ & + b'(\phi_{N,M}(\mu), u_{N,M}(\mu), u_{N,M}(\mu); \mu) \\ & - \lambda(\mu) m'(u_{N,M}(\mu), u_{N,M}(\mu); \mu) \\ & - \tfrac{1}{4} a_0'(\phi_{N,M}(\mu), \phi_{N,M}(\mu); \mu)\ ;\end{aligned} \tag{3.27}$$

where $\big(u_{N,M}(\mu), \phi_{N,M}(\mu), \lambda_{N,M}(\mu)\big) \in W^u_N \times W^\phi_N \times \mathbb{R}$ is the solution of

$$\begin{aligned}\tfrac{1}{2} a_0(u_{N,M}(\mu), v; \mu) \quad & a_1\big(u_{N,M}(\mu), v; g_{1M}(x;\mu)\big) \\ & - a_2\big(u_{N,M}(\mu), v; g_{2M}(x;\mu)\big) \\ & - a_3\big(u_{N,M}(\mu), v; g_{3M}(x;\mu)\big) \\ & + b(\phi_{N,M}(\mu), u_{N,M}(\mu), v; \mu) \\ & \qquad = \lambda_{N,M}(\mu) m(u_{N,M}(\mu), v; \mu), \quad \forall v \in W^u_N,\end{aligned} \tag{3.28}$$

$$\tfrac{1}{2} a_0(\phi_{N,M}(\mu), v; \mu) = b(u_{N,M}(\mu), u_{N,M}(\mu), v; \mu), \quad \forall v \in W^\phi_N, \tag{3.29}$$

$$m(u_{N,M}(\mu), u_{N,M}(\mu); \mu) = 1. \tag{3.30}$$

Here $g_{1M}(x;\mu)$, $g_{2M}(x;\mu)$, and $g_{3M}(x;\mu)$ are given by

$$\begin{aligned} g_{1M}(x;\mu) &= \sum_{m=1}^{M} \beta^{g_1}_m(\mu) q^{g_1}_m(x), \quad g_{2M}(x;\mu) = \sum_{m=1}^{M} \beta^{g_2}_m(\mu) q^{g_2}_m(x), \\ g_{3M}(x;\mu) &= \sum_{m=1}^{M} \beta^{g_3}_m q^{g_3}_m(x), \end{aligned} \tag{3.31}$$

where, for $j = 1, \ldots, M$,

$$\begin{aligned} &\sum_{m=1}^{M} q^{g_1}_m(t^{g_1}_j)\beta^{g_1}_m(\mu) = g_1(t^{g_1}_j;\mu), \quad \sum_{m=1}^{M} q^{g_2}_m(t^{g_2}_j)\beta^{g_2}_m(\mu) = g_2(t^{g_2}_j;\mu), \\ &\sum_{m=1}^{M} q^{g_3}_m(t^{g_3}_j)\beta^{g_3}_m(\mu) = g_3(t^{g_3}_j;\mu). \end{aligned} \tag{3.32}$$

The reduced basis approximation to the ground state energy of the hydrogen molecular system is then given by $\mathcal{E}_{N,M\mathrm{H}_2}(\mu) = 2s_{N,M}(\mu) + 1/\mu$ and the reduced basis approximation to the force exerted on the nuclei is given by $F_{N,M\mathrm{H}_2}(\mu) =$

$2\partial s_{N,M}(\mu)/\partial\mu - 1/\mu^2$. It remains to address the computational complexity of the reduced basis approximation.

3.2.2. *Offline-online procedure.* We first expand our reduced-basis approximations as

$$(3.33)\qquad u_{N,M}(\mu) = \sum_{j=1}^{N} u_{N,Mj}(\mu)\zeta_j, \quad \phi_{N,M}(\mu) = \sum_{j=1}^{N} \phi_{N,Mj}(\mu)\varsigma_j.$$

Inserting these representations and (3.31) into (3.28)–(3.30) yields

$$\begin{aligned}
\frac{1}{2}\sum_{j=1}^{N}\Bigg\{a_0(\zeta_j,\zeta_i;\mu) - \sum_{m=1}^{M}\big(&\beta_m^{g_1}(\mu)a_1(\zeta_j,\zeta_i;q_m^{g_1})\\
&+\beta_m^{g_2}(\mu)a_2(\zeta_j,\zeta_i;q_m^{g_2}) + \beta_m^{g_3}(\mu)a_3(\zeta_j,\zeta_i;q_m^{g_3})\big)\Bigg\}u_{N,Mj}(\mu)\\
&+\sum_{j=1}^{N}\sum_{k=1}^{N} b(\varsigma_j,\zeta_k,\zeta_i;\mu)\phi_{N,Mj}(\mu)u_{N,Mk}(\mu)\\
&\qquad = \lambda_{N,M}(\mu)\sum_{j=1}^{N} m(\zeta_j,\zeta_i;\mu)u_{N,Mj}(\mu), \quad 1 \le i \le N,\\
\frac{1}{2}\sum_{j=1}^{N} a_0(\varsigma_j,\varsigma_i;\mu)\phi_{N,Mj}(\mu) &= \sum_{j=1}^{N}\sum_{k=1}^{N} b(\zeta_j,\zeta_k,\varsigma_i;\mu)u_{N,Mj}(\mu)u_{N,Mk}(\mu), \quad 1 \le i \le N,\\
\sum_{j=1}^{N}\sum_{k=1}^{N} m(\zeta_j,\zeta_k;\mu)&u_{N,Mj}(\mu)u_{N,Mk}(\mu) = 1.
\end{aligned}$$

In a similar fashion as described in Section 2.2.2, this algebraic nonlinear system of $2N+1$ equations can be readily solved by the Newton's method for $u_{N,Mj}(\mu)$, $\phi_{N,Mj}(\mu)$, $1 \le j \le N$, and $\lambda_{N,M}(\mu)$.

The reduced basis outputs can then be calculated as

$$\begin{aligned}
(3.34)\quad s_{N,M}(\mu) &= \lambda_{N,M}(\mu)\\
&\quad - \frac{1}{2}\sum_{k=1}^{N}\sum_{j=1}^{N}\sum_{i=1}^{N}\phi_{N,Mk}(\mu)u_{N,Mj}(\mu)u_{N,Mi}(\mu)b(\varsigma_k,\zeta_j,\zeta_i;\mu),
\end{aligned}$$

$$\begin{aligned}
(3.35)\quad \frac{\partial s_{N,M}}{\partial\mu}(\mu) &= \sum_{j=1}^{N}\sum_{i=1}^{N} u_{N,Mj}(\mu)u_{N,Mi}(\mu)\\
&\times\Bigg(\frac{1}{2}a_0'(\zeta_j,\zeta_i;\mu) - \lambda(\mu)m'(\zeta_j,\zeta_i;\mu)\\
&\quad - \sum_{m=1}^{M}\big(\beta_m^{g_1\prime}(\mu)a_1(\zeta_j,\zeta_i;q_m^{g_1}) + \beta_m^{g_2\prime}(\mu)a_2(\zeta_j,\zeta_i;q_m^{g_2}) + \beta_m^{g_3\prime}(\mu)a_3(\zeta_j,\zeta_i;q_m^{g_3})\big)\\
&\qquad + \sum_{k=1}^{N}\phi_{N,Mk}(\mu)b'(\varsigma_k,\zeta_j,\zeta_i;\mu)\Bigg)\\
&- \frac{1}{4}\sum_{j=1}^{N}\sum_{i=1}^{N}\phi_{N,Mj}(\mu)\phi_{N,Mi}(\mu)a_0'(\varsigma_j,\varsigma_i;\mu),
\end{aligned}$$

where $\beta_m^{g_1\prime}(\mu)$, $\beta_m^{g_2\prime}(\mu)$, and $\beta_m^{g_3\prime}(\mu)$ are obtained from

$$\sum_{m=1}^{M} q_m^{g_1}(t_j^{g_1})\beta_m^{g_1\prime}(\mu) = g_1'(t_j^{g_1};\mu), \qquad \sum_{m=1}^{M} q_m^{g_2}(t_j^{g_2})\beta_m^{g_2\prime}(\mu) = g_2'(t_j^{g_2};\mu),$$

$$\sum_{m=1}^{M} q_m^{g_3}(t_j^{g_3})\beta_m^{g_3\prime}(\mu) = g_3'(t_j^{g_3};\mu).$$

Furthermore, we note from (3.17)–(3.21) that

$$a_0(\zeta_j,\zeta_i;\mu) = \int_{\Omega^1\cup\Omega^3} \nabla\zeta_j\nabla\zeta_i x_2 + \frac{3}{\mu}\int_{\Omega^2}\frac{\partial\zeta_j}{\partial x_1}\frac{\partial\zeta_i}{\partial x_1}x_2 + \frac{\mu}{3}\int_{\Omega^2}\frac{\partial\zeta_j}{\partial x_2}\frac{\partial\zeta_i}{\partial x_2}x_2 \tag{3.36}$$

$$a_0(\varsigma_j,\varsigma_i;\mu) = \int_{\Omega^1\cup\Omega^3} \nabla\varsigma_j\nabla\varsigma_i x_2 + \frac{3}{\mu}\int_{\Omega^2}\frac{\partial\varsigma_j}{\partial x_1}\frac{\partial\varsigma_i}{\partial x_1}x_2 + \frac{\mu}{3}\int_{\Omega^2}\frac{\partial\varsigma_j}{\partial x_2}\frac{\partial\varsigma_i}{\partial x_2}x_2 \tag{3.37}$$

$$b(\varsigma_k,\zeta_j,\zeta_i;\mu) = \int_{\Omega^1\cup\Omega^3}\varsigma_k\zeta_j\zeta_i x_2 + \frac{\mu}{3}\int_{\Omega^2}\varsigma_k\zeta_j\zeta_i x_2 \tag{3.38}$$

$$m(\zeta_j,\zeta_i;\mu) = \int_{\Omega^1\cup\Omega^3}\zeta_j\zeta_i x_2 + \frac{\mu}{3}\int_{\Omega^2}\zeta_j\zeta_i x_2 \tag{3.39}$$

$$a_1(\zeta_j,\zeta_i;q_m^{g_1}) = \int_{\Omega^1} q_m^{g_1}\zeta_j\zeta_i \tag{3.40}$$

$$a_2(\zeta_j,\zeta_i;q_m^{g_2}) = \int_{\Omega^2} q_m^{g_2}\zeta_j\zeta_i \tag{3.41}$$

$$a_3(\zeta_j,\zeta_i;q_m^{g_3}) = \int_{\Omega^3} q_m^{g_3}\zeta_j\zeta_i \tag{3.42}$$

$$a_0'(\zeta_j,\zeta_i;\mu) = -\frac{3}{\mu^2}\int_{\Omega^2}\frac{\partial\zeta_j}{\partial x_1}\frac{\partial\zeta_i}{\partial x_1}x_2 + \frac{1}{3}\int_{\Omega^2}\frac{\partial\zeta_j}{\partial x_2}\frac{\partial\zeta_i}{\partial x_2}x_2 \tag{3.43}$$

$$a_0'(\varsigma_j,\varsigma_i;\mu) = -\frac{3}{\mu^2}\int_{\Omega^2}\frac{\partial\varsigma_j}{\partial x_1}\frac{\partial\varsigma_i}{\partial x_1}x_2 + \frac{1}{3}\int_{\Omega^2}\frac{\partial\varsigma_j}{\partial x_2}\frac{\partial\varsigma_i}{\partial x_2}x_2 \tag{3.44}$$

$$b'(\varsigma_k,\zeta_j,\zeta_i;\mu) = \frac{1}{3}\int_{\Omega^2}\varsigma_k\zeta_j\zeta_i x_2 \tag{3.45}$$

$$m'(\zeta_j,\zeta_i;\mu) = \frac{1}{3}\int_{\Omega^2}\zeta_j\zeta_i x_2. \tag{3.46}$$

It should now be clear that we can develop an efficient offline-online procedure for the rapid evaluation of $s_{N,M}(\mu)$ and $\partial s_{N,M}(\mu)/\partial\mu$ for each μ in $\mathcal{D}$.

The operation count of the online stage is essentially the predominant Newton update component: at each Newton iteration, we first assemble the right-hand side and form the Jacobian matrix at a cost of $O(2N^3)$ (for a linear problem the scaling would be $O(2N^2)$ — the quadratic terms introduces another factor of N); we then invert the Jacobian matrix at cost $O\big((2N+1)^3\big)$. Note that we solve for $\beta_m^{g_1}(\mu)$, $\beta_m^{g_2}(\mu)$, $\beta_m^{g_3}(\mu)$, $\beta_m^{g_1\prime}(\mu)$, $\beta_m^{g_2\prime}(\mu)$, $\beta_m^{g_3\prime}(\mu)$ at cost $O(6M^2)$ by appealing to the triangular property of B^M and form the *affine* parameter-dependent quantities and their partial derivatives at cost $O(3MN^2+N^3)$ *before* pursuing the Newton steps. In summary, the operation count of the online stage is $O(6M^2+K(2N+1)^3)$, where

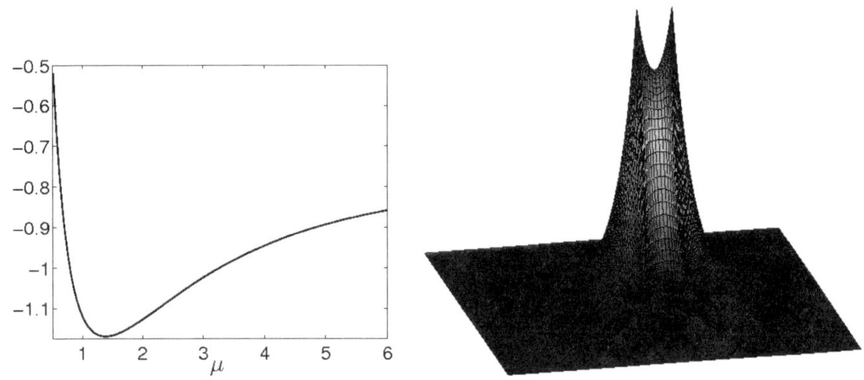

(a) (b)

FIGURE 4. (a) $\mathcal{E}_{\mathrm{H}_2}$ as a function of μ, and (b) $u(\mu)$ at the equilibrium internuclear separation $\mu = 1.4$.

K is the number of Newton iterations. The online complexity is thus *independent* of $\mathcal{N}$.

3.3. Numerical results. We first present in Figure 4(a) the finite element calculation of the ground state energy of the hydrogen molecular system $\mathcal{E}_{\mathrm{H}_2}$ as a function of the internuclear separation μ for the truth linear finite element approximation of dimension $\mathcal{N} = 8601$. The minimum binding energy is -1.1687 and the equilibrium internuclear separation, μ, is 1.4; this is in good agreement with the calculated binding energy of -1.1745 and the calculated equilibrium internuclear separation of 1.4 given in [33]. Figure 4(b) shows the wavefunction $u(\mu)$ at the equilibrium internuclear separation.

We next introduce a parameter sample $\Xi_{\mathrm{Test}} \subset \mathcal{D}$ of size 111 and define

$$\varepsilon^{s}_{N,M} = \max_{\mu \in \Xi_{\mathrm{Test}}} \frac{|s(\mu) - s_{N,M}(\mu)|}{|s(\mu)|},$$

$$\varepsilon^{ds}_{N,M} = \max_{\mu \in \Xi_{\mathrm{Test}}} \frac{|\partial s(\mu)\partial\mu - \partial s_{N,M}(\mu)/\partial\mu|}{|\partial s(\mu)/\partial\mu|},$$

and

$$\varepsilon^{\lambda}_{N,M} = \max_{\mu \in \Xi_{\mathrm{Test}}} \frac{|\lambda(\mu) - \lambda_{N,M}(\mu)|}{|\lambda(\mu)|}.$$

We present in Figure 5 $\varepsilon^{s}_{N,M}$ as a function of N and M. We observe very rapid convergence of the reduced basis approximation. Again, we note that the "plateau" in the curves for M fixed and the "drop" in the $N \to \infty$ asymptotes as M increases: for fixed M the error in our coefficient function approximation $g_M(x;\mu)$ to $g(x;\mu)$ will ultimately dominate at large N; increasing M renders the coefficient function approximation more accurate, which in turn leads to the drops in the asymptotic error. We tabulate in Table 1 $\varepsilon^{s}_{N,M}$, $\varepsilon^{ds}_{N,M}$, and $\varepsilon^{\lambda}_{N,M}$ for different values of N and M. We observe that the reduced basis outputs converge very rapidly and that, as expected, $s_{N,M}(\mu)$ and $\lambda_{N,M}(\mu)$ converge faster than $\partial s_{N,M}/\partial\mu(\mu)$. Indeed, it is

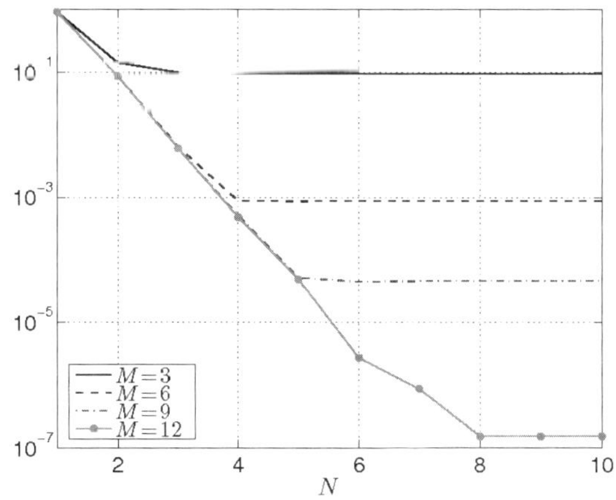

FIGURE 5. Convergence of the maximum error $\varepsilon^s_{N,M}$ as a function of N and M.

TABLE 1. Maximum relative errors for different values of (N, M) for the hydrogen molecule example.

N	M	$\varepsilon^s_{N,M}$	$\varepsilon^{ds}_{N,M}$	$\varepsilon^{\lambda}_{N,M}$
2	2	1.17 E -01	1.18 E -00	1.72 E -01
4	4	3.94 E -02	5.86 E -01	3.88 E -02
6	6	8.76 E -04	4.29 E -02	1.11 E -03
8	8	6.37 E -05	6.98 E -03	6.17 E -05
10	10	8.30 E -06	1.07 E -03	1.07 E -05
12	12	1.54 E -07	3.96 E -06	4.75 E -07

clear from Figure 5 that we can further reduce the errors in Table 1 by using larger M.

We now present in Table 2 the online computational times to calculate $\partial s_{N,M}/\partial\mu(\mu)$ as a function of N and M; the values are normalized with respect to the computational time for the direct calculation of the truth approximation output $\partial s/\partial\mu(\mu)$. We achieve significant computational savings: for a relative accuracy of close to 0.1 percent (corresponding to $N = M = 10$ in Table 1) in the output, the online saving is more than a factor of 5000 relative to the Finite Element Method. Of course, when competing with more efficient quantum chemistry methods particularly developed for solution of the RHF problems, the computational savings will be not really that great. Nevertheless, the reduced basis approach does provide an attractive alternative to the existing methods, in particular for problems with moving nuclei (this section) and problems that are periodic (the solid state models considered in the next section).

Finally, we apply our reduced basis approximation to the real-time molecular dynamics (MD) simulation of the hydrogen molecular system by using the velocity Verlet time integration scheme [10, 17, 19]. We present the time evolution of the internuclear separation and velocity in Figure 6 for small initial distance $R_0 = 1.0$

TABLE 2. Online computational times (normalized with respect to the time to solve for $\partial s(\mu)/\partial\mu$) for the hydrogen molecule example.

N	M	Online time for $\partial s_{N,M}(\mu)/\partial\mu$	(Online) time for $\partial s(\mu)/\partial\mu$
2	2	1.26 E –05	1
4	4	3.98 E –05	1
6	6	8.07 E –05	1
8	8	1.18 E –04	1
10	10	1.82 E –04	1
12	12	2.44 E –04	1

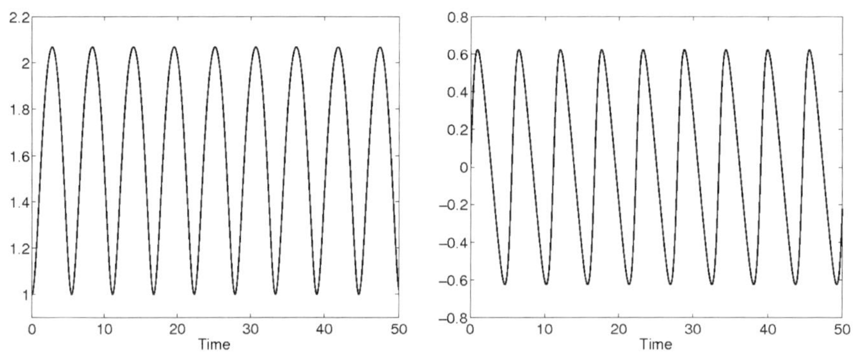

(a) (b)

FIGURE 6. MD simulation of the H_2 system for $R_0 = 1.0$: (a) internuclear separation and (b) velocity.

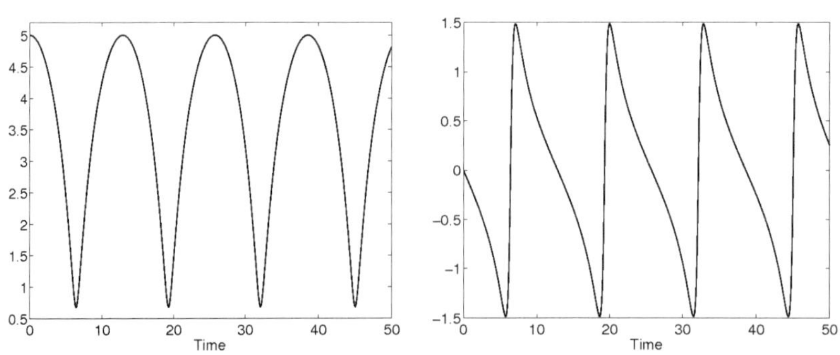

(a) (b)

FIGURE 7. MD simulation of the H_2 system for $R_0 = 5.0$: (a) internuclear separation and (b) velocity.

near the equilibrium internuclear separation and in Fig. 7 for large initial distance $R_0 = 5.0$ far from the equilibrium internuclear separation. In both cases, we take $N = M = 10$ and use a constant timestep of 0.02 for 2500 time steps. We see that in the first case the two nuclei interact with higher frequency and smaller magnitude than in the second case; and that in the first case the response is relatively harmonic whereas anharmonic effects dominate in the second case. The crucial new ingredient is fast output evaluations that allow us to perform a real-time MD simulation of the hydrogen molecular system: the "ab initio" molecular dynamics results are obtained online in less than 53 seconds on a Pentium 1.73 GHz laptop. (The same simulation would take 73 hours with (admittedly sub-optimal) finite element approximation.)

4. 1-D Kohn Sham Equation

4.1. Problem formulation. We shall now consider the parametrization of an one-dimensional periodic quantum problem[4]: we determine the ground state energy of a model periodic system with lattice parameter μ and hence unit cell $\Omega_o(\mu) \equiv]-\mu/2, \mu/2]$ based on the spinless Density Functional Theory [6, 8, 22, 25]. We further assume that a single nucleus of charge Z lies at the center of the cell and the number of electrons per nucleus is n_e, with $n_e = Z$ for charge neutrality. This model is rudimentary in the understanding of crystalline solids and has been studied in [31] and more recently [5].

This problem offers several opportunities for us to exercise the methodology described in previous sections: the parametrization procedure allows for efficient implementation of the *online-offline* computational framework; the empirical interpolation procedure allows for coefficient function approximation of the nonlinear terms; and the reduced basis procedure enables significant reduction in model size and computational cost in the *online* stage. More importantly, this problem introduces the computational challenges imposed by multiple electrons, and allows us to introduce in Section 4.2 a new ingredient that enables us to construct an efficient reduced basis space and ultimately a smaller reduced basis problem.

For simplicity, the parameter space consists of μ only; each new Z constitutes a new problem in our reduced basis approximation. Some possible applications from studying a system with varying μ includes the determination of forces exerted on the nuclei when the structure is deformed, and the characterization of the nonlinear behavior of the elasticity constant.

4.1.1. *Energy statement.* The equilibrium ground state energy of the above model can be obtained by solving a minimization problem for $\widehat{\mathbf{u}}_o([Z, \mu^*]) \equiv (u_{o1}, \ldots, u_{on_e})$, where [3, 7, 17, 18]

$$\widehat{\mathbf{u}}_o([Z,\mu]) = \arg\inf_{\widehat{\mathbf{w}}_o} \left\{ E_o(\widehat{\mathbf{w}}_o \equiv (w_{o1}, \ldots, w_{on_e}); [Z,\mu]), w_{oi} \in Y_o, \int_{\Omega_o(\mu)} w_{oi} w_{oj} = \delta_{ij}, 1 \le i, j \le n_e \right\}, \tag{4.1}$$

$$\mu^*(Z) = \arg\inf_{\mu} \{ \mathcal{E}_o(\widehat{\mathbf{u}}_o([Z,\mu]); [Z,\mu]); \mu > 0 \}; \tag{4.2}$$

[4]For a more general treatment of 3-D periodic quantum systems, work is currently being pursued to exploit existing planewave codes [11] to efficiently construct the reduced basis spaces [26].

here $Y_o \equiv H^1_{\text{per}}(\Omega_o(\mu))$ is the space of μ-periodic functions in $H^1(\mathbb{R})$; $\delta_{ij} = \{1 \text{ if } i = j, 0 \text{ otherwise}\}$; and u_{oi} is the Kohn–Sham orbital associated with the ith electron. (In practice, as in previous sections, Y_o is in fact our "truth" approximation, in this case a linear finite element approximation; a planewave (Fourier spectral) "truth" approximation is also possible.) The electronic energy $E_o(\widehat{\mathbf{w}}_o; [Z, \mu])$ is defined as

$$
\begin{aligned}
E_o(\widehat{\mathbf{w}}_o; [Z,\mu]) = {} & C_w \sum_{i=1}^{n_e} \int_{\Omega_o(\mu)} (\nabla w_{oi})^2 - Z \sum_{i=1}^{n_e} \int_{\Omega_o(\mu)} G_o {w_{oi}}^2 \\
& + \frac{1}{2} C_c \int_{\Omega_o(\mu)} \int_{\Omega_o(\mu)} \left(\sum_{i=1}^{n_e} w_{oi}^2(y_1) \right) \\
& \times G_o(y_1 - y_2) \left(\sum_{j=1}^{n_e} w_{oj}^2(y_2) \right) dy_1 \, dy_2 \\
& - C_x \sum_{i=1}^{n_e} \int_{\Omega_o(\mu)} \left(\sum_{j=1}^{n_e} w_{oj}^2 \right)^{4/3} w_{oi}^2,
\end{aligned} \tag{4.3}
$$

where we have used the X-α approximation to approximate the exchange-correlation term. Here, y denotes a point in $\Omega_o(\mu)$; C_w, C_c, and C_x are model constants (currently $C_w = 0.5$, $C_c = 1$ and $C_x = 0.7386$); and the periodic Green's function $G_o(\cdot; \mu)\colon \Omega_o(\mu) \to \mathbb{R}$ satisfies $-\triangle G_o = \{\delta(y) - 1/|\Omega_o(\mu)|\}$, $\int_{\Omega_o(\mu)} G_o = 0$, where $\triangle$ is the Laplacian operator, $\delta(y)$ is the Dirac delta distribution, and $|\Omega_o(\mu)| = \mu$ is the length of $\Omega_o(\mu)$. The total energy $\mathcal{E}_o(\widehat{\mathbf{w}}_o; [Z, \mu])$ — our output of interest — is then given by

$$
\mathcal{E}_o(\widehat{\mathbf{w}}_o; [Z,\mu]) = E_o(\widehat{\mathbf{w}}_o; [Z,\mu]) + \frac{Z^2}{2}\eta, \tag{4.4}
$$

where η is the nuclear — nuclear correction term given by $\eta = \lim_{y \to 0}\{G_o(y; \mu) - |y|/2\} = \pi^2 \mu / 12$.

Finally, the equivalent Euler–Lagrange equations for the constrained minimization problem (4.1) is given by:

$$
C_w \Delta u_{oi} - G_o u_{oi} - \phi_o u_{oi} + C_x \left(\sum_{j=1}^{n_e} u_{oj}^2 \right)^{1/3} u_{oi} = \lambda_{oi} u_{oi}, \quad 1 \le i \le n_e, \tag{4.5}
$$

$$
-\Delta \phi_o - C_c \left[\frac{Z}{\Omega_o(\mu)} - \sum_{j=1}^{n_e} u_{oj}^2 \right] = 0, \tag{4.6}
$$

$$
\int_{\Omega_o(\mu)} u_{oi} u_{oj} = \delta_{ij}, \quad 1 \le i < j < n_e \tag{4.7}
$$

$$
\int_{\Omega_o(\mu)} \phi_o = 0. \tag{4.8}
$$

Here, $\phi_o = -C_c \int_{\Omega_o(\mu)} (\sum_{j=1}^{n_e} u_{oj}^2) G_o(y_1 - y_2) dy_2$ is simply the Hartree potential [7] with a normalization of $\int_{\Omega_o(\mu)} \phi_o = 0$. Note that (4.5) contains a nonlinear term similar to Section 2.3; but now we have an eigenvalue problem for which we need to determine n_e eigenfunctions (or orbitals) and eigenvalues.

4.1.2. *Parametrized abstract formulation.* From the weak form of the Euler Lagrange equations (4.5)–(4.8), we derive the equivalent parametrized abstract formulation. We first define an affine geometric mapping, $\mathcal{G}(\mu)$, from $\Omega_o(\mu)$ to $\Omega \equiv]-\frac{1}{2}, \frac{1}{2}]$, which can be expressed as $x = \mathcal{G}(y;\mu) \equiv (y/\mu)$. We further define $u_i(\mu) = u_{oi} \circ \mathcal{G}^{-1}(\cdot;\mu)$, $\phi(\mu) = (\phi_o/\mu) \circ \mathcal{G}^{-1}(\cdot;\mu)$ and $G(\mu) = (G_o/\mu) \circ \mathcal{G}^{-1}(\cdot;\mu)$. Then, $\mathbf{u}([Z,\mu]) \equiv \big(\widehat{\mathbf{u}}([Z,\mu]), \phi([Z,\mu]), \widehat{\lambda}([Z,\mu]), \tau([Z,\mu])\big) \in \mathcal{Y} \equiv \big((Y)^{n_e} \times Y \times \mathbb{R}^{n_e(n_e+1)/2} \times \mathbb{R}\big)$ satisfies

$$\mathcal{A}(\mathbf{u}([Z,\mu]), \mathbf{v}; G; [Z,\mu]) = 0, \quad \forall \mathbf{v} \in \mathcal{Y}, \tag{4.9}$$

where $Y \equiv H^1_{\text{per}}(\Omega)$ is the space of 1-periodic functions in $H^1(\mathbb{R})$ (more precisely, an associated "truth" approximation subspace) with the associated inner product $(w,v)_Y \equiv \int_\Omega \nabla w \cdot \nabla v + \int_\Omega wv$ and norm $\|\cdot\| = (\cdot,\cdot)_Y^{1/2}$; $\widehat{\mathbf{u}}([Z,\mu]) \equiv (u_i, 1 \le i \le n_e)$; $\widehat{\lambda}([Z,\mu]) \equiv (\lambda_{ij}, 1 \le i \le j \le n_e)$; G satisfies $-\Delta G = \delta(x) - 1$, $\int_\Omega G = 0$; and $\mathcal{A}$ is defined as

$$\begin{aligned}
(4.10)\quad & \mathcal{A}(\mathbf{w} \equiv (\widehat{\mathbf{w}}, s, \hat{\sigma}, \kappa), \mathbf{v} \equiv (\widehat{\mathbf{v}}, \varsigma, \widehat{\varphi}, \varpi); t; [Z,\mu]) \\
&\equiv \sum_{i=1}^{n_e} \Bigg[\theta_1 a_0(w_i, v_i) + \theta_2 a_2(w_i, t, v_i) + \theta_3\, a_2(w_i, s, v_i) + \theta_5 a^{\text{nl}}\Bigg(w_i, \sum_{j=1}^{n_e} w_j^2, v_i\Bigg) \\
&\qquad + \theta_4 \sigma_{ii} a_1(w_i, v_i) + \theta_4 \sum_{j=1}^{n_e} \sigma_{ij} a_1(w_j, v_i)\Bigg] \\
&\quad + \sum_{i=1}^{n_e}\sum_{j=i}^{n_e} \varphi_{ij}\{\beta_1 a_1(w_i, w_j) + \beta_2 \delta_{ij}\} \\
&\quad + \Bigg[\alpha_1 a_0(s, \varsigma) + \alpha_2 \sum_{j=1}^{n_e} a_2(w_j, w_j, \varsigma) + \alpha_3 l(\varsigma) + \kappa\alpha_4\, l(\varsigma)\Bigg] + \varpi\, l(s).
\end{aligned}$$

Here, $a_0(w,v) \equiv \int_\Omega \nabla w \nabla v$, $a_1(w,v) \equiv \int_\Omega wv$, $a_2(w,s,v) \equiv \int_\Omega wsv$, $a^{\text{nl}}(w,t,v) \equiv \int_\Omega w t^{1/3} v$, and $l(w) \equiv \int_\Omega w$ for any $w \in Y$, $v \in Y$, $s \in Y$, and non-negative $t \in Y$. For prescribed C_w, C_x, and C_c[5] $\theta(Z,\mu) = \{C_w/\mu, -Z\mu^2, -\mu^2, -\mu/2, -C_x\frac{4}{3}\mu\}$, $\alpha(Z,\mu) = \{1, C_c\mu, -C_c Z, \mu\}$ and $\beta(Z,\mu) = \{\mu, -1\}$.

The total energy $\mathcal{E}(\mathbf{u}([Z,\mu]); G; [Z,\mu])$ is then given by

$$\begin{aligned}
(4.11)\quad \mathcal{E}(\mathbf{w} \equiv (\widehat{\mathbf{w}}, s, \hat{\sigma}, \kappa); G; [Z,\mu]) = \sum_{i=1}^{n_e} \Bigg[& \theta_1 a_0(w_i, w_i) + \theta_2 a_2(w_i, G, w_i) \\
& + \frac{1}{2}\theta_3 a_2(w_i, s, w_i) + \frac{3}{4}\theta_5 a_1^{\text{nl}}(w_i, \sum_{j=1}^{n_e} w_j^2, w_i)\Bigg] + \frac{Z^2}{2}\eta,
\end{aligned}$$

for $\eta = \pi^2\mu/12$ as defined earlier.

4.2. Reduced basis formulation.

4.2.1. *Reduced basis spaces.* Proceeding from Sections 2.2 and 3.2, it is not immediately clear how we should construct our reduced basis space for this multi-electron problem — there are now n_e components to the solution. A simplest (and perhaps naive) approach would be as follows: we introduce nested sample sets $S^u_{N^u_{\text{naive}}} = \{\mu^u_1, \ldots, \mu^u_{N^u_{\text{naive}}}\}, 1 \le N^u_{\text{naive}} \le N^u_{\text{naive max}}$; the reduced basis space is then given by $W^{u,\text{naive}}_{n_e \times N^u_{\text{naive}}} = \text{span}\{u_i([Z,\mu^u_n]), 1 \le i \le n_e, 1 \le n \le N^u_{\text{naive}}\}$,

[5] For our numerical results, we have used $C_w = 0.5$, $C_x = 0.7386$ and $C_c = 1$.

$1 \leq N^u_{\text{naive}} \leq N^u_{\text{naive}\max}$. For any μ, we now find an approximation to $u_i([Z,\mu])$, $1 \leq i \leq n_e$, in $W^{u,\text{naive}}_{n_e \times N^u_{\text{naive}}}$. The space $W^{u,\text{naive}}_{n_e \times N^u_{\text{naive}}}$ is sufficiently rich that the approximation will be good, and it provides sufficient degrees of freedom to honor the orthonormality constraints. However, the number of basis functions required is $N^u_{\text{naive}} \times n_e$; the size of our problem will increase linearly with n_e and, depending on N^u_{naive}, it can rapidly become unacceptably large. In addition, degeneracy can also become an issue.

This bring us to the final ingredient in our method, the vector reduced basis space. This approximation attempts to exploit (through the reduced basis space) the inherent orthogonality properties between the solutions $u_i([Z,\mu])$, $1 \leq i \leq n_e$ for a given μ, and their common smoothness: we introduce nested sample sets $S^u_{N^u} = \{\mu^u_1, \ldots, \mu^u_{N^u}\}$, $1 \leq N^u \leq N^u_{\max}$ and define the associated nested reduced basis spaces as

$$W^u_{N^u} = \text{span}\{\widehat{\mathbf{u}}([Z,\mu^u_n]), 1 \leq n \leq N^u\}, \quad 1 \leq N^u \leq N^u_{\max}, \tag{4.12}$$

$$= \text{span}\{\widehat{\zeta}_n, 1 \leq n \leq N^u\}, \quad 1 \leq N^u \leq N^u_{\max}; \tag{4.13}$$

where $\widehat{\mathbf{u}}([Z,\mu^u_n]) \equiv \big(u_1([Z,\mu^u_n]), \ldots, u_{n_e}([Z,\mu^u_n])\big)$ are the solutions of (4.10) at $\mu = \mu^u_n$ for a given Z; and $\hat{\boldsymbol{\zeta}} \equiv (\zeta_1, \ldots, \zeta_{n_e})$ are basis functions obtained after $\widehat{\mathbf{u}}([Z,\mu^u_n])$, $1 \leq n \leq N^u$ are aligned and pseudo-orthogonalized; these two preprocessing steps will lead to smaller N^u and better stability in the resulting discrete system, and will be described in the following paragraphs. Then, an approximation of $\widehat{\mathbf{u}}$ in W^u_N is given by $\widehat{\mathbf{u}}_{N,M}(\mu) = \sum_{n=1}^{N^u} \psi_n(\mu)\widehat{\zeta}_n$ — the ith component of $\widehat{\mathbf{u}}_{N,M}(\mu)$ is given by $u_{N,Mi}(\mu) = \sum_{n=1}^{N^u} \psi_n(\mu)(\zeta_i)_n$, $1 \leq i \leq n_e$.

We shall now describe the preprocessing steps alluded to earlier. First, we observe that the solutions $\widehat{\mathbf{u}}$ obtained based on a numerical algorithm such as the Arnoldi Method [32] are usually presented in ascending order of their respective eigenvalues, λ_{ii}, and not according to any particular structure of the orbitals identified by the form of the solutions $u_i([Z,\mu])$, $1 \leq i \leq n_e$. This, in addition to fact that the orbitals are only equivalent up to a sign, leads to discontinuities in $u_i([Z,\mu])$, $1 \leq i \leq n_e$ in the parameter space, as shown in Fig. 8. Referring to Fig. 8, we can identify three types of discontinuities: (1) sign switching (as demonstrated by u_1, u_4, and u_5); (2) mode crossing, where u_2 at $\mu = 4.5$ and 5.5 are a smooth transition of u_3 at $\mu = 1.5$, 2.5 and 3.5, and vice versa; and (3) mode entering, where there are more than n_e forms of orbitals appearing in our solutions as μ varies (not exhibited by the solutions in Fig. 8).

The alignment process seeks to remove the first two discontinuities by employing the following greedy algorithm: given a pre-sorted space $W^u_{N^u} \equiv \text{span}\{\widehat{\zeta}^s_n, 1 \leq n \leq N^u\}$ where $\hat{\zeta}^s_n$, $1 \leq n \leq N^u$ are the sorted basis functions of $\widehat{\mathbf{u}}_n$, $1 \leq n \leq N^u$, we wish to add $\widehat{\mathbf{u}}([Z,\mu^u_{N^u+1}])$ to $W^u_{N^u}$ to form $W^u_{N^u+1}$. We first select a $\hat{\zeta}^s_n \in W^u_{N^u}$ such that $\mu^u_n \in S^u_{N^u}$ is closest to $\mu^u_{N^u+1}$. We compute $e^+_{w,j} = \|(\zeta^s_1)_n + u_j(\mu^u_{N^u+1})\|_Y$ and $e^-_{w,j} = \|(\zeta^s_1)_n - u_j(\mu^u_{N^u+1})\|_Y$ for $1 \leq j \leq n_e$; we then determine $j^* \equiv \arg\min_{j,1\leq j\leq n_e}\{e^+_{w,j}, e^-_{w,j}\}$. If $e^-_{w,j^*} > e^+_{w,j^*}$, then $(\zeta^s_1)_{N^u+1} = -u_{j^*}(\mu^u_{N^u+1})$; otherwise $(\zeta^s_1)_{N^u+1} = u_{j^*}(\mu^u_{N^u+1})$. This is then repeated for all $(\zeta^s_i)_{N^u+1}$, for $i = 2, \ldots, n_e$. First part of the algorithm associates $u_j(\mu)$ to the correct orbital and

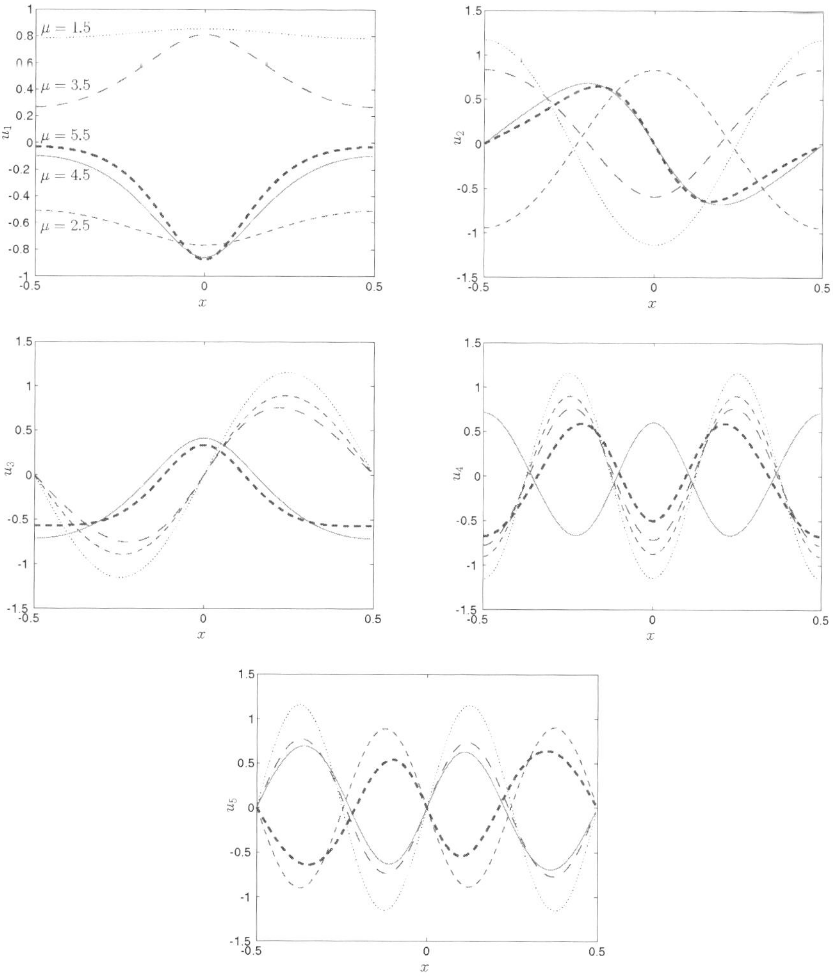

FIGURE 8. Solutions of $\widehat{\mathbf{u}}(\mu)$ at $\mu = 1.5, 2.5, 3.5, 4.5$, and 5.5, before the alignment process.

the second part of the algorithm remove the sign variation in ζ_i^s. The results are sets of orbitals that vary smoothly with μ, as shown in Fig. 9.[6]

The sorted bases are then pseudo-orthogonalized: given a space $W_{N^u}^u = \operatorname{span}\{\widehat{\zeta}_n, 1 \leq n \leq N^u\}$ where $\widehat{\zeta}_n$ are the pseudo-orthogonalized basis functions

[6]The 3rd type of the discontinuities — mode entering — has to be handled differently. We may allow this discontinuity to exist although the resulting approximation is suboptimal. Alternatively, one could expand the number of orbitals taken into account, i.e. we may wish to construct a reduced basis space given by $W_{N^u}^{u,\text{augment}} \equiv \text{span } \{\widehat{\mathbf{u}}([Z, \mu_n^u]) \equiv (u_1([Z, \mu_n^u]), \ldots, u_{n_{e,\text{augment}}}([Z, \mu_n^u])),\ 1 \leq n \leq N^u\}$, where $n_{e,\text{augment}} > n_e$ and $n_{e,\text{augment}}$ is sufficiently large such that the discontinuity resulting from mode entering does not appear within our parameter space of interest. Thanks to the vectorial nature of $W_{N^u}^{u,\text{augment}}$, this will not lead to significant increase in the N^u required.

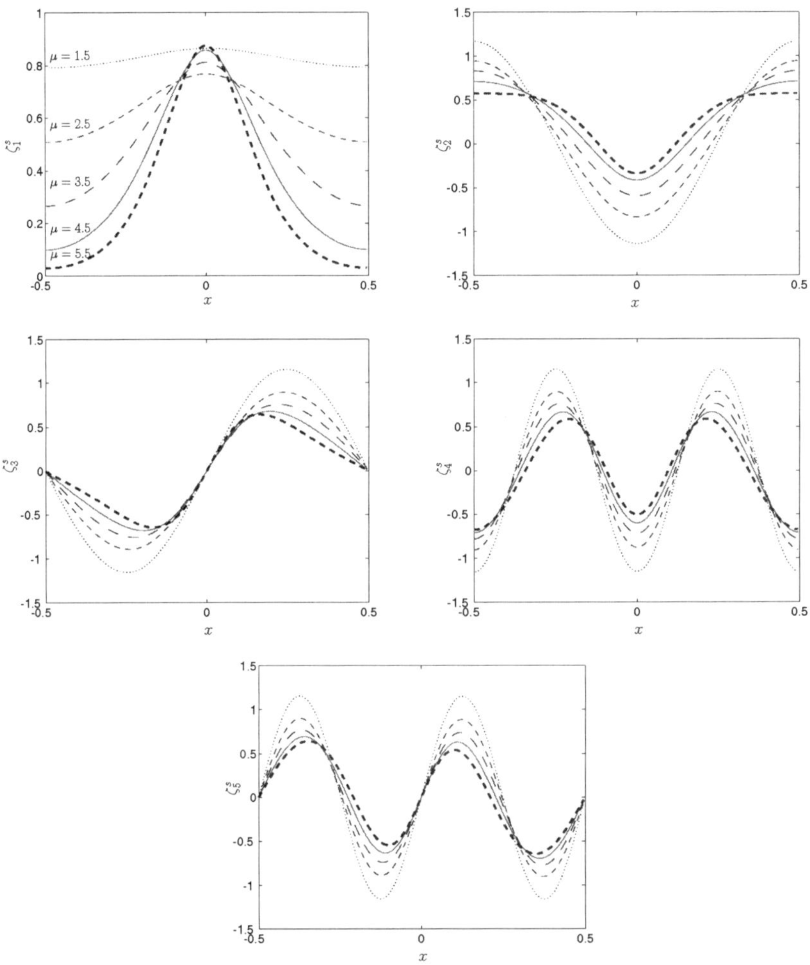

FIGURE 9. $\widehat{\zeta}^s$ at $\mu = 1.5, 2.5, 3.5, 4.5$, and 5.5, after the alignment process.

of $\widehat{\zeta}_n^s$, $1 \le n \le N^u$, and the next member of the space, $\widehat{\zeta}^s([Z, \mu_{n+1}^u])$, we first compute $\widehat{\mathbf{b}} = \widehat{\zeta}^s - \sum_{n=1}^{N^u} \nu_n \widehat{\zeta}_n$, where $\nu = \arg\min_{\nu \in \mathbb{R}^{N^u}} \sum_{i=1}^{n_e} \|\zeta_i^s - \sum_{n=1}^{N^u} \nu_n (\zeta_i)_n\|_Y^2$; the new basis, pseudo-orthogonalized, is then given by $\widehat{\zeta}_{N^u+1} = \widehat{\mathbf{b}} / (\sum_{i=1}^{n_e} \|b_i\|_Y^2 / n_e)^{1/2}$ and $W_{N^u+1}^u = W_{N^u}^u + \widehat{\zeta}_{N^u+1}$.

We may similarly define for ϕ the nested sample sets $S_{N^\phi}^\phi = \{\mu_1^\phi, \dots, \mu_{N^\phi}^\phi\}$ and the associated reduced basis space $W_{N^\phi}^\phi = \text{span}\{\phi([Z, \mu_n^\phi]), 1 \le n \le N^\phi\} = \text{span}\{\chi_n, 1 \le n \le N^\phi\}$, $1 \le N^\phi \le N_{\max}^\phi$. The χ_n, $1 \le n \le N^\phi$ are obtained by orthonormalizing $\phi([Z, \mu_n^\phi])$, $1 \le n \le N^\phi$ relative to the $(\cdot;\cdot)_Y$ inner product.

Finally (although in actual practice, initially), we construct the $S_{N^u}^u$ and $S_{N^\phi}^\phi$ based on the greedy selection process. As described in Section 2.2.1, we may employ different forms of error measures. Here, we first construct $S_{N^\phi}^\phi$ (and correspondingly $W_{N^\phi}^\phi$) based on the greedy selection process using the projection error as defined by (2.12); we wish to isolate the behavior of the space $W_{N^u}^u$. The $S_{N^u}^u$ (and correspondingly $W^u_{N^u}$) is then constructed based on the greedy selection process using the reduced basis error defined as

$$\frac{(\sum_{i=1}^{n_e} \|u_i([Z,\mu]) - u_{N,Mi}([Z,\mu])\|_Y^2)^{1/2}}{(\sum_{i=1}^{n_e} \|u_i([Z,\mu])\|_Y^2)^{1/2}}.$$

The approximations by which $\widehat{\mathbf{u}}_{N,M}([Z,\mu])$ are obtained will now be elaborated.

4.2.2. *The approximations.* In our reduced basis approximation, the equilibrium ground state of the resulting neutral structure for a particular Z $(= n_e)$ is given by $\widehat{\mathbf{u}}_{N,M}([Z,\mu]) \equiv \big(u_{N,M1}([Z,\mu]),\ldots,u_{N,Mn_e}([Z,\mu])\big)$, where

$$\widehat{\mathbf{u}}_{N,M}([Z,\mu]) = \arg\inf_{\widehat{\mathbf{w}}}\Big\{E_{N,M}(\widehat{\mathbf{w}} \equiv (w_1,\ldots,w_{n_e});[Z,\mu]),\quad w_i \in W^u_N, \mu\int_\Omega w_i^2 = 1, 1 \le i \le n_e\Big\}, \tag{4.14}$$

$$\mu^*(Z) = \arg\inf_{\mu}\{\mathcal{E}_{N,M}(\widehat{\mathbf{u}}_{N,M}([Z,\mu]);[Z,\mu]);\mu > 0\}. \tag{4.15}$$

Here, $\widehat{\mathbf{u}}_{N,M}([Z,\mu])$ is obtained by solving the following Euler Lagrange equations: find $\mathbf{u}_{N,M}([Z,\mu]) \equiv \big(\widehat{\mathbf{u}}_{N,M}([Z,\mu]),\phi_{N,M}([Z,\mu]),\hat{\lambda}_{N,M}([Z,\mu]),\tau_{N,M}([Z,\mu])\big) \in \mathcal{Y}_N \equiv (W^u_{N^u} \times W^\phi_N \times \mathbb{R}^{n_e} \times \mathbb{R})$ such that

$$\mathcal{A}_M(\mathbf{u}_{N,M}([Z,\mu]),\mathbf{v};G;[Z,\mu]) = 0, \quad \forall \mathbf{v} \equiv (\hat{\mathbf{v}},\varsigma,\hat{\varphi},\varpi) \in \mathcal{Y}_N, \tag{4.16}$$

where $\hat{\lambda}_{N,M}([Z,\mu]) \equiv ((\lambda_{N,M})_{ii}, 1 \le i \le n_e)$ and

$$\begin{aligned}
&\mathcal{A}_M(\mathbf{w} \equiv (\hat{\mathbf{w}},s,\hat{\sigma},\kappa),\mathbf{v} \equiv (\hat{\mathbf{v}},\varsigma,\hat{\varphi},\varpi);t;[Z,\mu]) \\
&\equiv \sum_{i=1}^{n_e}\Bigg[\theta_1 a_0(w_i,v_i) + \theta_2 a_2(w_i,t,v_i) + \theta_3\ a_2(w_i,s,v_i) \\
&\qquad + \theta_5 a^{\mathrm{nl},M}\Bigg(w_i,\sum_{j=1}^{n_e} w_j^2, v_i\Bigg) + 2\theta_4\sigma_{ii}a_1(w_i,v_i)\Bigg] \\
&\quad + \sum_{i=1}^{n_e}\varphi_{ii}\{\beta_1 a_1(w_i,w_i)\beta_2\delta_{ii}\} \\
&\quad + \Bigg[\alpha_1 a_0(s,\varsigma) + \alpha_2\sum_{j=1}^{n_e} a_2(w_j,w_j,\varsigma) + \alpha_3 l(\varsigma) + \kappa\alpha_4 l(\varsigma)\Bigg] + \varpi l(s).
\end{aligned} \tag{4.17}$$

Then, $E_{N,M}(\mathbf{u}_{N,M}([Z,\mu]);G;[Z,\mu])$, the reduced basis approximation for the electronic energy $E(\mathbf{u}([Z,\mu]);G;[Z,\mu])$, is given by

$$\begin{aligned}
&E_{N,M}(\mathbf{w} \equiv (\hat{\mathbf{w}},s,\hat{\sigma},\kappa);G;[Z,\mu]) \\
&= \sum_{i=1}^{n_e}\Bigg[\theta_1 a_0(w_i,w_i) + \theta_2 a_2(w_i,G,w_i) + \frac{1}{2}\theta_3\ a_2(w_i,s,w_i) \\
&\qquad + \frac{3}{4}\theta_5\ a^{\mathrm{nl},M}\Bigg(w_i,\sum_{j=1}^{n_e} w_j^2, w_i\Bigg)\Bigg]
\end{aligned} \tag{4.18}$$

and $\mathcal{E}_{N,M}(\mathbf{u}_{N,M}([Z,\mu]);G;[Z,\mu])$, the reduced basis approximation to the total energy $\mathcal{E}(\mathbf{u}([Z,\mu]);G;[Z,\mu])$, is given by $E_{N,M}(\mathbf{u}_{N,M}([Z,\mu]);G;[Z,\mu]) + (Z^2\eta(\mu)/2)$.

When compared to (4.1) and (4.10), we have made two approximations. First, we approximate $a^{\mathrm{nl}}(w_i, \sum_{j=1}^{n_e} w_j^2, v_i)$ by $a^{\mathrm{nl},M}(w_i,\sum_{j=1}^{n_e} w_j^2, v_i) \equiv \int_\Omega g^w_{Mi} v$, where

g^u_{Mi} is an empirical interpolation approximation to $g_i(\widehat{\mathbf{u}}) \equiv u_i(\sum_{j=1}^{n_e} u_j^2)^{1/3}$. Thus, we require n_e empirical interpolation approximations. Secondly, we only impose the constraints $\mu \int_\Omega u^2_{N,Mi} = 1$, $1 \le i \le n_e$. Note that in the original minimization problem (4.9), the constraints are $\mu \int_\Omega u_i u_j = \delta_{ij}$, $1 \le i < j \le n_e$; we have thus assumed that the orthogonality of the components in $\widehat{\mathbf{u}}_{N,M}$ will be approximately satisfied by construction (implicit to our space $W^u_{N^u}$) in (4.16). We note that $\int_\Omega u_{N,Mi} u_{N,Mj}$, $1 \le i < j \le n_e$ can be bounded: let $(w,v)_{L^2} = a_1(w,v) = \int_\Omega wv$ and $\|\cdot\|_{L^2} = \sqrt{(\cdot,\cdot)_{L^2}}$; then

$$
\begin{aligned}
(4.19)\qquad (u_{N,Mi}, u_{N,Mj})_{L^2} &= (u_{N,Mi} - u_i, u_{N,Mj})_{L^2} + (u_i, u_{N,Mj} - u_j)_{L^2} \\
&\qquad + (u_i, u_j)_{L^2} \\
&\le \|u_{N,Mi} - u_i\|_{L^2} + \|u_{N,Mj} - u_j\|_{L^2},
\end{aligned}
$$

since $\|u_i\|_{L^2} = \|u_{N,Mj}\|_{L^2} = 1$ and $(u_i, u_j)_{L^2} = 0$, for all $1 \le i < j \le n_e$. Therefore,

$$
(4.20)\qquad (u_{N,Mi}, u_{N,Mj})_{L^2} \le \sum_{n=1}^{n_e} \|u_n - u_{N,Mn}\|_{L^2}.
$$

From above, we can conclude that as $\widehat{\mathbf{u}}_{N,M} \to \widehat{\mathbf{u}}$, $\int_\Omega u_{N,Mi} u_{N,Mj} \to 0$, for $1 \le i < j \le n_e$.

However, note that for N^u sufficiently large, we can represent any member of our (finite-dimensional) truth approximation space, presuming the linear independence of the snapshots. But, clearly (4.14) is not equivalent to (4.1) due to the absence of the orthogonality constraints. Hence, to be consistent, as N^u increases, we should also systematically add in orthogonality constraints, for example by adding one orthogonality constraint for every two additional basis functions. This will be considered in future work.

Finally, since $\int_\Omega \chi_n = 0$, $1 \le n \le N$, $\int_\Omega \phi_{N,M}$ is perforce zero; our discrete (nonlinear) algebraic system will thus have an actual dimension of $N^u + N^\phi + n_e$.

4.2.3. *Solution method.* Our truth approximations are obtained based on the finite element formulation with $\mathcal{N} = 400$. The resulting discrete equation is solved through self-consistent field procedure [4, 6]: the nonlinearity is tackled based on a fixed point method in which a diagonalization of the algebraic system for (4.10) is performed within each iteration. As shown in Fig. 9, the solutions in $\widehat{\mathbf{u}}$ exhibit considerable variation with respect to μ.

However, the fixed point method is not suited for our reduced basis approximation given by (4.16). We thus resort to Newton's method as described in Section 2.2.2, and, to assist convergence, we exploit a homotopy procedure in $\varepsilon \in [0,1]$ and $\mu_k = \mu_i + \varepsilon(\mu - \mu_i)$, where μ_k is the μ at k intermediate homotopy step and μ_i is the initial $\mu_{k=0}$ at start of the homotopy procedure — usually chosen to be the closest $\mu^u \in S^u_{N^u}$ to μ. The online complexity is then $O\big((N^\phi)^3 + n_e N^\phi (N^u)^2 + (N^u)^3 + n_e N^u M\big)$ per Newton iteration.

4.3. Numerical results. We consider μ in the interval $[1.5, 5.5]$. We introduce a parameter test sample $\Xi_{\rm Test}$ of size 161 and choose M, the dimension of the empirical interpolation space, such that ε_M is less than 10^{-10}. We then define the following

$$
(4.21)\qquad \varepsilon^u_{N,M} = \max_{\mu \in \Xi_{\rm Test}} \frac{\big(\sum_{i=1}^{n_e} \|u_i([Z,\mu]) - u_{N,Mi}([Z,\mu])\|^2_Y\big)^{1/2}}{\big(\sum_{i=1}^{n_e} \|u_i([Z,\mu])\|^2_Y\big)^{1/2}},
$$

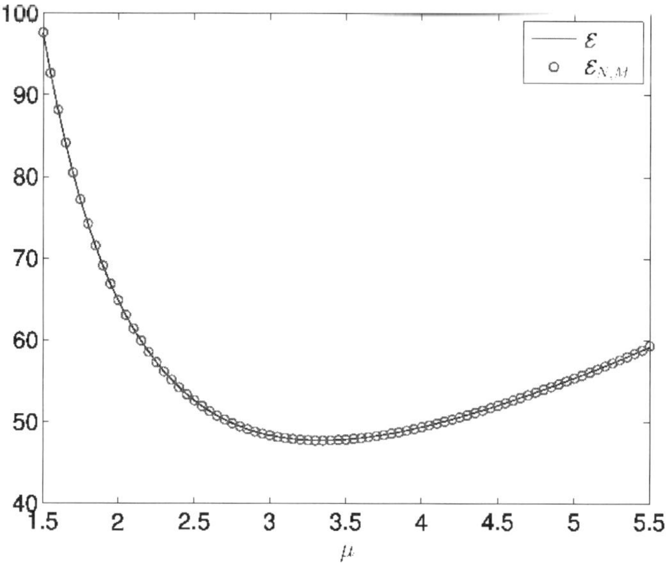

FIGURE 10. Comparison between $\mathcal{E}$ and $\mathcal{E}_{N,M}$ for $n_e = 5$, $1.5 < \mu < 5.5$; there is no discernable difference between the two.

$$\varepsilon^{\phi}_{N,M} = \max_{\mu \in \Xi_{\text{Test}}} \frac{\|\phi([Z,\mu]) - \phi_{N,M}([Z,\mu])\|_Y}{\|\phi([Z,\mu])\|_Y}, \tag{4.22}$$

$$\varepsilon^{\mathcal{E}}_{N,M} = \max_{\mu \in \Xi_{\text{Test}}} |\mathcal{E}_N([Z,\mu]) - \mathcal{E}([Z,\mu])|/|\mathcal{E}([Z,\mu])| \tag{4.23}$$

$$\varepsilon^{\text{ortho}}_{N,M} = \max_{\mu \in \Xi_{\text{Test}}} \max_{1 \le i < j \le n_e} (u_{N,Mi}([Z,\mu]), u_{N,Mj}([Z,\mu]))_{L^2}. \tag{4.24}$$

where $\varepsilon^{u}_{N,M}$, $\varepsilon^{\phi}_{N,M}$ and $\varepsilon^{\mathcal{E}}_{N,M}$ are respectively the maximum error in the reduced basis approximation of $\widehat{\mathbf{u}}$, ϕ and $\mathcal{E}$ within a given sample Ξ_{Test}; $\varepsilon^{\text{ortho}}_{N,M}$ is a measure of non-compliance in the orthogonality constraints; and $(w,v)_{L^2} = a_1(w,v) = \int_\Omega wv$. We present the results for the above quantities with increasing N^u for $n_e = 5$ in Table 3. Here, N^ϕ is 5 and M is 12. We observe a monotonic decrease in $\varepsilon^{u}_{N,M}$, $\varepsilon^{\phi}_{N,M}$ and $\varepsilon^{\mathcal{E}}_{N,M}$. We only require $N^u = 10$ to reduce $\varepsilon^{u}_{N,M}$ and $\varepsilon^{\mathcal{E}}_{N,M}$ to 9.76×10^{-6} and 3.05×10^{-10} respectively. In addition, $\varepsilon^{\mathcal{E}}_{N,M}$ is approximately the square of $\varepsilon^{u}_{N,M}$ (and $\varepsilon^{\phi}_{N,M}$), indicating that $\mathcal{E}$ can in fact be approximated very accurately with very few basis functions by $\mathcal{E}_{N,M}$; there is virtually no noticeable distinction between the two as shown in Fig. 10.

From Table 3, we also observe that $\varepsilon^{\text{ortho}}_{N,M}$ converges very rapidly with N^u. This is perhaps not surprising. From (4.20), we obtain

$$\varepsilon^{\text{ortho}}_{N,M} \le \max_{\mu \in \Xi_{\text{Test}}} \sum_{n=1}^{n_e} \|u_n([Z,\mu]) - u_{N,Mn}([Z,\mu])\|_{L^2}. \tag{4.25}$$

TABLE 3. Variations of the reduced basis errors $\varepsilon^u_{N,M}$, $\varepsilon^{\mathcal{E}}_{N,M}$, $\varepsilon^{\phi}_{N,M}$ and $\varepsilon^{\text{ortho}}_{N,M}$ with N^u. Here, $n_e = 5$ and $1.5 \le \mu \le 5.5$.

N^u	$\varepsilon^u_{N,M}$	$\varepsilon^{\mathcal{E}}_{N,M}$	$\varepsilon^{\phi}_{N,M}$	$\varepsilon^{\text{ortho}}_{N,M}$
5	7.9044 E −2	4.6557 E −04	1.4647 E + 0	5.1756 E −14
6	4.5693 E −2	3.5279 E −05	1.2839 E −1	3.6342 E −3
7	2.1383 E −4	1.3947 E −09	1.0334 E −3	2.1783 E −5
8	9.8819 E −5	8.8168 E −10	3.7635 E −4	1.0686 E −5
9	9.7602 E −6	3.0509 E −10	3.8463 E −5	8.9840 E −7

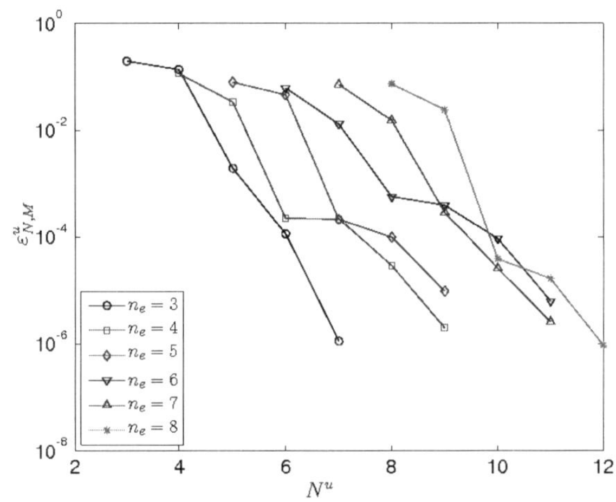

FIGURE 11. Convergence of the reduced basis error $\varepsilon^u_{N,M}$ for $3 \le n_e \le 8$. The μ range is $[1.5, 5.5]$.

This implies that as $\widehat{\mathbf{u}}_{N,M}([Z,\mu]) \to \widehat{\mathbf{u}}([Z,\mu])$, $\varepsilon^{\text{ortho}}_{N,M} \to 0$ and since $\varepsilon^u_{N,M}$ decreases with increasing N^u, so does $\varepsilon^{\text{ortho}}_{N,M}$ [7]. The error $\varepsilon^{\text{ortho}}_{N,M}$ can certainly be reduced further by the systematic inclusion of orthogonality constraints, as argued in Section 4.2.2.

The above results can be extended to other cases of $n_e > 1$. In Fig. 11, we show the convergence results of the error $\varepsilon^u_{N,M}$ with respect to N^u for $3 \le n_e \le 8$. We observe that the error decreases monotonically with increasing N^u and with just 12 basis functions in $W^u_{N^u}$, an error of $\varepsilon^u_{N,M} < 10^{-5}$ can be obtained for the $n_e = 8$ case. In addition, the number of basis functions required for each case is only slightly higher than n_e; N^u scales approximately as $n_e + C$, where C is a small integer. The naive approach which does not exploit the vectorial nature of the reduced basis space (as described at beginning of Section 4.2.1) will require

[7] At $N = n_e$, the apparent "good orthogonality" is deceiving — the size of N is too small and the reduced basis solution gravitates towards a single basis, of which the orthogonality is obviously satisfied.

$N^u_{\text{naive}} \times n_e$ basis functions, and hence is much more sensitive to the number of electrons.

References

1. B.O. Almroth, P. Stern, and F.A. Brogan, *Automatic choice of global shape functions in structural analysis*, AIAA J. **16** (1978), 525–528.
2. M. Barrault, N. C. Nguyen, Y. Maday, and A.T. Patera, *An "empirical interpolation" method: application to efficient reduced-basis discretization of partial differential equations*, C. R. Acad. Sci. Paris Sér. I Math. **339** (2004), 667–672.
3. X. Blanc, *A mathematical insight into ab initio simulations of the solid phase*, Mathematical Models and Methods for Ab Initio Quantum Chemistry, Lecture Notes in Chem., vol.74, Springer, Berlin, 2000, pp. 133–158.
4. ———, *Geometry optimization for crystals in Thomas–Fermi type theories of solids*, Comm. Partial Differential Equations **26** (2001), no. 3-4, 651–696.
5. X. Blanc and C. Le Bris, *Periodicity of the infinite-volume ground state of a one-dimensional quantum model*, Nonlinear Anal. **48** (2002), no. 6, 791–803.
6. E. Cancès, M. Defranceschi, W. Kutzelnigg, C. Le Bris, and Y. Maday, *Computational quantum chemistry: a primer*, Handbook Numer. Anal., vol. 10, North-Holland, Amsterdam, 2003, pp. 3–270.
7. I. Catto, C. Le Bris, and P.-L. Lions, *Recent mathematical results on the quantum modeling of crystals*, Mathematical Models and Methods for Ab Initio Quantum Chemistry, Lecture Notes in Chem., vol. 74, Springer, Berlin, 2000, pp. 95–119.
8. R.M. Dreizler and E.K.U. Gross, *Density functional theory: an approach to the quantum many-body problem*, Springer-Verlag, Berlin, Heidelberg, New York, 1991.
9. J.P. Fink and W.C. Rheinboldt, *On the error behavior of the reduced basis technique for nonlinear finite element approximations*, Z. Angew. Math. Mech. **63** (1983), 21–28.
10. D. Frenkel and B. Smit, *Understanding molecular simulation*, 2nd ed., Academic Press, New York, 2001.
11. X. Gonze, J.M. Beuken, R. Caracas, F. Detraux, M. Fuchs, G.M. Rignanese, L. Sindic, M. Verstraete, G. Zerah, and F. Jollet, *First-principles computation of material properties: the ABINIT software project*, Comput. Materials Sci. **25** (2002), no. 2, 478–492.
12. M.A. Grepl, Y. Maday, N.C. Nguyen, and A.T. Patera, *Efficient reduced-basis treatment of nonaffine and nonlinear partial differential equations*, M2AN Math. Model. Numer. Anal., 2006 (to appear).
13. M.A. Grepl, N.C. Nguyen, K. Veroy, A.T. Patera, and G.R. Liu, *Certified rapid solution of partial differential equations for real-time parameter estimation and optimization*, 2nd Sandia Workshop of PDE-Constrained Optimization: Towards Real-Time and On-Line PDE-Constrained Optimization, SIAM Comput. Sci. Engr. Book Series, 2006 (to appear).
14. M.D. Gunzburger, *Finite element methods for viscous incompressible flows: a guide to theory, practice, and algorithms*, Academic Press, Boston, 1989.
15. W.L. Hehre, L. Radon, P.V.R. Schleyer, and J. A. Pople, *Ab initio molecular orbital theory*, Wiley, New York, 1986.
16. K. Ito and S.S. Ravindran, *A reduced-order method for simulation and control of fluid flows*, J. Comput. Phys. **143** (1998), no. 2, 403–425.
17. C. Le Bris, *Computational chemistry from the perspective of numerical analysis*, Acta Numer. **14** (2005), 363–444.
18. C. Le Bris and P.-L. Lions, *From atoms to crystals: a mathematical journey*, Bull. Amer. Math. Soc. (N.S.) **42** (2005), no. 3, 291–363.
19. B. Leimkuhler and S. Reich, *Simulating Hamiltonian dynamics*, Cambridge Monogr. Appl. Comput. Math., no. 14, Cambridge University Press, Cambridge, 2005.
20. L. Machiels, Y. Maday, I.B. Oliveira, A.T. Patera, and D.V. Rovas, *Output bounds for reduced-basis approximations of symmetric positive definite eigenvalue problems*, C. R. Acad. Sci. Paris Sér. I Math. **331** (2000), no. 2, 153–158.
21. Y. Maday, A.T. Patera, and G. Turinici, *Global* a priori *convergence theory for reduced-basis approximation of single-parameter symmetric coercive elliptic partial differential equations*, C. R. Acad. Sci. Paris Sér. I Math. **335** (2002), no. 3, 289–294.

22. R.M. Martin, *Electronic structure: basic theory and practical methods*, Cambridge University Press, Cambridge, 2004.
23. N.C. Nguyen, K. Veroy, and A.T. Patera, *Certified real-time solution of parametrized partial differential equations*, Handbook of Materials Modeling (S. Yip, ed.), Springer, 2005, pp. 1523–1558.
24. A.K. Noor and J.M. Peters, *Reduced basis technique for nonlinear analysis of structures*, AIAA J. **18** (1980), no. 4, 455–462.
25. R.G. Parr and Y. Weitao, *Density-functional theory of atoms and molecules*, Intern. Ser. Monogr. Chem., vol. 16, Oxford University Press, U.K., 1994.
26. G.S.H. Pau, *Reduced-basis method for quantum models of periodic solids*, Ph.D thesis, MIT (in progress).
27. J.S. Peterson, *The reduced basis method for incompressible viscous flow calculations*, SIAM J. Sci. Stat. Comput **10** (1989), no. 4, 777–786, 1989.
28. T.A. Porsching, *Estimation of the error in the reduced basis method solution of nonlinear equations*, Math. Comp. **45** (1985), no. 172, 487–496.
29. C. Prud'homme, D. Rovas, K. Veroy, Y. Maday, A.T. Patera, and G. Turinici, *Reliable real-time solution of parametrized partial differential equations: reduced-basis output bound methods*, J. Fluids Engr. **124** (2002), no. 1, 70–80.
30. A. Quarteroni and A. Valli, *Numerical approximation of partial differential equations*, 2nd ed., Springer-Verlag Berlin Heidelberg , 1997.
31. C. Radin, *Periodicity of classical ground states*, Phys. Rev. Lett. **51** (1983), no. 8, 621–622.
32. Y. Saad, *Numerical methods for large eigenvalue problems: theory and Algorithms*, John Wiley and Sons, New York, 1992.
33. J.S. Sims and S.A. Hagstrom, *High precision variational calculations for the Born–Oppenheimer energies of the ground state of the hydrogen molecule*, J. Chem. Phys. **124** (2006), no. 9, 094101–094108.
34. A. Szabo and N.S. Ostlund, *Modern quantum chemistry: introduction to advanced electronic structure theory*, Dover Publications, New York, 1996.
35. K. Veroy and A.T. Patera, *Certified real-time solution of the parametrized steady incompressible Navier–Stokes equations; rigorous reduced-basis* a posteriori *error bounds*, Internat. J. Numer. Methods Fluids **47** (2005), 773–788.
36. K. Veroy, C. Prud'homme, D.V. Rovas, and A.T. Patera, A posteriori *error bounds for reduced-basis approximation of parametrized noncoercive and nonlinear elliptic partial differential equations*, Proceedings of the 16th AIAA Computational Fluid Dynamics Conference (Orlando), AIAA paper 2003–3847, 2003.
37. K. Veroy, D. Rovas, and A.T. Patera, A posteriori *error estimation for reduced-basis approximation of parametrized elliptic coercive partial differential equations: "convex inverse" bound conditioners*, ESAIM Control Optim. Calc. Var. **8** (2002), 1007–1028.

CERMICS, ÉCOLE NATIONALE DES PONTS ET CHAUSSÉES, 6 & 8 AVENUE BLAISE PASCAL, 77455 MARNE LA VALLÉE CEDEX 2, FRANCE AND INRIA ROCQUENCOURT, MICMAC PROJECT, DOMAINE DE VOLUCEAU, B.P. 105, 78153 LE CHESNAY CEDEX, FRANCE.
E-mail address: cances@cermics.enpc.fr

CERMICS, ÉCOLE NATIONALE DES PONTS ET CHAUSSÉES, 6 & 8 AVENUE BLAISE PASCAL, 77455 MARNE LA VALLÉE CEDEX 2, FRANCE AND INRIA ROCQUENCOURT, MICMAC PROJECT, DOMAINE DE VOLUCEAU, B.P. 105, 78153 LE CHESNAY CEDEX, FRANCE.
E-mail address: lebris@cermics.enpc.fr

LABORATOIRE JACQUES-LOUIS LIONS, UNIVERSITÉ PIERRE ET MARIE CURIE, BOÎTE COURRIER 187, 75252 PARIS CEDEX, FRANCE.
E-mail address: maday@ann.jussieu.fr

DEPARTMENT OF AERONAUTICS AND ASTRONAUTICS, MASSACHUSETTS INSTITUTE OF TECHNOLOGY, ROOM 37-435, 77 MASSACHUSETTS AVENUE, CAMBRIDGE, MA 02139-4307, USA.
E-mail address: cuongng@mit.edu

MECHANICAL ENGINEERING, MASSACHUSETTS INSTITUTE OF TECHNOLOGY, ROOM 3-266, CAMBRIDGE, MA 02139-4307, USA.
E-mail address: patera@mit.edu

MASSACHUSETTS INSTITUTE OF TECHNOLOGY, ROOM 3-264, CAMBRIDGE, MA 02139-4307, USA.
E-mail address: gpau@mit.edu

Centre de Recherches Mathématiques
CRM Proceedings and Lecture Notes
Volume **41**, 2007

Some Fundamental Mathematical Properties in Atomic and Molecular Quantum Mechanics

Goong Chen, Zhonghai Ding, Alain Perronnet, Marlan O. Scully, Ruihua Xie, and Zhigang Zhang

Abstract. The theoretical basis for modeling atomic and molecular systems containing a few particles of electrons and nuclei is the Schrödinger equation and the Born–Oppenheimer separation. The objective of this paper is to review some fundamental mathematical properties for such systems in order to facilitate contemporary interdisciplinary research in quantum computing, quantum control and chemical physics. The following are the main topics addressed in the paper:

(1) derivations of the virial theorem and cusp conditions;
(2) discussion of H_2^+ (hydrogen molecular ion) orbitals;
(3) related examples and illustrations

1. Introduction

During the past decade, we have seen rapid advances in the multidisciplinary research at the interfaces of chemistry, physics and information science. For example, in molecular chemistry, new femtosecond and attosecond laser applications have opened up new frontiers [2]. In quantum computing, coherent excited atomic and molecular states are utilized for quantum information processing [5]. What is at work may be largely attributed to the interplay between quantum optics and quantum chemistry.

At a fundamental level, the center of this research is molecular modeling and computing. We are thus motivated to reexamine certain problems in mathematical physics and molecular chemistry, based on the multi-particle Schrödinger equation. But such an equation lives in a very high dimensional space (which is the focal topic of the Conference) and is rather analytically intractable in general. Numerous methods have been devised in modern computational chemistry: diffusion Monte Carlo methods, coupled cluster expansions, configuration interactions, density function

2000 *Mathematics Subject Classification.* Primary: 35J10; Secondary: 81Q05.

This research is supported in part by ONR Grant Nos. N00014-03-1-0693 and N00014-04-1-0336, TITF initiative from Texas A&M University, and a Robert A. Welch Foundation grant No. A-1261.

This is the final form of the paper.

theory, etc. Despite these successful, intensely computational methods, there remains the need to understand the underlying mathematics in some relatively simple way. In this connection, the following is widely identified as some of the issues of importance:

(1) *scaling and other invariant properties* of the Hamiltonian H and the solutions;
(2) solution behaviours near the *singularities* of the Hamiltonian operator H, called the *proximal*, or *short-range*, or *coalescing* conditions;
(3) *long-range*, or *large-separation*, or *ionization*, or *dissociation* solution behaviours when one or several of the particles H move to infinity (i.e., $|\mathbf{R}_K| \to \infty$, $|\boldsymbol{r}_k| \to \infty$, cf. (2.2) below);
(4) *correlation relations* between any two particles;
(5) special choices of *trial basis* wave functions, including *configuration interactions.*

To address all (1)–(5) above in depth would require a few hundred pages of a book to do it. Here in this paper, we will provide partial answers to issues (1) and (2) by

(i) presenting the *virial theorem* as a consequence of scaling, in Section 3.1;
(ii) deriving the coalescing conditions in terms of *cusp conditions*; Section 3.2.

Some study of issues (3), (4) and (5) may be found in a long tutorial paper [4] and in the book [20], for example.

The remainder of this paper is organized as follows: In Section 2, we introduce the Schrödinger equation model with Coulomb potentials and the Born–Oppenheimer separation.

In Section 3, we prove the quantum virial theorem and derive the cusp conditions, the main mathematical properties as we have just indicated above.

In Section 4, classical solutions to the H_2^+ hydrogen molecular ion problems and two-center molecular orbitals are discussed.

In the last section, Section 5, cusps of some wave functions are illustrated, and some numerical results are also discussed, for the special case of 3-body problems.

2. The Schrödinger Equation for Atoms and Molecules and the Born–Oppenheimer Separation

We start off the technical discussion with a general problem model. The underlying theoretical basis for problems involving a few particles in atomic and molecular physics is the many-particle Schrödinger equation for the electrons and nuclei, which provides satisfactory explanations of the chemical, electromagnetic and spectroscopic properties of the atoms and molecules. Assume that the system under consideration has N_1 nuclei with masses M_K and charges eZ_K, e being the electron charge, for $K = 1, 2, \ldots, N_1$, and $N_2 = \sum_{K=1}^{N_1} Z_K$ is the number of electrons. The position vector of the Kth nucleus will be denoted as $\mathbf{R}_K$, while that of the kth electron will be $\boldsymbol{r}_k$, for $k = 1, 2, \ldots, N_2$. Let m be the mass of the electron. The Schrödinger equation for the overall system is given by

$$\text{(2.1)} \quad \mathrm{H}\Psi(\mathbf{R}, \boldsymbol{r}) = \left[-\sum_{K=1}^{N_1} \frac{\hbar^2}{2M_K} \nabla_K^2 - \sum_{k=1}^{N_2} \frac{\hbar^2}{2m} \nabla_k^2 + V(\mathbf{R}, \boldsymbol{r}) \right] \Psi(\mathbf{R}, \boldsymbol{r}) = E\Psi(\mathbf{R}, \boldsymbol{r}),$$

where H is the Hamiltonian (for the spatial wave function Ψ) and

$$\mathbf{R} = (\mathbf{R}_1, \mathbf{R}_2, \ldots, \mathbf{R}_{N_1}), \quad \boldsymbol{r} = (\boldsymbol{r}_1, \boldsymbol{r}_2, \ldots, \boldsymbol{r}_{N_2}), \tag{2.2}$$

with $\mathbf{R}_K = (X_K, Y_K, Z_K)$ and $\boldsymbol{r}_k = (x_k, y_k, z_k) \in \mathbb{R}^3$,

$$V(\mathbf{R}, \boldsymbol{r}) \equiv -\sum_{K=1}^{N_1}\sum_{k=1}^{N_2} \frac{Z_K e^2}{|\mathbf{R}_K - \boldsymbol{r}_k|} + \frac{1}{2}\sum_{\substack{k,k'=1\\ k\neq k'}}^{N_2} \frac{e^2}{|\boldsymbol{r}_k - \boldsymbol{r}_{k'}|} + \frac{1}{2}\sum_{\substack{K,K'=1\\ K\neq K'}}^{N_1} \frac{Z_K Z_{K'} e^2}{|\mathbf{R}_K - \mathbf{R}_{K'}|}$$

and

$$\nabla_K^2 = \frac{\partial^2}{\partial X_K^2} + \frac{\partial^2}{\partial Y_K^2} + \frac{\partial^2}{\partial Z_K^2}, \quad \nabla_k^2 = \frac{\partial^2}{\partial x_k^2} + \frac{\partial^2}{\partial y_k^2} + \frac{\partial^2}{\partial z_k^2},$$

being Laplacian operators, for $1 \le K \le N_1$ and $1 \le k \le N_2$. E is the energy level of the system. The terms involving the Laplacians are the *kinetic energy*, while the remaining term V in the brackets in (2.1) is the *potential energy* containing the Coulomb attracting and repelling potentials.

Equation (2.1) is a *perfect* "High-Dimensional PDEs" model for this volume and the theme of the conference.

The mathematical foundation of (2.1) was established by T. Kato in a series of important papers published (mostly) in the 1950s [13–15]. One of the primary mathematical difficulties associated with (2.1) is the singularities associated with the Coulomb potentials. Kato proved that the wave function Ψ is Hölder continuous, with bounded first order (weak) partial derivatives. However, these first-order partial derivatives are discontinuous at the singularities of the Coulomb potentials. Thus, (2.1) cannot have classical C^∞ solutions. Kato [14] went on to derive his famous *cusp conditions* at the singularities. (Such cusp conditions will be derived in Section 3.1 and visualized in Section 5.)

The full wave-mechanical model (2.1) is often too complex for practical purposes of studying molecular problems. Born and Oppenheimer [3] provide a reduced order model by separation, permitting a particularly accurate decoupling of the motions of the electrons and the nuclei. The main idea is to assume that Ψ in (2.1) takes the form of a product

$$\Psi(\mathbf{R}, \boldsymbol{r}) = G(\mathbf{R})F(\mathbf{R}, \boldsymbol{r}). \tag{2.3}$$

Substituting (2.3) into (2.1), we obtain

$$\begin{aligned} G(\mathbf{R})&\left\{\left[-\sum_{k=1}^{N_2}\frac{\hbar^2}{2m}\nabla_k^2 - \sum_{K=1}^{N_1}\sum_{k=1}^{N_2}\frac{Z_K e^2}{|\mathbf{R}_K - \boldsymbol{r}_k|} - \frac{1}{2}\sum_{\substack{k,k'=1\\ k\neq k'}}^{N_2}\frac{e^2}{|\boldsymbol{r}_k - \boldsymbol{r}_{k'}|}\right]F(\mathbf{R}, \boldsymbol{r})\right\} \\ &+ \left\{\left[-\sum_{K=1}^{N_1}\frac{\hbar^2}{2M_K}\nabla_K^2 + \frac{1}{2}\sum_{\substack{K,K'=1\\ K\neq K'}}^{N_1}\frac{Z_K Z_{K'} e^2}{|\mathbf{R}_K - \mathbf{R}_{K'}|}\right]G(\mathbf{R})\right\}F(\mathbf{R}, \boldsymbol{r}) \\ &+ \mathcal{T}_1 + \mathcal{T}_2 = EG(\mathbf{R})F(\mathbf{R}, \boldsymbol{r}), \end{aligned} \tag{2.4}$$

where

$$\mathcal{T}_1 \equiv -G(\mathbf{R}) \sum_{K=1}^{N_1} \frac{\hbar^2}{M_K} \nabla_K G(\mathbf{R}) \cdot \nabla_K F(\mathbf{R}, \boldsymbol{r}), \tag{2.5}$$

$$\mathcal{T}_2 \equiv -G(\mathbf{R}) \sum_{K=1}^{N_1} \frac{\hbar^2}{2M_K} \nabla_K^2 F(\mathbf{R}, \boldsymbol{r}). \tag{2.6}$$

These two terms, $\mathcal{T}_1$ and $\mathcal{T}_2$ are called *non-adiabatic corrections* (or *couplings*) in molecular physics. The essential step in the Born–Oppenheimer separation consists in dropping $\mathcal{T}_1$ and $\mathcal{T}_2$ in (2.4). This leads to the separation of the electronic wave function

$$\left(-\frac{\hbar^2}{2m} \sum_{k=1}^{N_2} \nabla_k^2 + \frac{1}{2} \sum_{\substack{k,k'=1 \\ k \neq k'}}^{N_2} \frac{e^2}{|\boldsymbol{r}_k - \boldsymbol{r}_{k'}|} - \sum_{K=1}^{N_1} \sum_{k=1}^{N_2} \frac{Z_K e^2}{|\mathbf{R}_K - \boldsymbol{r}_k|} \right) F(\mathbf{R}, \boldsymbol{r}) \\ = E_e(\mathbf{R}) F(\mathbf{R}, \boldsymbol{r}). \tag{2.7}$$

Equation (2.7) constitutes the basis of the quantum chemistry "software" that is available commercially. It is a static nuclei equation, where the nuclear momenta are zero. Equation (2.7) alone is deficient in making the total angular momentum conservative, but (2.5) and (2.6) can correct this deficiency. We also have a second equation for the nuclear wave function

$$\left[-\sum_{K=1}^{N_1} \frac{\hbar^2}{2M_K} \nabla_K^2 + \frac{1}{2} \sum_{\substack{K,K'=1 \\ K \neq K'}}^{N_1} \frac{Z_K Z_{K'} e^2}{|\mathbf{R}_K - \mathbf{R}_{K'}|} + E_e(\mathbf{R}) \right] G(\mathbf{R}) = E G(\mathbf{R}), \tag{2.8}$$

where $E_e(\mathbf{R})$ is the constant of separation. This equation (2.8) describes nuclear dynamics in the adiabatic approximation and is part of the new frontier in scientific research — quantum dynamics.

In "typical" molecules, the time scale for the valence electrons to orbit about the nuclei is about once every 10^{-15} sec. (and that of the inner-shell electrons is even smaller, about attoseconds, i.e., 10^{-18} sec.), that of the molecular vibration is about once every 10^{-14} sec., and that of the molecule rotation is every 10^{-12} sec. This difference of time scale is what make $\mathcal{T}_1$ and $\mathcal{T}_2$ in (2.5) and (2.6) negligible, as the electrons move so fast that they can instantaneously adjust their motions with respect to the vibration and rotation movements of the slower and much heavier nuclei.

The equation (2.7) for the electronic wave function $F(\mathbf{R}, \boldsymbol{r})$ will be the primary subject of study in this paper. It will be convenient here for us to use the *atomic units*, where the unit of length is the radius of "the first Bohr orbit" $a = \hbar^2(me^2)$, i.e., we set $\hbar^2(me^2) = 1$. Then (2.7) can be rewritten as

$$\left(-\frac{1}{2} \sum_{k=1}^{N_2} \nabla_k^2 + \frac{1}{2} \sum_{\substack{k \neq k' \\ k,k'=1}}^{N_2} \frac{1}{r_{kk'}} - \sum_{K=1}^{N_1} \sum_{k=1}^{N_2} \frac{Z_K}{r_{Kk}} \right) \Psi(\boldsymbol{r}) = E \Psi(\boldsymbol{r}), \tag{2.9}$$

where

$$r_{kk'} \equiv |\boldsymbol{r}_k - \boldsymbol{r}_{k'}|, \quad r_{Kk} = |\mathbf{R}_K - \boldsymbol{r}_k|.$$

The unit of energy E will be given in hartrees. We call equation (2.9) the Schrödinger–Born–Oppenheimer model.

Closed-form solutions to (2.9) are hard to come by in general. What is known today is the following:

(i) $\boldsymbol{N_1 = 1, N_2 = 1}$:
This is the case of hydrogen atom H (or H-like ions with a single nucleus and electron) whose solutions are known *explicitly* in closed form, available in most quantum mechanics books.

(ii) $\boldsymbol{N_1 = 2, N_2 = 1}$:
This is the H_2^+ (or H_2^+-like) two-centered molecular ion, whose solutions are separable and expressible as an infinite series of products of special functions in prolate spheroidal coordinates, where coefficients of the series are *not explicitly* given. Such are the two renowned classic solutions due to Hylleraas and Jaffé, see [4], e.g. Since such wave functions constitute the basis functions of the important class of 2-center *molecular orbitals*, we discuss them in some detail in Section 4.

(iii) $\boldsymbol{N_1 \geq 3, N_2 = 1}$:
This one-electron multi-centered molecular ion has an *analytic* solution due to Shibuya and Wulfman [21] in terms of integral equations on the unit hypersphere of the 4-dimensional momentum space.

In our study here, the special cases of 3-*body problems*, due to their simplicity, will be used as examples:

(a) $N_1 = 2, N_2 = 1$, the case (ii) above of H_2^+-like molecular ions, cf. (3.5);
(b) $N_1 = 1, N_2 = 2$, the case of He or He-like isoelectronic cases, cf. (3.6).

See Section 5.

3. Fundamental Properties of the Schrödinger–Born–Oppenheimer Model

3.1. Basic analytical properties (I): the virial theorem. We assume that the reader is familiar with the Dirac bra-ket notation. For generality, in this subsection we treat the full molecular Schrödinger model equation (2.1). The readers can easily see that by setting $N_1 = 0$ and $\hbar^2(me^2) = 1$, the treatment will go through for the reduced-order Schrödinger–Born–Oppenheimer model (2.7). The underlying Hilbert space here for equation (2.1) is $\mathcal{H}_0 \equiv L^2(\mathbb{R}^{3(N_1+N_2)})$. Associated with the Hamiltonian operator, we also need the Hilbert space $\mathcal{H}_1 \equiv W^{1,2}(\mathbb{R}^{3(N_1+N_2)})$, where $W^{1,2}$ is the Sobolev space of L^2-functions with square summable weak first-order partial derivatives.

For $\psi_1, \psi_2 \in \mathrm{H}^1(\mathbb{R}^{3(N_1+N_2)})$, define their usual L^2-product, kinetic-energy product and potential-energy product, in sequential order, by the following: for

$\mathcal{R} = (\mathbf{R}, \boldsymbol{r}) \in \mathbb{R}^{3(N_1+N_2)}$, with $\mathbf{R} \in \mathbb{R}^{3N_1}$ and $\boldsymbol{r} \in \mathbb{R}^{3N_2}$,

$$\langle \psi_1 | \psi_2 \rangle = \int_{\mathbb{R}^{3(N_1+N_2)}} \psi_1^*(\mathcal{R}) \psi_2(\mathcal{R}) d\mathcal{R},$$

$$\langle \psi_1, \psi_2 \rangle_{\text{kin}} = \int_{\mathbb{R}^{3(N_1+N_2)}} \left\{ \sum_{K=1}^{N_1} \frac{\hbar^2}{2M_K} \nabla_K \psi_1^* \cdot \nabla_K \psi_2 \right. \tag{3.1}$$
$$\left. + \sum_{k=1}^{N_2} \frac{\hbar^2}{2m} \nabla_k \psi_1^* \cdot \nabla_k \psi_2 \right\} d\mathcal{R},$$

$$\langle \psi_1, \psi_2 \rangle_{\text{pot}} = \int_{\mathbb{R}^{3(N_1+N_2)}} \psi_1^*(\mathcal{R}) V(\mathcal{R}) \psi_2(\mathcal{R}) d\mathcal{R},$$

where ψ_1^* denotes the complex conjugate of ψ_1. Then we have

$$\langle \psi_1, \psi_2 \rangle_{\text{kin}} + \langle \psi_1, \psi_2 \rangle_{\text{pot}} = \langle \psi_1 | \text{H} | \psi_2 \rangle.$$

where H is the Hamiltonian as given in (2.1). In this notation one can formulate a minimax variational principle for (2.1): consider

$$\min_{\substack{\langle \psi | \psi \rangle = 1 \\ \psi \in \mathcal{H}_1}} \langle \psi | \text{H} | \psi \rangle = E.$$

The (unique) solution Ψ_0 attaining the minimum of $\langle \Psi | \text{H} | \Psi \rangle$ is called the *ground state* and the associated value $E_0 \equiv \langle \Psi_0 | \text{H} | \Psi_0 \rangle$ is the corresponding ground state energy. *Excited states* Ψ_k with successively higher energy levels may be obtained sequentially through

$$\begin{cases} \min \langle \Psi | \text{H} | \psi \rangle \equiv E_k = \langle \Psi_k | \text{H} | \Psi_k \rangle, \text{ where } \Psi \text{ is subject to the constraints} \\ \langle \Psi | \Psi \rangle = 1, \langle \Psi | \Psi_0 \rangle = \langle \Psi | \Psi_1 \rangle = \cdots = \langle \Psi | \Psi_{k-1} \rangle = 0, \text{ for } k = 1, 2, 3, \ldots. \end{cases} \tag{3.2}$$

(The above is equivalent to the following:

$$E_0 = \min_{\substack{\langle \psi | \psi \rangle = 1 \\ \psi \in \mathcal{H}_1}} \langle \psi | \text{H} | \psi \rangle;$$

$$E_k = \max_{V_{k+1}} \min_{\substack{\langle \psi | \psi \rangle = 1 \\ \psi \in V_n \cap \mathcal{H}_1}} \langle \psi | \text{H} | \psi \rangle;$$

$V_n \equiv$ any n-dimensional subspace of $\mathcal{H}_1, n = 1, 2, \ldots$).

Let ψ_n satisfy (2.1):

$$\text{H}\psi_n = E_n \psi_n. \tag{3.3}$$

For any such eigenstate ψ_n, a useful analytic property satisfied by ψ_n is the virial theorem:

$$\langle \psi_n, \psi_n \rangle_{\text{kin}} = -\tfrac{1}{2} \langle \psi_n, \psi_n \rangle_{\text{pot}}, \tag{3.4}$$

i.e., the kinetic energy that ψ_n has is equal to negative one-half of its potential energy. This profound fact follows from the basic observation that with respect to scaling

$$\mathcal{R} \mapsto \text{k}\mathcal{R}; \quad \mathcal{R} = (\mathbf{R}, \boldsymbol{r}) \in \mathbb{R}^{3(N_1+N_2)} \quad \text{as in (2.2)},$$

the kinetic energy term scales as

$$-\sum_K \frac{\hbar^2}{2M_K} \nabla_K^2 - \sum_k \frac{\hbar^2}{2m} \nabla_k^2 \longmapsto \frac{1}{\text{k}^2} \left(-\sum_K \frac{\hbar^2}{2M_K} \nabla_K^2 - \sum_k \frac{\hbar^2}{2m} \nabla_k^2 \right),$$

while the potential energy, as the *inverse power law*, scales as

$$V(\mathrm{k}\boldsymbol{\mathcal{R}}) \longmapsto \frac{1}{\mathrm{k}} V(\boldsymbol{\mathcal{R}}).$$

Even though the proof of the virial theorem is given in several quantum mechanics books, a good number of researchers still seems to be unaware of the simple fact (3.4). Since the proof is quite elegant, we include it below.

Theorem 3.1 (Virial). *Let* H *be the Hamiltonian operator and let* ψ_n *be an eigenstate of* H *satisfying* (3.3). *Then* (3.4) *holds.*

PROOF. For any observable $\mathcal{O}$ (i.e., a self-adjoint operator), the commutator $[\mathrm{H}, \mathcal{O}] = \mathrm{H}\mathcal{O} - \mathcal{O}\mathrm{H}$ and ψ_n satisfies

$$\begin{aligned}\langle \psi_n | [\mathrm{H}, \mathcal{O}] | \psi_n \rangle &= \langle \psi_n | \mathrm{H}\mathcal{O} - \mathcal{O}\mathrm{H} | \psi_n \rangle \\ &= \langle E_n \psi_n | \mathcal{O} | \psi_n \rangle - \langle \psi_n | \mathcal{O} | E_n \psi_n \rangle \\ &= E_n^* \langle \psi_n | \mathcal{O} | \psi_n \rangle - E_n \langle \psi_n | \mathcal{O} | \psi_n \rangle \\ &= (E_n - E_n) \langle \psi_n | \mathcal{O} | \psi_n \rangle = 0.\end{aligned}$$

Now, choose

$$\begin{aligned}\mathcal{O} &= \sum_{K=1}^{N_1} \mathbf{R}_K \cdot p_K + \sum_{k=1}^{N_2} \boldsymbol{r}_k \cdot p_k \quad \left(p_j \equiv \frac{\hbar}{i} \nabla_j = \text{momentum, for } j = K \text{ and } k \right) \\ &\equiv \frac{\hbar}{i} \left[\sum_{K=1}^{N_1} \mathbf{R}_K \cdot \nabla_K + \sum_{k=1}^{N_2} \boldsymbol{r}_k \cdot \nabla_k \right].\end{aligned}$$

(This operator $\mathcal{O}$ is actually the "infinitesimal generator" of the scaling transformation if we ignore its factor $\hbar/i$. This may be viewed as follows. Let $\boldsymbol{\mathcal{R}} = (\mathbf{R}, \boldsymbol{r}) \in \mathbb{R}^{3(N_1+N_2)}$. Then $\mathcal{O} = \boldsymbol{\mathcal{R}} \cdot \nabla_{\boldsymbol{\mathcal{R}}}$, where $\nabla_{\boldsymbol{\mathcal{R}}}$ is the gradient operator in $\mathbb{R}^{3(N_1+N_2)}$. For any differentiable function f,

$$\frac{d}{d\mathrm{k}} f(\mathrm{k}\boldsymbol{\mathcal{R}})|_{\mathrm{k}=1} = \boldsymbol{\mathcal{R}} \cdot \nabla_{\boldsymbol{\mathcal{R}}} f(\boldsymbol{\mathcal{R}}) = \mathcal{O} f(\boldsymbol{\mathcal{R}}).$$

In particular, if f is invariant under scaling, i.e., $f(\mathrm{k}\boldsymbol{\mathcal{R}}) = f(\boldsymbol{\mathcal{R}})$ for any $\mathrm{k} \in \mathbb{R}$, then $\mathcal{O} f(\boldsymbol{\mathcal{R}}) = 0$.) Then a straightforward calculation shows that

$$[\mathrm{H}, \mathcal{O}] = \frac{\hbar}{i} \left[-\sum_{K=1}^{N_1} \frac{\hbar^2}{M_K} \nabla_K^2 - \sum_{k=1}^{N_2} \frac{\hbar^2}{m} \nabla_k^2 - \sum_{K=1}^{N_1} \mathbf{R}_K \cdot \nabla_K V - \sum_{k=1}^{N_2} \boldsymbol{r}_k \cdot \nabla_k V \right].$$

Therefore

$$\begin{aligned}0 &= \langle \psi_n | [\mathrm{H}, \mathcal{O}] | \psi_n \rangle \\ &= \frac{\hbar}{i} \left[2 \langle \psi_n, \psi_n \rangle_{\mathrm{kin}} - \left\langle \psi_n \left| \sum_K \mathbf{R}_K \cdot \nabla_K V + \sum_k \boldsymbol{r}_k \cdot \nabla_k V \right| \psi_n \right\rangle \right].\end{aligned}$$

But

$$\sum_K \mathbf{R}_K \cdot \nabla_K V + \sum_k \boldsymbol{r}_k \cdot \nabla_k V = -V(\mathbf{R}, \boldsymbol{r})$$

as V satisfies the inverse power law. Hence

$$\begin{aligned}2 \langle \psi_n, \psi_n \rangle_{\mathrm{kin}} + \langle \psi_n | V | \psi_n \rangle = 0 \\ \langle \psi_n, \psi_n \rangle_{\mathrm{kin}} = -\tfrac{1}{2} \langle \psi_n, \psi_n \rangle_{\mathrm{pot}}.\end{aligned}$$

The above arguments actually need to be more rigorized as eigenfunctions ψ_n in general lack regularity for the operator $[\mathrm{H}, \mathcal{O}]$. This may be done by using mollifiers. □

The virial theorem is useful in checking the accuracy and quality of numerical solutions — the farther the ratio of the kinetic energy over the potential energy deviates from $-\frac{1}{2}$, the less reliable the numerical solutions become.

Infinitesimal generators $\mathcal{O}$ from other transformations (such as rotation) may also be used to derive additional identities for wave functions. See [20, pp. 18–19], e.g.

3.2. Basic analytical properties (II): the cusp conditions. As a preparation for the subsequent material of the paper, we state the two types of 3-body problem under study here: *the* H_2^+*-like molecular ion*

$$-\frac{1}{2}\nabla^2\Psi - \left(\frac{Z_a}{r_a} + \frac{Z_b}{r_b} - \frac{Z_a Z_b}{R}\right)\Psi = E\Psi, \tag{3.5}$$

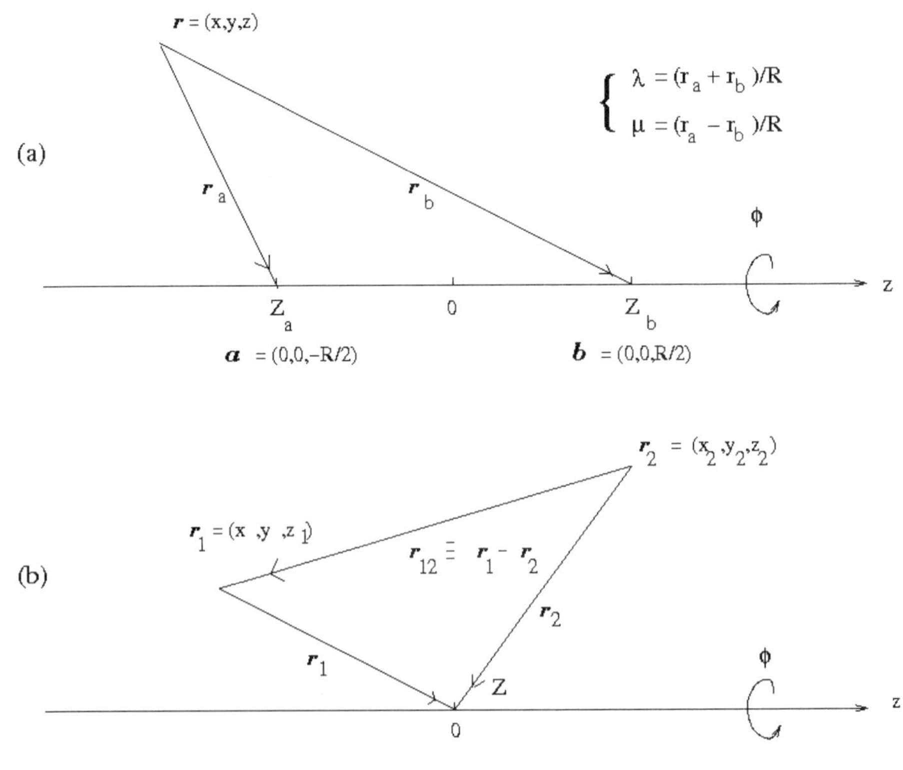

FIGURE 1. Geometry of the 3-body problem: (a) one electron-two nuclei H_2^+-like molecular ion, and (b) two electron — one nucleus He atom (or He-like ion). In part (a), the variables λ and μ, together with the azimuth angle ϕ, constitute the prolate spheroidal coordinate system.

($\nabla^2 = \partial^2/\partial x^2 + \partial^2/\partial y^2 + \partial^2/\partial z^2$, R = internuclear separation; $Z_a > 0$ and $Z_b > 0$ are positive charges, respectively, of nuclei at $\boldsymbol{a}$ and $\boldsymbol{b}$, cf. Figure 1 (a)). Equation (3.5) is separable, whose solutions are sketched in Section 4; and the *the helium* (He) *atom* (*or* He-*like ion*)

$$-\frac{1}{2}(\nabla_1^2 + \nabla_2^2)\Psi - \left(\frac{Z}{r_1} + \frac{Z}{r_2} - \frac{1}{r_{12}}\right)\Psi = E\Psi \tag{3.6}$$

($\nabla_j^2 = \partial^2/\partial x_j^2 + \partial^2/\partial y_j^2 + \partial^2/\partial z_j^2$, $j = 1, 2$, $Z > 0$ is the nuclear charge at origin, cf. Figure 1 (b)).

We wish to derive the cusp conditions for (3.5) and (3.6). Instead of treating (3.5) and (3.6) directly, let us consider the following slightly more general form of the Schrödinger equation for a 2-particle system

$$(\widehat{\mathrm{H}} - E)\psi = 0, \tag{3.7}$$

where

$$\widehat{\mathrm{H}} = -\frac{1}{2m_1}\nabla_1^2 - \frac{1}{2m_2}\nabla_2^2 - \frac{Z_a}{r_{1a}} - \frac{Z_b}{r_{1b}} - \frac{Z_a}{r_{2a}} - \frac{Z_b}{r_{2b}} + \frac{q_1 q_2}{r_{12}} + \frac{Z_a Z_b}{R}, \tag{3.8}$$

cf. the geometry and notation in Figure 1. The operator $\widehat{\mathrm{H}}$ has five sets of singularities, at

$$r_{1a} = 0, \quad r_{1b} = 0, \quad r_{2a} = 0, \quad r_{2b} = 0 \quad \text{and} \quad r_{12} = 0. \tag{3.9}$$

Note that these five sets of singularities are not mutually exclusive. It is possible to have, e.g.,

$$r_{1a} = 0, \quad r_{2a} = 0, \quad r_{12} = 0.$$

But it is not possible to have, e.g., $r_{1a} = 0$ and $r_{1b} = 0$ simultaneously.

In order for the wave function ψ to satisfy the eigenvalue problem (3.7) at the singularities (3.9), the kinetic energy operators $-\nabla_1^2/2m_1$ and $-\nabla_2^2/2m_2$, after acting on ψ, must produce terms that *exactly cancel* those singularity terms in the potential in order to give us back just a constant E times ψ, because the wave function ψ is *bounded* everywhere in space, including the points where the nuclei are located, without exception. From the numerical viewpoint, one can see that, if the cusp conditions are not satisfied, then there is some *unboundedness* at the singularities (3.9) which can affect the accuracy in numerical computations. Conversely, if the cusp conditions are satisfied, this normally improves the numerical accuracy.

We also need, as preparation, to introduce the *spherical means* of a function. Given a point $\boldsymbol{r}_0 \in \mathbb{R}^3$, the spherical means of a function $u(\boldsymbol{r})$ at $\boldsymbol{r}_0$ on the sphere with radius ρ is defined to be

$$u_{av,\rho}(\boldsymbol{r}_0) \equiv \frac{1}{4\pi}\iint_{\mathcal{S}_1} u(\boldsymbol{r}_0 + \rho\boldsymbol{\nu})\, d\omega, \tag{3.10}$$

where $\mathcal{S}_\rho$ is the sphere with radius ρ centered at $\boldsymbol{r}_0$, here $\rho = 1$; $d\omega = \sin\theta\, d\theta\, d\phi$, where ω represents all the angular variables; $\boldsymbol{\nu} = (\nu_1, \nu_2, \nu_3)$, $\nu_1^2 + \nu_2^2 + \nu_3^2 = 1$; $\boldsymbol{\nu}$ is the unit outward pointing normal vector on $\mathcal{S}_1$. It is easy to see that if $u(\boldsymbol{r})$ is continuous in a neighbourhood of $\boldsymbol{r}_0$, then the spherical means just converge to the pointwise value:

$$\lim_{\rho\to 0} u_{av,\rho}(\boldsymbol{r}_0) = u(\boldsymbol{r}_0).$$

Theorem 3.2 (Cusp conditions). *Let ψ be a generalized solution to (3.7) corresponding to a spectral value E. Then the following hold in the pointwise sense:*

(i) *the electron-nucleus cusp conditions at a and b:*

$$\begin{cases} \left.\dfrac{\partial\psi}{\partial r_{1a}}\right|_{r_{1a}=0} = -m_1 Z_a \cdot \psi|_{r_{1a}=0}, & \textit{if } r_{2a}\cdot r_{12}\neq 0; \\ \left.\dfrac{\partial\psi}{\partial r_{1b}}\right|_{r_{1b}=0} = -m_1 Z_b \cdot \psi|_{r_{1b}=0}, & \textit{if } r_{2b}\cdot r_{12}\neq 0; \\ \left.\dfrac{\partial\psi}{\partial r_{2a}}\right|_{r_{2a}=0} = -m_2 Z_a \cdot \psi|_{r_{2a}=0}, & \textit{if } r_{1a}\cdot r_{12}\neq 0; \\ \left.\dfrac{\partial\psi}{\partial r_{2b}}\right|_{r_{2b}=0} = -m_2 Z_b \cdot \psi|_{r_{2b}=0}, & \textit{if } r_{1b}\cdot r_{12}\neq 0; \end{cases} \tag{3.11}$$

(ii) *the interelectronic (or electron-electron) cusp condition*

$$\left.\frac{\partial\psi}{\partial r_{12}}\right|_{r_{12}=0} = \frac{m_1 m_2}{m_1+m_2} q_1 q_2 \cdot \psi|_{r_{12}=0}, \quad \textit{if} \quad r_{1a}\cdot r_{2a}\neq 0 \quad \textit{or} \quad r_{1b}\cdot r_{2b}\neq 0. \tag{3.12}$$

PROOF. Since the proofs of (3.11) and (3.12) are similar, we will prove the case of (3.12) as it contains more complication. We first use the center-of-mass (CM) coordinate system of two electrons by making the change of variables

$$\begin{cases} \boldsymbol{S} = \dfrac{m_1\boldsymbol{r}_1 + m_2\boldsymbol{r}_2}{m_1+m_2} & (\boldsymbol{S} \text{ is the CM coordinate}), \\ \boldsymbol{r}_{12} = \boldsymbol{r}_1 - \boldsymbol{r}_2 & (\boldsymbol{r}_{12} \text{ is the relative coordinate}), \end{cases} \tag{3.13}$$

cf. Fig. 2. With respect to the new variables $\boldsymbol{S}\in\mathbb{R}^3$, $\boldsymbol{r}_{12}\in\mathbb{R}^3$, the Laplacians are written as $\nabla^2_{\boldsymbol{S}}$ and $\nabla^2_{\boldsymbol{r}_{12}}$, respectively.

Then equation (3.7) becomes

$$(\widehat{\mathrm{H}} - E)\psi = 0, \tag{3.14}$$

where

$$\begin{aligned} \widehat{\mathrm{H}} = \frac{1}{2M}\nabla^2_{\boldsymbol{S}} - \frac{1}{2\mu}\nabla^2_{\boldsymbol{r}_{12}} &- \frac{Z_a}{|\mathbf{R}_a - m_1/(m_1+m_2)\boldsymbol{r}_{12}|} \\ &- \frac{Z_b}{|\mathbf{R}_b - m_1/(m_1+m_2)\boldsymbol{r}_{12}|} - \frac{Z_a}{|\mathbf{R}_a + m_2/(m_1+m_2)\boldsymbol{r}_{12}|} \\ &- \frac{Z_b}{|\mathbf{R}_b + m_2/(m_1+m_2)\boldsymbol{r}_{12}|} + \frac{q_1q_2}{r_{12}} + \frac{Z_aZ_b}{R}. \end{aligned} \tag{3.15}$$

(Details of the change of variables may be found in [4, pp. 208–210], e.g.) Note that the Laplacian $\nabla^2_{\boldsymbol{r}_{12}}$ may be written as

$$\nabla^2_{12} = \frac{1}{r_{12}^2}\frac{\partial}{\partial r_{12}}\left(r_{12}^2\frac{\partial}{\partial r_{12}}\right) - \frac{L^2}{r_{12}^2}, \tag{3.16}$$

where L^2 is the Legendrian (i.e., the angular momentum operator) associated with the angular variables of $\boldsymbol{r}_{12}$ (θ,ϕ are spherical coordinates):

$$L^2 \equiv \frac{1}{\sin\theta}\frac{\partial}{\partial\theta}\left(\sin\theta\frac{\partial}{\partial\theta}\right) + \frac{1}{\sin^2\theta}\frac{\partial^2}{\partial\phi^2}; \quad 0\le\theta\le\pi,\ 0\le\phi\le 2\pi.$$

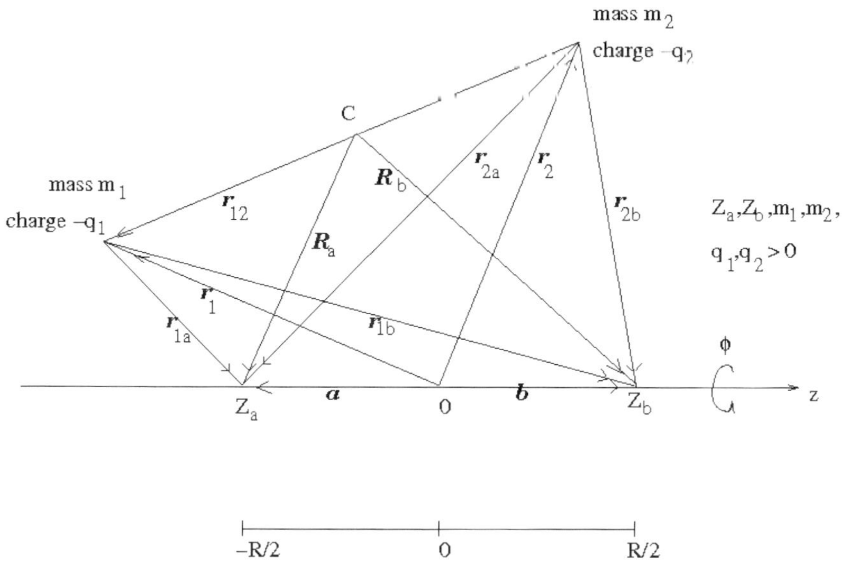

FIGURE 2. Two nuclei, with positive charges Z_a and Z_b, are located at $\mathbf{R}_1 \equiv \boldsymbol{a} = (0, 0, -R/2)$ and $\mathbf{R}_2 \equiv \boldsymbol{b} = (0, 0, R/2)$ on the z-axis. the internuclear separation is R. Two negatively charged particles with masses m_1 and m_2 and charges $-q_1$ and q_2, respectively, are located at $\boldsymbol{r}_1 = (x_1, y_1, z_1)$ and $\boldsymbol{r}_2 = (x_2, y_2, z_2)$. The vector from the first electron to the nucleus Z_a is $\boldsymbol{r}_{1a}$, while that from the second electron to the nucleus Z_b is $\boldsymbol{r}_{2b}$. The vectors $\boldsymbol{r}_{1b}$ and $\boldsymbol{r}_{2a}$ are defined similarly. C is the origin of the center-of-mass coordinates. The relative coordinate vector is $\boldsymbol{r}_{12} = \boldsymbol{r}_1 - \boldsymbol{r}_2$. The vectors $\mathbf{R}_a$, $\mathbf{R}_b$ are defined by

$$\mathbf{R}_a = \frac{m_1}{m_1 + m_2}\boldsymbol{r}_{12} + \boldsymbol{r}_{1a}, \quad \mathbf{R}_b = -\frac{m_1}{m_1 + m_2}\boldsymbol{r}_{12} + \boldsymbol{r}_{2b}.$$

As far as the radial variable r_{12} is concerned in (3.15) and (3.16), the equation is *regular singular* at $r_{12} = 0$:

$$(3.17) \qquad -\frac{1}{2\mu}\frac{1}{r_{12}^2}\frac{\partial}{\partial r_{12}}\left(r_{12}^2\frac{\partial \psi}{\partial r_{12}}\right) + \left(-\frac{1}{2\mu}\frac{L^2}{r_{12}^2} + \frac{q_1 q_2}{r_{12}}\right)\psi.$$

Therefore near $r_{12} \neq 0, \psi$ admits a power series solution

$$(3.18) \quad \psi(\boldsymbol{r}_{12}, \boldsymbol{S}) = r_{12}^{\nu}\left[\psi(0, \boldsymbol{S}) + \left.\frac{\partial \psi}{\partial r_{12}}\right|_{r_{12}=0} \cdot C_1(\omega, S) r_{12} + \frac{1}{2!}\left.\frac{\partial^2 \psi}{\partial r_{12}^2}\right|_{r_{12}=0} \cdot C_2(\omega, S) r_{12}^2 + \cdots\right],$$

where ω represents the spherical angular variables θ and ϕ of the vector $\boldsymbol{r}_{12}$, and ν is the usual power determined from the *indicial equation* associated with any ordinary differential equation near a regular-singular point (here $r_{12} = 0$):

$$(3.19) \qquad \nu(\nu + 1) - \ell(\ell + 1) = 0, \quad \ell = 0, 1, 2, \ldots,$$

where we have used the eigenvalues $-\ell(\ell+1)$ for the Legendrian operator Λ^2 (whose eigenfunctions are the spherical harmonics $Y_{\ell m}$, $m = -\ell, -\ell+1, \ldots, -1, 0, 1, 2, \ldots, \ell$). Equation (3.19) has roots

$$\nu = \ell \quad \text{or} \quad \nu = -(\ell+1), \qquad \ell = 0, 1, 2, \ldots .$$

The second root $\nu = -(\ell+1)$ is immediately ruled out as its causes (3.18) to be discontinuous at $r_{12} = 0$, violating the property that the wave function is (Hölder) continuous everywhere in space. Thus $\nu = \ell$, or nonnegative integer. Therefore, we can just write (3.18) as

$$\psi(\boldsymbol{r}_{12}, \boldsymbol{S}) = \psi(\boldsymbol{0}, \boldsymbol{S}) + r_{12} \frac{\partial \psi(r_{12}, \omega, \boldsymbol{S})}{\partial r_{12}}\bigg|_{r_{12}=0} + \frac{r_{12}^2}{2!} \frac{\partial^2 \psi(r_{12}, \omega, \boldsymbol{S})}{\partial r_{12}^2}\bigg|_{r_{12}=0} + \cdots (\text{terms } \mathcal{O}(r_{12}^3)), \tag{3.20}$$

with the understanding that the first ℓ terms in the above are zero if $\ell \geq 1$.

For $\widehat{\mathrm{H}}$ given in the form (3.15), we now integrate $\widehat{\mathrm{H}}\Psi$ over a small 3-dimensional ball B_{ρ_0} with radius ρ_0 centered at $r_{12} = 0$, for any $\boldsymbol{S} \in \mathbb{R}^3$, $\boldsymbol{S} \neq \boldsymbol{0}$:

$$\begin{aligned} \iiint_{B_{\rho_0}} \bigg\{ & -\frac{1}{2M} \nabla_S^2 \psi(\boldsymbol{r}_{12}, \boldsymbol{S}) - \frac{1}{2\mu} \nabla_{12}^2 \psi(\boldsymbol{r}_{12}, \boldsymbol{S}) - \frac{Z_a}{|\mathbf{R}_a - m_1/(m_1+m_2)\boldsymbol{r}_{12}|} \psi(\boldsymbol{r}_{12}, \boldsymbol{S}) \\ & - \frac{Z_b}{|\mathbf{R}_b - m_1/(m_1+m_2)\boldsymbol{r}_{12}|} \psi(\boldsymbol{r}_{12}, \boldsymbol{S}) - \frac{Z_a}{|\mathbf{R}_a - m_2/(m_1+m_2)\boldsymbol{r}_{12}|} \psi(\boldsymbol{r}_{12}, \boldsymbol{S}) \\ & - \frac{Z_b}{|\mathbf{R}_b + m_2/(m_1+m_2)\boldsymbol{r}_{12}|} \psi(\boldsymbol{r}_{12}, \boldsymbol{S}) + \frac{q_1 q_2}{r_{12}} \psi(\boldsymbol{r}_{12}, \boldsymbol{S}) \\ & + \frac{Z_a Z_b}{R} \psi(\boldsymbol{r}_{12}, \boldsymbol{S}) \bigg\} \, d\boldsymbol{r}_{12} = 0. \end{aligned}$$

Note that as $\rho_0 \to 0$, the integrals of all the terms above vanish (by their continuity at $r_{12} = 0$) as $\mathbf{R}_a \neq \mathbf{0}$ and $\mathbf{R}_b \neq \mathbf{0}$ due to the assumptions that $r_{1a} \cdot r_{2a} \neq 0$ and $r_{1b} \cdot r_{2b} \neq 0$, except possibly those of

$$-\frac{1}{2\mu} \nabla_{12}^2 \psi(\boldsymbol{r}_{12}, \boldsymbol{S}), \qquad \frac{q_1 q_2}{r_{12}} \psi(\boldsymbol{r}_{12}, \boldsymbol{S}).$$

Thus, we need only consider

$$\lim_{\rho_0 \to 0} \iiint_{B_{\rho_0}} \left[-\frac{1}{2\mu} \nabla_{12}^2 \psi(\boldsymbol{r}_{12}, \boldsymbol{S}) + \frac{q_1 q_2}{r_{12}} \psi(\boldsymbol{r}_{12}, \boldsymbol{S}) \right] d\boldsymbol{r}_{12} = 0. \tag{3.21}$$

Apply the Gauss Divergence Theorem to the first term of the integral to get

$$\begin{aligned} \iiint_{B_{\rho_0}} \left(-\frac{1}{2\mu}\right) \nabla_{12}^2 \psi(\boldsymbol{r}_{12}, \boldsymbol{S}) d\boldsymbol{r}_{12} &= -\frac{1}{2\mu} \iint_{S_{\rho_0}} \frac{\partial \psi}{\partial r_{12}}(\boldsymbol{r}_{12}, \boldsymbol{S}) dS_{\rho_0} \\ &= -\iiint_{B_{\rho_0}} \frac{q_1 q_2}{r_{12}} \psi(\boldsymbol{r}_{12}, \boldsymbol{S}) d\boldsymbol{r}_{12}. \end{aligned} \tag{3.22}$$

But $dS_{\rho_0} = \rho_0^2 \, d\omega$; thus

$$\begin{aligned} \iint_{S_{\rho_0}} \frac{\partial \psi}{\partial r_{12}} dS_{\rho_0} = \iint_{S_1} \frac{\partial \psi}{\partial r_{12}} \, d\omega_1 \cdot \rho_0^2 &= \rho_0^2 \iint_{S_1} \frac{\partial \psi(\boldsymbol{r}_{12}, \boldsymbol{S})}{\partial r_{12}} \, d\omega \\ &= 4\pi \rho_0^2 \left[\frac{\partial \psi(\boldsymbol{0}, \boldsymbol{S})}{\partial r_{12}} \right]_{av, \rho_0} \end{aligned} \tag{3.23}$$

where $\mathcal{S}_1 = \mathcal{S}_{\rho_0}|_{\rho_0=1}$. By using spherical coordinates, $d\boldsymbol{r}_{12} = r_{12}^2\, d\omega\, dr_{12}$, we have

$$\iiint_{B_{\rho_0}} \frac{q_1 q_2}{r_{12}} \psi(\boldsymbol{r}_{12}, \boldsymbol{S}) d\boldsymbol{r}_{12} = \int_0^{\rho_0} \left[\iint_{\mathcal{S}_\rho} \frac{q_1 q_2}{r_{12}} \psi(\boldsymbol{r}_{12}, \boldsymbol{S})\, d\omega \right] r_{12}^2 dr_{12}$$

$$= q_1 q_2 \int_0^{\rho_0} \left\{ \iint_{\mathcal{S}_\rho} [\psi(\boldsymbol{0}, \boldsymbol{S}) + \varepsilon(\boldsymbol{r}_{12}, \boldsymbol{S})]\, d\omega \right\} r_{12}\, dr_{12}, \tag{3.24}$$

where $\varepsilon(\boldsymbol{r}_{12}, \boldsymbol{S}) \to 0$ as $\boldsymbol{r}_{12} \to \boldsymbol{0}$. This follows from the fact that $\psi(\boldsymbol{r}_{12}, \boldsymbol{S})$ is continuous at $\boldsymbol{r}_{12} = \boldsymbol{0}$ for any $\boldsymbol{S}$.

$$(\text{Continuing from the above}) \equiv q_1 q_2 \left[\psi(\boldsymbol{0}, \boldsymbol{S}) \int_0^{\rho_0} r_{12}\, dr_{12} + \tilde{\varepsilon}(\rho_0, \boldsymbol{S}) 4\pi \frac{\rho_0^2}{2} \right]$$

$$= q_1 q_2 \psi(\boldsymbol{0}, \boldsymbol{S}) 4\pi \cdot \frac{\rho_0^2}{2} + 2\pi \rho_0^2 \tilde{\varepsilon}(\rho_0, \boldsymbol{S}).$$

From (3.21), (3.22), (3.23) and (3.24), we now get

$$-\frac{1}{2\mu} \rho_0^2 \left[\frac{\partial \psi(\boldsymbol{0}, \boldsymbol{S})}{\partial r_{12}} \right]_{av, \rho_0} = -\frac{1}{2} \rho_0^2 q_1 q_2 \psi(\boldsymbol{0}, \boldsymbol{S}) + \frac{\rho_0^2}{2} \tilde{\varepsilon}(\rho_0, \boldsymbol{S}).$$

Dividing all the terms above by ρ_0^2 and let $\rho_0 \to 0$, we have $\tilde{\varepsilon}(\rho_0, \boldsymbol{S}) \to 0$ and, therefore,

$$\lim_{\rho_0 \to 0} \left[\frac{\partial \psi(\boldsymbol{0}, \boldsymbol{S})}{\partial r_{12}} \right]_{av, \rho_0} = \frac{\partial \psi(\boldsymbol{0}, \boldsymbol{S})}{\partial r_{12}} = \mu q_1 q_2 \psi(\boldsymbol{0}, \boldsymbol{S}), \text{for all } \boldsymbol{S} \in \mathbb{R}^3, \boldsymbol{S} \neq \boldsymbol{0}. \tag{3.25}$$

Similarly, one can derive the electron-nucleus cusp condition at, e.g., $r_{1a} = 0$, for (3.7) to be the following:

$$\lim_{\rho_0 \to 0} \left[\frac{\partial \psi(\boldsymbol{r}_1, \boldsymbol{r}_2)}{\partial r_{1a}} \right]_{av, \rho_0} = \frac{\partial \psi(\boldsymbol{r}_1, \boldsymbol{r}_2)}{\partial r_{1a}} = -m_1 Z_a \psi(\boldsymbol{r}_1, \boldsymbol{r}_2)|_{r_{1a}=0}, \tag{3.26}$$

$$\text{for all } \boldsymbol{r}_2, \boldsymbol{r}_{2a} \neq \boldsymbol{0}. \quad \square$$

Corollary 3.1. *Assume that $m_1 = m_2$ and $q_1 = q_2 = -1$ in (3.7). Let $(\boldsymbol{S}, \boldsymbol{r}_{12})$ denote the CM and relative coordinates and ω_{12} denote the angular variables of the vector $\boldsymbol{r}_{12}$. Let ψ be a nontrivial solution of (3.7) such that its local Taylor expansion near $r_{12} = 0$ is of the form*

$$\psi(\boldsymbol{r}_1, \boldsymbol{r}_2) = C_0(\boldsymbol{S}) + C_1(\boldsymbol{S}) r_{12} + \mathcal{O}(r_{12}^2), \quad \textit{for } r_{12} \textit{ small}, \tag{3.27}$$

where $C_0(\boldsymbol{S})$ and $C_1(\boldsymbol{S})$ are independent of ω_{12} while the remainder satisfies $\mathcal{O}(r_{12}^2) = C_3(\boldsymbol{r}_{12}, \boldsymbol{S}) r_{12}^2$ for some bounded function C_3 which depends on $\boldsymbol{r}_{12}$ and $\boldsymbol{S}$. Then $C_1(\boldsymbol{S}) = \frac{1}{2} C_0(\boldsymbol{S})$ for all $\boldsymbol{S}$ and, consequently,

$$\psi(\boldsymbol{r}_1, \boldsymbol{r}_2) = C_0(\boldsymbol{S}) \big(1 + \tfrac{1}{2} r_{12}\big) + \mathcal{O}(r_{12}^2). \tag{3.28}$$

PROOF. We substitute the RHS of (3.27) into (3.25). There is no need to take the spherical mean on the left of (3.25) anymore as the dominant term on the RHS of (3.28) does not depend on the angular variables of $\boldsymbol{r}_{12}$. We therefore obtain

$$C_1(\boldsymbol{S}) = \mu q_1 q_2 C_0(\boldsymbol{S}) = \frac{1}{2} C_0(\boldsymbol{S}), \text{ as } \mu = \frac{m_1 m_2}{m_1 + m_2} = \frac{1}{2}, \quad q_1 = q_2 = -1.$$

Hence

$$\begin{aligned}\psi(\boldsymbol{r}_1,\boldsymbol{r}_2) &= C_0(\boldsymbol{S}) + \tfrac{1}{2}C_0(\boldsymbol{S})r_{12} + \mathcal{O}(r_{12}^2)\\ &= C_0(\boldsymbol{S})\big[1+\tfrac{1}{2}r_{12}\big] + \mathcal{O}(r_{12}^2),\end{aligned}$$

which is (3.27). □

Similarly, we can obtain the cusp conditions at $r_{1a}=0$, $r_{1b}=0$, $r_{2a}=0$ and $r_{2b}=0$, as given in the following.

Corollary 3.2. *Let ψ be either a nontrivial solution of (3.7) or a trial wave function for (3.7). Let r_{1a} and ω_{1a} denote, respectively, the radial and angular variables of the vector $\boldsymbol{r}_{1a}$. Assume that for r_{1a} sufficiently small, ψ satisfies the Taylor expansion*

(3.29) $$\psi(\boldsymbol{r}_1,\boldsymbol{r}_2) = C_0(\boldsymbol{r}_2) + C_1(\boldsymbol{r}_2)r_{1a} + \mathcal{O}(r_{1a}^2), \quad r_{1a} \text{ small},$$

for some functions C_0 and C_1 depending on $\boldsymbol{r}_2$ only, where the remainder satisfies $\mathcal{O}(r_{1a}^2) = C_3(\boldsymbol{r}_1,\boldsymbol{r}_2)r_{1a}^2$ for some bounded function C_3 depending on $\boldsymbol{r}_1$ and $\boldsymbol{r}_2$. Then

(3.30) $$\psi(\boldsymbol{r}_1,\boldsymbol{r}_2) = C_0(\boldsymbol{r}_2)(1-m_1Z_ar_{1a}) + \mathcal{O}(r_{1a}^2), \quad r_{1a} \text{ small},$$

where $\mathcal{O}(r_{1a}^2) = C_3(\boldsymbol{r}_1,\boldsymbol{r}_2)r_{1a}^2$ for some smooth function C_3.

PROOF. The kinetic energy term $-1/(2m_1)\nabla_1^2$, after the translation

$$z_1 \to z_1 + \frac{R}{2}, \quad \text{cf. Figure 2 for notation,}$$

becomes $-1/(2m_1)\nabla^2_{\boldsymbol{r}_{1a}}$ which is centered at $(x_1,y_1,z_1)=(0,0-R/2)$. Now apply the Hamiltonian

(3.31) $$\widehat{\mathrm{H}} = -\frac{1}{2m_1}\nabla^2_{\boldsymbol{r}_{1a}} - \frac{Z_a}{r_{1a}} + \left[-\frac{1}{2m_2}\nabla_2^2 - \frac{Z_b}{r_{1b}} - \frac{Z_a}{r_{2a}} - \frac{Z_b}{r_{2b}} + \frac{q_1q_2}{r_{12}} + \frac{Z_aZ_b}{R}\right]$$

to (3.29). We need only focus our attention locally near $r_{1a}=0$ and ignore the terms in the bracket of (3.31) as they have no effect on the singularity at $r_{1a}=0$. We obtain

$$\begin{aligned}(\widehat{\mathrm{H}}-E)\psi &= -\frac{1}{2m_1}\frac{1}{r_{1a}^2}\frac{\partial}{\partial r_{1a}}[r_{1a}^2\cdot C_1(\boldsymbol{r}_2)] - \frac{Z_a}{r_{1a}}[C_0(\boldsymbol{r}_2)+C_1(\boldsymbol{r}_2)r_{1a}]\\ &\qquad + \text{ higher order terms in } r_{1a}\\ &= -\frac{1}{2m_1}\frac{2C_1(\boldsymbol{r}_2)}{r_{1a}} - \frac{Z_a\cdot C_0(\boldsymbol{r}_2)}{r_{1a}} + \text{ higher order terms in } r_{1a}.\end{aligned}$$

To eliminate the singularity at $r_{1a}=0$ above, it is necessary that

$$\frac{C_1(\boldsymbol{r})}{m_1} + Z_aC_0(\boldsymbol{r}_2) = 0.$$

The above gives (3.30), as desired. □

Note that if we have the *coalescing of three particles*, say two electrons get close together near a nucleus, then the assumptions of Theorem 3.2 are violated and the cusp conditions as given therein no longer holds. Fock [7, 8] has shown that the expansion (3.18) (now with $\nu > 0$ therein) must be modified to a more sophisticated form incorporating logarithmic dependence on functions of several of the variables r_{12}, r_{1a}, r_{1b}, r_{2a}, r_{2b}, etc. Thus, one can see that the intrinsic optimal (Sobolev space) regularity properties of the wave functions can be quite complicated.

Graphics of wave functions featuring the electron-nucleus conditions will be displayed in Section 5.

4. The H_2^+-like molecular ion

As one of the 3-body problems, H_2^+ (-like) molecular ion, with two nuclei Z_a^+ and Z_b^+ sharing one covalent electron as depicted in Figure 1 (a), is the simplest molecule in the universe. There are significant mathematical interests in understanding the ground and excited states of such a basic molecule. In its general form of (2.1), H_2^+ is not separable. But the Schrödinger–Born–Oppenheimer model (3.5) is separable in prolate spheroidal coordinates. Define $\lambda = (r_a + r_b)/R$, $\mu = (r_a - r_b)/R$, $\phi =$ the usual azimuth or longitudinal angle about the z-axis, cf. the geometry in Figure 1 (a). From the triangle inequality, one can immediately deduce that $\lambda \geq 1$, $-1 \leq \mu \leq 1$. The transformation from prolate spheroidal to Cartesian coordinates is given by

$$x = \frac{R}{2}\sqrt{(\lambda^2-1)(1-\mu^2)}\cos\phi,\ y = \frac{R}{2}\sqrt{(\lambda^2-1)(1-\mu^2)}\sin\phi,\ z = \frac{R}{2}\lambda\mu.$$

Write Ψ in (3.5) as

$$\Psi = \Lambda(\lambda)M(\mu)e^{im\phi}; \quad \lambda \geq 1, -1 \leq \mu \leq 1, 0 < \phi < 2\pi, m = \text{ an integer.}$$

Then Λ and M separate, satisfying

$$\text{(4.1)} \quad \frac{d}{d\lambda}\left[(\lambda^2-1)\frac{d\Lambda}{d\lambda}\right] + \left[A + 2R_1\lambda - p^2\lambda^2 - \frac{m^2}{\lambda^2-1}\right]\Lambda = 0, \quad \left(R_1 \equiv \frac{R(Z_a+Z_b)}{2}\right),$$

$$\text{(4.2)} \quad \frac{d}{d\mu}\left[(1-\mu^2)\frac{dM}{d\mu}\right] + \left[-A - 2R_2\mu + p^2\mu^2 - \frac{m^2}{1-\mu^2}\right]M = 0, \quad \left(R_2 \equiv \frac{R(Z_a-Z_b)}{2}\right),$$

where two separation constants A and p^2 have been introduced, with $p^2 \equiv \frac{1}{2}(-R^2E + RZ_aZ_b)$, which are yet to be determined from an eigenvalue problem. Note that the two ODEs satisfied by Λ and M are similar, so they can be solved in a very similar way.

Two solutions to the Λ-equation above are well known:

(1) Jaffé's [12]:

$$\text{(4.3)} \quad \Lambda(\lambda) = e^{-p\lambda}(\lambda^2-1)^{|m|/2}(\lambda+1)^{\sigma}\sum_{n=0}^{\infty} g_n \cdot \left(\frac{\lambda-1}{\lambda+1}\right)^n, \left(\sigma \equiv \frac{R_1}{p} - |m| - 1\right).$$

This leads to a 3-term recurrence relation solvable and approximable by continued fractions:

$$\text{(4.4)} \quad \alpha_n g_{n-1} - \beta_n g_n + \gamma_n g_{n+1} = 0, \quad n = 0, 1, 2, \ldots,; \quad g_{-1} \equiv 0,$$

$\alpha_n = (n-1-\sigma)(n-1-\sigma-m)$, $\beta_n = 2n^2 + (4p-2\sigma)n - A + p^2 - 2p\sigma - (m+1)(m+\sigma)$, $\gamma_n = (n+1)(n+m+1)$;

$$\frac{\beta_0}{\gamma_0} = \cfrac{\alpha_1}{\beta_1 - \cfrac{\gamma_1\alpha_2}{\beta_2 - \cfrac{\gamma_2\alpha_3}{\beta_3 - \dots}}}. \tag{4.5}$$

(2) Hylleraas' [11]:

$$\Lambda(\lambda) = e^{-p(\lambda-1)}(\lambda^2-1)^{|m|/2}\sum_{n=0}^{\infty}\frac{c_n}{(m+n)!}L^m_{m+n}(x), \quad x \equiv 2p(\lambda-1), \tag{4.6}$$

where $L^m_{m+n}(x)$ is the associated Laguerre polynomial and c_n also satisfies a 3-term recurrence relation just as (4.5) as above (with g_n replaced by c_n), such that

$$\begin{aligned}\alpha_n &= (n-m)(n-m-1-\sigma), \beta_n = 2(n-m)^2 + 2(n-m)(2p-\sigma) \\ &\qquad - [A - p^2 + 2p\sigma + (m+1)(m+\sigma)], \\ \gamma_n &= (n+1)(n-2m-\sigma).\end{aligned}$$

For large λ, one can show that

$$\Lambda(\lambda) = a_0 e^{-p\lambda}\lambda^\beta \sum_{j=0}^{\infty} c_j\lambda^{-j}, \quad \beta \equiv -1 + R_1/p, \qquad p > 0. \tag{4.7}$$

Therefore the wave function Ψ decays exponentially in the radial direction.

The M-equation in (4.2) can be solved by first writing

$$M(\mu) = e^{\pm p\mu}\widetilde{M}(\mu), \quad -1 \le \mu \le 1, \tag{4.8}$$

in order to eliminate the $p^2\mu^2$ term in (4.2). We obtain

$$\begin{aligned}&[(1-\mu^2)\widetilde{M}']' \pm 2p(1-\mu^2)\widetilde{M}' \\ &\qquad + \left[(-2R_2 \mp 2p)\mu + (p^2 - A) - \frac{m^2}{1-\mu^2}\right]\widetilde{M} = 0.\end{aligned} \tag{4.9}$$

To simplify notation, let us just consider the case $M(\mu) = e^{-p\mu}\widetilde{M}(\mu)$, but note that for $M = e^{p\mu}\widetilde{M}(\mu)$, we need only make the changes of $p \to -p$ in (4.12) below. Write

$$M(\mu) = e^{-p\mu}\sum_{k=0}^{\infty} f_k P^m_{m+k}(\mu), \tag{4.10}$$

where $P^m_n(\mu)$ are the associated Legendre polynomials, and substitute (4.10) into (4.2). We obtain a 3-term recurrence relation

$$\alpha_n f_{n-1} - \beta_n f_n + \gamma_n f_{n+1} = 0; \quad n = 0, 1, 2, \dots; \ f_{-1} \equiv 0, \tag{4.11}$$

where

$$\begin{cases} \alpha_n = \dfrac{1}{2(m+n)-1}[-2nR_2 + 2pn(m+n)], \\ \beta_n = A - p^2 + (m+n)(m+n+1), \\ \gamma_n = \dfrac{2m+n+1}{2(m+n)+3}\{-2R_2 - 2p(m+n+1)\}, \end{cases} \tag{4.12}$$

and, consequently, again the continued fractions of the same form as (4.5). The continued fractions obtained here should be *coupled* with the continued fraction (4.5) for the variable μ to solve A and p.

In the *homonuclear* case, $R_2 = R(Z_a - Z_b)/2 = 0$, equation (4.2) reduces to

$$[(1-\mu^2)M']' + \left(-A + p^2\mu^2 - \frac{m^2}{1-\mu^2}\right)M = 0.$$

In this case, several optional representations of M can be used:

$$(4.13)\qquad \begin{cases} \text{(a) } M(\mu) = (1-\mu^2)^{|m|/2}\sum_{k=0}^{\infty} c_k\mu^{2k}, \\ \quad\;\; M(\mu) = (1-\mu^2)^{|m|/2}\sum_{k=0}^{\infty} c_k\mu^{2k+1}; \\ \\ \text{(b) } M(\mu) = (1-\mu^2)^{|m|/2}\sum_{k=0}^{\infty} c_k P_{m+2k}^m(\mu), \\ \quad\;\; M(\mu) = (1-\mu^2)^{|m|/2}\sum_{k=0}^{\infty} c_k P_{m+2k+1}^m(\mu); \\ \\ \text{(c) } M(\mu) = e^{\pm p\mu}(1-\mu^2)^{|m|/2}\sum_{k=0}^{\infty} c_k(1\mp\mu)^k. \end{cases}$$

Truncated solutions for $\Lambda(\lambda)M(\mu)e^{im\phi}$ can be used as *trial basis functions* in the modeling and computations of simple molecules, while Corollaries 3.1 and 3.2 can be applied to secure that the cusp conditions are satisfied. Such trial functions are called 2-*centered (molecular) orbitals*, in contrast to the (1-centered) *atomic orbitals* which originate from truncated wave functions of the hydrogen atom. A French school, especially exemplified in the work of Le Sech [16–18], has achieved admirable success in using 2-centered orbitals to compute diatomic molecules.

The following is a simple example.

Example 1 (2-centered orbitals satisfying the cusp conditions). Suppose we are modeling the ground state of a 1-electron homonuclear molecular ion whose Hamiltonian is given by

$$\mathrm{H} = -\frac{1}{2}\nabla^2 - \frac{Z}{r_a} - \frac{Z}{r_b}, \quad \left(\nabla^2 = \frac{\partial^2}{\partial x_1^2} + \frac{\partial^2}{\partial x_2^2} + \frac{\partial^2}{\partial x_3^2}\right),$$

where the geometry is given as in Figure 1 (a), with $Z_a = Z_b \equiv Z > 0$. Motivated by the Jaffé solution (4.3) and (4.13) (b) with $m = 0$, we consider a trial function

$$(4.14)\qquad \psi(\lambda,\mu) = e^{-\alpha\lambda}[1 + BP_2(\mu)],$$

where P_2 is the Legendre polynomial of degree 2 and α and B are two free parameters. Then the cusp conditions (Theorem 3.2) dictate that at $r_a = 0$ and $r_b = 0$,

$$(4.15)\qquad \alpha(1+B) + 3B = RZ(1+B).$$

On the other hand, if we are modeling the ground state of a 1-electron *heteronuclear* molecular ion with Hamiltonian

$$\mathrm{H} = -\frac{1}{2}\nabla^2 - \frac{Z_a}{r_a} - \frac{Z_b}{r_b}, \quad Z_a \not\equiv Z_b;\ Z_a, Z_b > 0,$$

then (4.14) is no longer appropriate. We need to use

$$\psi(\lambda,\mu) = e^{-\alpha\lambda}e^{\beta\mu}[1 + B_1P_1(\mu) + B_2P_2(\mu)],$$

where α, β, B_1, B_2 are free parameters yet to be determined. The cusp conditions at $r_a = 0$ and $r_b = 0$ then dictate that

$$(4.16)\quad \begin{cases} B_1 = -\dfrac{3(RZ_a+2\beta-RZ_b)}{2(3+4\alpha-2RZ_a-2RZ_b-RZ_a\alpha-RZ_a\beta+R^2Z_aZ_b-\alpha RZ_b+\beta RZ_b+\alpha^2-\beta^2)}, \\ B_2 = -\dfrac{(2\alpha^2+2\alpha-2RZ_a\alpha-2\alpha RZ_b-RZ_a+2\beta RZ_b-2\beta^2+2R^2Z_aZ_b-RZ_b-2RZ_a\beta)}{2(3+4\alpha-2RZ_a-2RZ_b-RZ_a\alpha-RZ_a\beta+R^2Z_aZ_b-\alpha RZ_b+\beta RZ_b+\alpha^2-\beta^2)}. \end{cases}$$

The derivations of (4.15) and (4.16) can be found in [4, Appendix H].

5. Additional Numerical Examples and Graphical Illustrations for 3-Body Problem

In Section 5.1, we use the finite element method (FEM) to carry out some 1-electron problems and illustrate the shape of cusps. In Section 5.2, we review some numerical results for the 2-electron helium problem obtained with the commercial Gaussian software.

5.1. Visualization of cusps from FEM solutions of H_2^+. As we noted earlier, the model equation (3.5) for H_2^+-like molecular ions has separable solutions — in prolate spheroidal coordinates. Such "exact" solutions provide values of E which can be approximated to any desirable degree of accuracy.

The numerical method to be employed in this subsection is the finite element method (FEM). FEM, based on the Rayleigh–Ritz and Galerkin principles, is a powerful numerical method for solving boundary value problems in mathematical physics. However, we have not yet seen many examples actually applying FEM to compute (3.5) in the literature. A notable exception is the paper by Yu and Bandrauk [23], where a time-dependent Schrödinger equation for the molecules H_2^+, H_2 and H_3^+ in intense laser fields is computed by FEM and Rabi oscillations are shown.

In this section, we utilize the property that analytic solution Ψ of (3.5) decays exponentially with respect to the λ-variable of the prolate spheroidal coordinates where (3.5) is separable, (4.7). Thus by truncating the entire space $\mathbb{R}^3$ to a ball with a reasonably large radius, we should still be able to capture the solutions with good numerical accuracy.

In truncating the unbounded space $\mathbb{R}^3$ to a ball B_{r_0} with radius r_0, we also need to impose some artificial boundary condition on the boundary ∂Br_0 of B_{r_0} (which is a sphere with radius r_0 centered at the origin). One may try to set

$$\Psi = 0 \text{ (Dirichlet)}, \quad \frac{\partial \Psi}{\partial n} = 0 \text{ (Neumann)} \quad \text{or} \quad \Psi + \alpha\frac{\partial \Psi}{\partial n} = 0;\ \alpha > 0 \text{ (Robin)},$$

on ∂B_{r_0}. We have found through numerical experiments that the outcomes of such are all similar with accuracy of 5 digits or more. A good choice for r_0 here is $r_0 = 40R$, with R being the internuclear distance. We also choose $\Psi = 0$ as the artificial boundary condition.

We choose a family of finite element approximation spaces V_{h_j} on B_{r_0}, where h_j signifies the mesh size parameter. V_{h_j} is defined by

$$(5.1)\qquad V_{h_j} = \text{span}\{\phi_{h_j}^{(k)}(x,y,z) \mid k = 1, 2, \ldots, n(h_j)\},$$

where the finite element basis functions $\phi_{h_j}^{(k)}$ satisfy certain smoothness and Lagrange (or Hermite) interpolation properties and are locally supported, with a total number of $n(h_j)$ independent basis of functions. An approximate solution for Ψ is represented as

$$(5.2)\quad \Psi_{h_j} = \sum_{k=1}^{n(h_j)} c_{k,j}\phi_{h_j}^{(k)}; \quad c_{1,j}, c_{2,j}, \ldots, c_{n(h_j),j} \text{ are coefficients to be computed.}$$

We compute the coefficients $c_{k,j}$ by requiring

$$\frac{1}{2}\langle \nabla\Psi_{h_j} \mid \nabla\phi_{h_j}^{(k)}\rangle - \left\langle \left(\frac{Z_a}{r_a} + \frac{Z_b}{r_b} - \frac{Z_a Z_b}{R} - E_{h_j}\right)\Psi_{h_j} \mid \phi_{h_j}^{(k)}\right\rangle = 0,$$
$$\text{for } 1 \le k \le n(h_j);\ \langle\ \mid\ \rangle \text{ is the } L^2\text{-product on } B_{r_0},\ \text{i.e., (3.1).}$$

The above, after numerical quadrature and the assembling of $n(h_j)$ linear equations in matrix form, yields an eigenvalue problem

$$M_j\boldsymbol{c} - E_{h_j}\boldsymbol{c} = \boldsymbol{0},$$

where for each given h_j, $\boldsymbol{c} = (c_{1,j}, c_{2,j}, \ldots, c_{n(h_j)})^T$, and M_j is an $n(h_j)\times n(h_j)$ matrix. We then compute the eigenvalues E_{h_j} and the eigenvector $\boldsymbol{c}$, which in turn via (5.2) yields an approximate wave function Ψ_{h_j}. The parameter h_j can be decreased in order to test whether the numerical results have any trend of convergence, which is confirmed in our case.

In our computation, we have chosen the finite element basis functions $\phi_{h_j}^k$ in (5.1) to be C^0 quadratic Lagrangian tetrahedral elements [6, pp. 176–178]; $n(h_j) = 11590$. There are a total of 11497 tetrahedral, 11558 internal nodes and 362

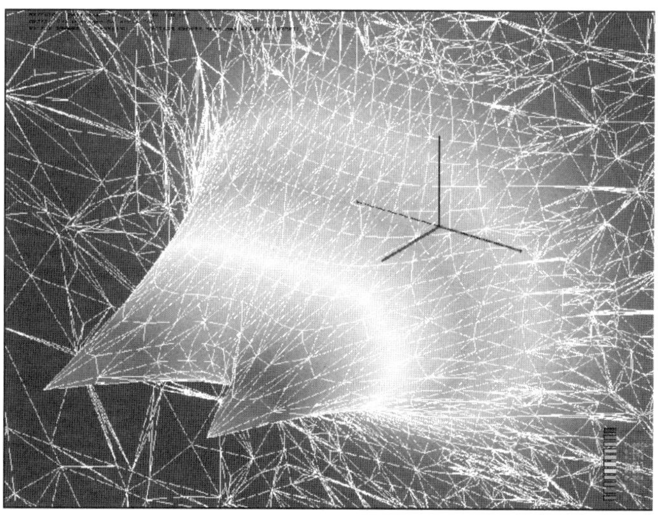

FIGURE 3. The profile of the ground state wave function $\Psi(0, y, z)$ for the molecular ion H_2^+ on the plane $x = 0$. Here $Z_a = Z_b = 1$ and $R = 2$ in equation (3.5). Two prominent cusps can be observed here. The numerical value of E is computed to be $E = -0.599468$, versus the exact value $E = -0.6026273$. The density distribution as shown by the electron wave function shows that the electron lies between the two nuclei, forming the molecular bond.

boundary nodes. Grid points are chosen with *finer resolution* near the two nuclei. The ground state of the hydrogen H_2^+ molecular ion (i.e., $Z_a = Z_b = 1$) is displayed in order to illustrate the cusps, in Figures 3 and 5. The FEM computer software, MEFISTO, developed by the third coauthor A. Perronnet of this article, is available

FIGURE 4. The profile of the ground state wave function $\Psi(x, 0, z)$ for the same molecular ion H_2^+ as in the preceding Figure 3, on the plane $y = 0$. In this view, cusps do not show up.

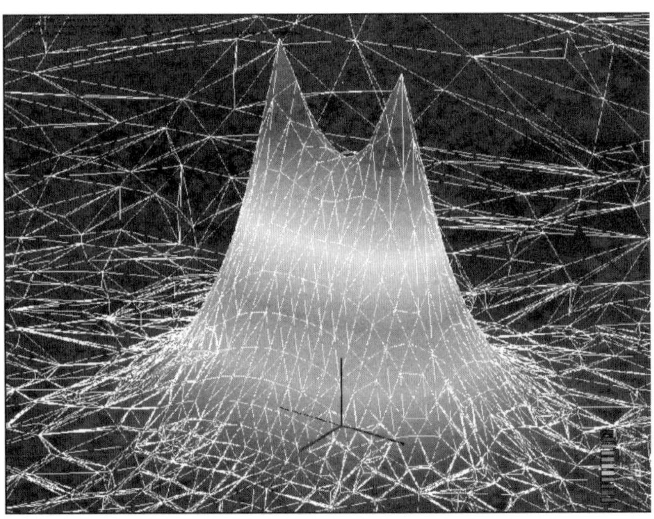

FIGURE 5. The profile of the ground state wave function $\Psi(x, y, 0)$ for the same molecular ion H_2^+ as in the preceding Figure 3, on the plane $z = 0$. Here two cusps are again prominent.

at: `http://www.ann.jussieu.fr/~perronnet/mefisto.gene.html`. More graphics, in *color*, can be viewed at: `http://www.ann.jussieu.fr/~perronnet/vp21/vp21.html`.

We remark that Theorem 3.1 is not applicable to the H_2^+ problem because the Coulomb potential in (3.5) is not homogeneous of degree -1 with respect to the x_1, x_2 and x_3 variables.

5.2. Computation of the helium atom by the Gaussian software. The He atom model (3.6) is a PDE in six dimensions (6D). Presently, we don't have our own FEM software package, nor do we know of any other FEM packages in the public domain, which can be used to compute PDEs in 6D without preprocessing such as SCF (self-consistent fields) in, e.g., [23]. Therefore, here we have to rely on a commercial package, [10].

The inner working of the Gaussian subroutines used by us from [10] contains the methods and ideas of Hartree–Fock, DFT (density function theory, as developed by Hohenberg, Kohn and Sham), LDA (local density approximation), etc. Interested readers are referred to [9, 22] for detailed explanations.

With Gaussian, an important approximation is made by assuming that the molecular orbital function $\psi_k(\boldsymbol{r})$ is contained in a finite dimensional linear space of Gaussian functions and can be written as a linear combination

$$\psi_k(\boldsymbol{r}) = \sum_{\mu=1}^{N} c_{k\mu}\chi_\mu, \tag{5.3}$$

where $c_{k\mu}$ are the molecular orbital expansion coefficients, while the functions $\chi_1, \ldots, \chi_N$ constitute a basis set and N is the number of basis functions. There are various basis sets available for different applications. Each basis function is a linear combination of several Gaussian-type functions, also called a primitive Gaussian.

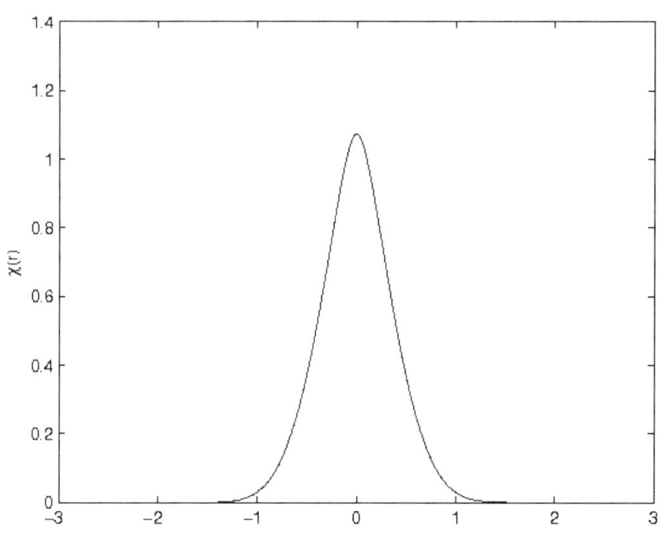

FIGURE 6. The curve of an s-type basis function in the basic set $6-311++\mathrm{G(3df,2pd)}$.

A Gaussian-type function has the following form [9, 10]:

$$g(\alpha, \boldsymbol{r}) = cx^m y^n z^l e^{-\alpha r^2}. \tag{5.4}$$

For example, an s-type basis function in the basis set $6-311++\mathrm{G(3df, 2pd)}$

$$\chi(r) = 0.02874e^{-98.1243r^2} + 0.2081e^{-14.7689r^2} + 0.8376e^{-3.3188r^2},$$

is shown in Figure 6. For this basis set, N is equal to 15 in (5.3). But in the actual computations, only the following 10 basis functions are involved:

$$\begin{aligned}
\chi_{1s} &= 0.0287452025e^{-98.1243r^2} \\
&\quad + 0.2080610181e^{-14.7689r^2} + 0.8376350728e^{-3.31883r^2}; \\
\chi_{2s} &= e^{-0.874047r^2}; \quad \chi_{3s} = e^{-0.244564r^2}; \quad \chi_{4s} = e^{-0.086r^2}; \\
\chi_{5px} &= xe^{-1.5r^2}; \quad \chi_{5py} = ye^{-1.5r^2}; \quad \chi_{5pz} = ze^{-1.5r^2}; \\
\chi_{6px} &= xe^{-0.375r^2}; \quad \chi_{6py} = ye^{-0.375r^2}; \quad \chi_{6pz} = ze^{-0.375r^2}.
\end{aligned}$$

Their linear combinations lead to the following orbitals:

$$\begin{aligned}
\psi_1 &= 0.27083\chi_{1s} + 0.47315\chi_{2s} + 0.38419\chi_{3s} + 0.03374\chi_{4s} \\
&= 0.00778506e^{-98.1243r^2} + 0.0563492e^{-14.7689r^2} + 0.226857e^{-3.31883r^2} \\
&\quad + 0.47315e^{-0.874047r^2} + 0.38419e^{-0.244564r^2} + 0.03374e^{-0.086r^2}; \\
\psi_2 &= -0.10498\chi_{1s} - 0.10381\chi_{2s} - 0.88678\chi_{3s} + 1.56653\chi_{4s} \\
&= -0.00301767e^{-98.1243r^2} - 0.0218422e^{-14.7689r^2} - 0.0879349e^{-3.31883r^2} \\
&\quad - 0.10381e^{-0.874047r^2} - 0.88678e^{-0.244564r^2} + 1.56653e^{-0.086r^2}; \\
\psi_3 &= -0.03868\chi_{5pz} + 1.02164\chi_{6pz} \\
&= -0.03868ze^{-1.5r^2} + 1.02164ze^{-0.375r^2}; \\
\psi_6 &= -0.16865\chi_{1s} - 1.31965\chi_{2s} + 2.49893\chi_{3s} - 1.25673\chi_{4s} \\
&= -0.00484788e^{-98.1243r^2} - 0.0350895e^{-14.7689r^2} - 0.141267e^{-3.31883r^2} \\
&\quad - 1.31965e^{-0.874047r^2} + 2.49893e^{-0.244564r^2} - 1.25673e^{-0.086r^2}.
\end{aligned}$$

Four (unnormalized) wave functions of He are obtained:

(i) ground state:

$$\phi_0 = \psi_1(\vec{r}_1)\psi_1(\vec{r}_2); \tag{5.5}$$

(ii) first excited state:

$$\phi_1 = 0.71694\big(\psi_1(\vec{r}_1)\psi_2(\vec{r}_2) - \psi_2(\vec{r}_1)\psi_1(\vec{r}_2)\big) + 0.07264\big(\psi_1(\vec{r}_1)\psi_6(\vec{r}_2) - \psi_6(\vec{r}_1)\psi_1(\vec{r}_2)\big);$$

(iii) second excited state:

$$\phi_2 = 0.68571\psi_1(\vec{r}_1)\psi_2(\vec{r}_2) - 0.07283\psi_1(\vec{r}_1)\psi_6(\vec{r}_2);$$

(iv) third excited state:

$$\phi_3 = \psi_1(\vec{r}_1)\psi_3(\vec{r}_2) - \psi_3(\vec{r}_1)\psi_1(\vec{r}_2).$$

TABLE 1. Values of E of four eigenstates of the helium atom. The "exact values" are those listed by NIST [19], while those obtained by using the Gaussian (code name B3LYP/6311++G(3df,2pd)) are listed on the rightmost column. Note that the Schrödinger–Born–Oppenheimer He-model (3.6) itself does not predict the spin state singlet or triplet, as Ψ in (3.6) models only the spatial part of the wave function. Nevertheless, the prediction of the spin state can often be done with the help of Hund's rules [1].

States	E: Exact (NIST)	E: Gaussian B3LYP/6311++G(3df,2pd)
$\Psi^{(1)}$: $1s1s\ ^1S$ (singlet)	-2.90358	-2.9135439
$\Psi^{(2)}$: $1s2s\ ^3S$ (triplet)	-2.17522	-2.227579805
$\Psi^{(3)}$: $1s2s\ ^1S$ (singlet)	-2.14596	-2.1602840072
$\Psi^{(4)}$: $1s2p\ ^3P$ (triplet)	-2.13316	-1.83958833258

Even though the helium atom model (3.6) satisfies the assumptions of Theorem 3.1 and, thus, (3.4) holds and for each eigenstate we should have

$$\text{kinetic energy} + \text{potential energy} = \text{kinetic energy- } 2 \text{ kinetic energy} = E,$$

i.e.,

$$\text{kinetic energy} \ = -E.$$

But for the ground state of He in (5.5) and Table 1, we actually computed its kinetic energy to be 1.089702169, i.e.,

$$1.089702169 \neq 2.9135439,$$

and their deviation is quite significant. This really leaves something for us to think about.

Comparison between exact values and numerical values of E can be found in Table 1.

Summary

In this paper we have reviewed some basic mathematical properties for quantum mechanical equations governing atoms and molecules. We hope this material will help the intuition and facilitate interdisciplinary research in molecular modeling and computation.

References

1. P.W. Atkins, *Molecular quantum mechanics*, 2nd ed., Oxford University Press, Oxford, 1983.
2. A.D. Bandrauk, M.C. Delfour, and C. Le Bris, (eds.), *Quantum control: mathematical and numerical challenges* (Montréal, 2002), CRM Proceedings & Lecture Notes, vol. 33, Amer. Math. Soc., Providence, R.I., 2003.
3. M. Born and J.R. Oppenheimer, *Quantum theory of the molecules*, Ann. Physik **84** (1927), 457–484.
4. G. Chen, S.A. Chin, Y. Dou, K.T. Kapale, M. Kim, A. Svidzinsky, K. Urtekin, H. Xiong, and M.O. Scully, *The two electron molecular bond revisited — from Bohr orbits to two center orbits* (H.H. Stroke, ed.), Adv. Atom. Mol. Opt. Phys., vol. 51, Elsevier-Academic Press, San Diego, 2005, pp. 93–238.

5. G. Chen, D.A. Church, B.-G. Englert, C. Henkel, B. Rohwedder, M.O. Scully, and M.S. Zubairy, *Quantum computing devices: principles, designs and analysis*, Chapman and Hall/CRC, Boca Raton, 2006.
6. G. Chen and J. Zhou, *Boundary element methods*, Comput. Math. Appl., Academic Press, Ltd., London, 1992.
7. V.A. Fock, *On the Schrödinger equation for the helium atom*, Izv. Akad. Nauk. **18** (1954), 161–172.
8. ______, *On the Schrödinger equation for the helium atom,* I *and* II, K. Norske Vidensk. Selsk. Fosh. **31** (1958), 138–152.
9. J.B. Foresman, and Æ. Frisch, *Exploring chemistry with electronic structure methods*, Gaussian, Inc., Pittsburgh, 1996.
10. Æ. Frisch, M.J. Frisch, and G.W. Trucks, *Gaussian* 03 *user's reference*, Gaussian, Inc., Carnegie, 2003.
11. E.A. Hylleraas, *Über die Elektronenterme des Wasserstoffmoleküls*, Z. Physik **71** (1931), 739–763.
12. G. Jaffé, *Zur Theorie des Wasserstoffmoleklions*, Z. Physik **87** (1934), 535–544.
13. T. Kato, *Integration of the equations of evolution in a Banach space*, J. Math. Soc. Japan **5** (1953), 208–234.
14. ______, *On the eigenfunctions of many-particle systems in quantum mechanics*, Comm. Pure Appl. Math. **10** (1957), 151–177.
15. ______, *Perturbations of continuous spectra by trace class operators*, Proc. Japan Acad. **33** (1957), 260–264.
16. C. Le Sech, *Determination of highly excited states of diatomic-molecular ions using exact* H_2^+*-like orbitals*, Phys. Rev. A **51** (1995), R2668–R2671.
17. C. Le Sech, A.M. Jorus, and M. Aubert-Frecon, *The two electron molecular bond revisited: from Bohr orbits to two-center orbitals*, J. Chem. Phys. **75** (1981), 2932–2934.
18. C. Le Sech and A. Sarsa, *Simple analysis of correlation in few-body Coulomb systems: application in the diffusion Monte Carlo method*, Phys. Rev. A **63** (2001), 022501.
19. NIST (Nat. Inst. of Standards Techno.), `http://physics.nist.gov/cgi-bin/ASD/energy1.pl`.
20. S.H. Patil and K.T. Tang, *Asymptotic methods in quantum mechanics, applications to atoms, molecules and nuclei*, Springer-Verlag, Berlin, 2000.
21. T. Shibuya and C.E. Wulfman, *Molecular orbitals in momentum space*, Proc. Roy. Soc. Ser. A **286** (1965), 376–389.
22. J.M. Thijssen, *Computational physics*, Cambridge University Press, Cambridge, 1999.
23. H. Yu and A.D. Bandrauk, *3-D finite element method for time dependent Schrödinger equation of molecules in laser fields*, J. Chem. Phys. **192** (1995), 1257–1265.

Department of Mathematics and Institute for Quantum Studies, Texas A&M University, College Station, TX 77843, USA.
E-mail address: `gchen@math.tamu.edu`

Department of Mathematical Sciences, University of Nevada, Las Vegas, Las Vegas, NV 89154-4020, USA.
E-mail address: `dingz@unlv.nevada.edu`

Laboratoire J.-L. Lions, Université Pierre et Marie Curie, Paris Cedex 05, France.
E-mail address: `perronnet@ann.jussieu.fr`

Department of Physics and Institute for Quantum Studies, Texas A&M University, College Station, TX 77843; and Departments of Chemistry, Mechanical and Aerospace Engineering, Princeton University, Princeton, NJ, 08544, USA.
E-mail address: `scully@physics.tamu.edu`

Department of Physics and Institute for Quantum Studies, Texas A&M University, College Station, TX 77843, USA.
E-mail address: `rhxie04@tamu.edu`

Department of Mathematics, Texas A&M University, College Station, TX 77843, USA.
E-mail address: `zgzhang@math.tamu.edu`

Centre de Recherches Mathématiques
CRM Proceedings and Lecture Notes
Volume **41**, 2007

Sparse Tensor-Product Fokker–Planck-Based Methods for Nonlinear Bead-Spring Chain Models of Dilute Polymer Solutions

Pascal Delaunay, Alexei Lozinski, and Robert G. Owens

ABSTRACT. In this paper we present a spectral method for the solution of a high-dimensional Fokker–Planck equation describing the evolution of the configuration probability density function for a dilute solution of idealized polymer chains. The numerical method draws on ideas found in the literature on sparse grids and allows solutions to be obtained at much lower cost, for comparable accuracy, than using a full tensor-product of one-dimensional basis functions. Numerical results for homogeneous shear flow serve to illustrate the attractive features of the new method.

1. Introduction

The dimensionless equation of conservation of linear momentum and of mass for an incompressible polymeric solution may be written

$$Re\frac{D\boldsymbol{v}}{Dt} - \beta\nabla^2\boldsymbol{v} + \nabla p = \nabla\cdot\boldsymbol{\tau}, \tag{1.1}$$

$$\nabla\cdot\boldsymbol{v} = 0, \tag{1.2}$$

where $\boldsymbol{v}$ denotes the velocity, p the pressure, Re a Reynolds number and β is a solvent to total viscosity ratio. The tensor $\boldsymbol{\tau}$ accounts for the contribution of the dissolved polymer to the Cauchy stress tensor and has traditionally been determined as the solution to a closed-form differential or integral equation. In acknowledgement of the unphysical (and, therefore, undesirable) assumptions that may be implied in modelling the elastic stress tensor of a complex fluid as the solution to a closed-form equation a more recent trend, in the so-called "micro-macro" approach [13], has been the computation of $\boldsymbol{\tau}$ from a coarse-grained kinetic theory model for the polymer conformations. Although most papers in the micro-macro rheological literature have been concerned with the solution of stochastic differential equations

---

2000 *Mathematics Subject Classification.* Primary: 65M70, 74S25; Secondary: 74A60.
The second author was supported for the duration of this project by CTI project 6437.1 IWS-IW.
This is the final form of the paper.

(see, for example, [12, 14, 19]), there is growing representation by advocates of numerical methods specially adapted to the solution of an equivalent Fokker–Planck equation [5–9, 17, 23, 27].

Our own recent work has been directed towards developing Fokker–Planck-based high-order methods for both dilute [4, 15] and concentrated [16] polymer solutions that are at the same time cheaper and more accurate than stochastic techniques. The number of configuration space dimensions in all our papers to date has been low (≤ 3), however, and the cost advantage of Fokker–Planck-based methods is quickly lost once high-dimensional simulations are attempted. This is because of the so-called "curse of dimension": standard full tensor-product bases for the probability density function (pdf) ψ lead to computational effort that can grow exponentially with total dimension. Higher-order configuration spaces are encountered in bead-spring chain models of polymer solutions, for example.

Our aim in this paper is to present a suitably inexpensive numerical method for the determination of the joint probability density function ψ for the internal configurations of a bead-spring chain that Fokker–Planck-based numerical methods will become competitive with stochastic techniques for (at least some) configuration dimensions exceeding 3. Recently, Venkiteswaran and Junk [24, 25] have employed a quasi-Monte Carlo method [18] for the simulation of the Fokker–Planck equation for a dilute solution of polymers modelled by a bead-spring chain and undergoing simple shear flow. Monte Carlo methods are a well-known approach for overcoming the curse of dimension in high dimensional systems since the error estimate in terms of the number of nodes M (say) behaves like $\sqrt{M}$, independently of the dimension. Although the provable error estimate of Venkiteswaran and Junk [25] was no better than for a standard Monte Carlo method, their new method manifested less variance in the results and, for fixed accuracy, the authors achieved an improvement in the computational time over the simple Monte Carlo method. Another way of attempting to overcome the curse of dimension in the present context is by using sparse tensor-product spaces [22, 28] in order to represent ψ. In this class of methods a high-dimensional basis is derived from a one-dimensional basis by forming tensor products of univariate formulae whose indices lie in a simplex. An extensive and recent review of sparse grid methods may be found in the survey paper of Bungartz and Griebel [3]. The advantages of using a sparse grid approach are seen in the cost savings made when compared with the use of traditional full tensor-product bases for the dependent variables, without compromising on the order of accuracy. For example, whereas the number of degrees of freedom on a full tensor product finite element mesh of mesh size h in d dimensions grows like $O(h^{-d})$, that associated with a sparse tensor product only grows like $O(h^{-1}|\log h|^{d-1})$. At the same time, the order of accuracy in the H^1 norm of both methods is $O(h^p)$ for finite elements of degree $p \geq 1$ and sufficiently smooth approximated functions [26].

In Section 2 we begin by explaining how the multi-dimensional Fokker–Planck equation is derived from the bead-spring micro-mechanical model. The eigenfunctions of the Fokker–Planck operator for a single dumbbell are computed using a Galerkin spectral method in configuration space and we then employ both full and sparse tensor products of these eigenfunctions to represent the configuration pdf ψ for the full chain. Exactly how this is done is described in Section 3. An implicit stochastic algorithm is detailed in Section 4, and in the same section we explain how the elastic stress contribution to the total Cauchy stress is calculated on the basis of the solutions to the stochastic and equivalent Fokker–Planck equations. The advantages of using a sparse tensor product basis for ψ rather than the usual full tensor product for such functions is showcased in Section 5 by comparing the costs of the two approaches against that of the Brownian dynamics simulation for simple shear flow at various shear rates.

2. Micromechanical Model and Governing Equations

The micromechanical model used to represent a single linear polymer chain in this paper consists of $(d+1)$ beads, each having the same mass m and joined consecutively by d nonlinear massless springs, as shown in Figure 1. The equations of motion for the bead having position vector $\boldsymbol{r}_j$ $(j = 2, \ldots, d)$ may be written

$$m\frac{d^2\boldsymbol{r}_j}{dt^2} = \zeta\left(\boldsymbol{v}(\boldsymbol{r}_j) - \frac{d\boldsymbol{r}_j}{dt}\right) + \boldsymbol{F}(\boldsymbol{r}_{j+1} - \boldsymbol{r}_j) - \boldsymbol{F}(\boldsymbol{r}_j - \boldsymbol{r}_{j-1}) + \boldsymbol{B}_j, \tag{2.1}$$

whereas for the beads at the ends the chain, (2.1) simplifies to

$$m\frac{d^2\boldsymbol{r}_1}{dt^2} = \zeta\left(\boldsymbol{v}(\boldsymbol{r}_1) - \frac{d\boldsymbol{r}_1}{dt}\right) + \boldsymbol{F}(\boldsymbol{r}_2 - \boldsymbol{r}_1) + \boldsymbol{B}_1, \tag{2.2}$$

$$m\frac{d^2\boldsymbol{r}_{d+1}}{dt^2} = \zeta\left(\boldsymbol{v}(\boldsymbol{r}_{d+1}) - \frac{d\boldsymbol{r}_{d+1}}{dt}\right) - \boldsymbol{F}(\boldsymbol{r}_{d+1} - \boldsymbol{r}_d) + \boldsymbol{B}_{d+1}. \tag{2.3}$$

In (2.1)–(2.3) ζ is a friction coefficient, $\boldsymbol{v}(\boldsymbol{r}_j)$ is the velocity of the solvent at the point with position vector $\boldsymbol{r}_j$, $\boldsymbol{F}(\boldsymbol{r}_{j+1} - \boldsymbol{r}_j) = Hf(|\boldsymbol{r}_{j+1} - \boldsymbol{r}_j|)(\boldsymbol{r}_{j+1} - \boldsymbol{r}_j)$ (for some spring constant H and spring law f) is the force acting on the jth bead due to tension in the spring of the jth segment and $\boldsymbol{B}_j$ denotes the Brownian forces acting on the jth bead.

The acceleration terms in (2.1)–(2.3) may be neglected (see [10] for a justification) and no account is taken of any segment-segment or bead-bead interaction apart from that which exists through the tension in the segment springs. In the

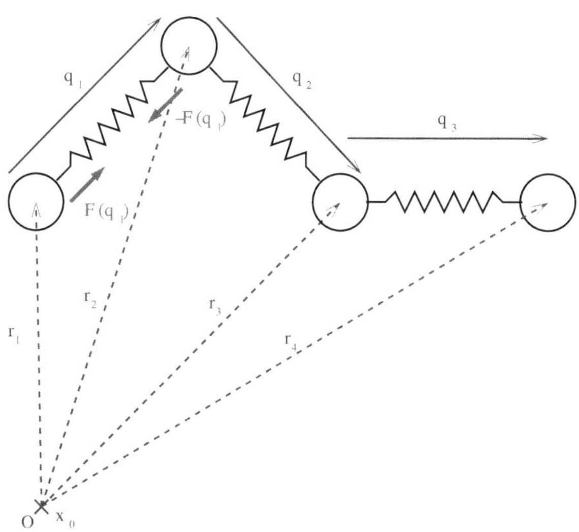

FIGURE 1. Schematic of a bead-spring chain of $d = 3$ segments.

present paper for simplicity, the chains are supposed to live on a plane and the vectors $\boldsymbol{r}_j$, $\boldsymbol{B}_j$ etc. are hence two-dimensional. Note that simulations of bead-spring chains evolving in the three-dimensional physical space should be feasible by the approach analogous to that described below (cf. [4] for the case of a single dumbbell model).

The Brownian forces are approximated by a Wiener process:

$$\boldsymbol{B}_j dt = \sqrt{2k_B T\zeta} d\boldsymbol{W}_j, \tag{2.4}$$

where k_B is Boltzmann's constant, T is the absolute temperature and $\boldsymbol{W}_j(t) = \big(W_{j,1}(t), W_{j,2}(t)\big)$ is a Gaussian process in two-dimensions with

$$\langle W_{j,k}(t)\rangle = 0, \quad \langle W_{j,k}(t)W_{j',k'}(t')\rangle = \delta_{jj'}\delta_{kk'}\min(t,t'). \tag{2.5}$$

Introducing the jth connector vector

$$\boldsymbol{q}_j = \boldsymbol{r}_{j+1} - \boldsymbol{r}_j, \quad j = 1, \dots, d, \tag{2.6}$$

and the position vector of the centre of mass

$$\boldsymbol{r}_c = \frac{1}{d+1}\sum_{j=1}^{d+1} \boldsymbol{r}_j, \tag{2.7}$$

we see that the equations of motion (2.1)–(2.3) may be rewritten in the form

$$d\boldsymbol{q}_j(t) = \left[\boldsymbol{v}(\boldsymbol{r}_{j+1}) - \boldsymbol{v}(\boldsymbol{r}_j) - \frac{1}{\zeta}\sum_{k=1}^{d} A_{jk}\boldsymbol{F}(\boldsymbol{q}_k)\right] dt + \sqrt{\frac{4k_BT}{\zeta}} d\boldsymbol{W}^q_j(t), \tag{2.8}$$

$$d\boldsymbol{r}_c(t) = \frac{1}{d+1}\sum_{j=1}^{d+1} \boldsymbol{v}(\boldsymbol{r}_j)\, dt + \sqrt{\frac{2k_BT}{\zeta(d+1)}} d\boldsymbol{W}^c(t), \tag{2.9}$$

where $\boldsymbol{W}^q_j(t) = \boldsymbol{W}_{j+1} - \boldsymbol{W}_j/\sqrt{2}$ and $\boldsymbol{W}^c(t) = 1/\sqrt{d+1}\sum_{j=1}^{d+1} \boldsymbol{W}_j(t)$ both have the same distribution as $\boldsymbol{W}_j$, and $\boldsymbol{A}$ is the Rouse matrix

$$\boldsymbol{A} = \begin{pmatrix} 2 & -1 & & (0) \\ -1 & \ddots & \ddots & \\ & \ddots & \ddots & -1 \\ (0) & & -1 & 2 \end{pmatrix}.$$

Suppose that the flow is homogeneous (i.e., $\nabla \boldsymbol{v}$ is a constant matrix so that we are permitted to write $\boldsymbol{v}(\boldsymbol{r}_{j+1}) - \boldsymbol{v}(\boldsymbol{r}_j) = \nabla\boldsymbol{v}\cdot\boldsymbol{q}_j$). Then, after scaling $\boldsymbol{q}_k$ with $l_0 = \sqrt{k_BT/H}$ it may be shown (see for example, [19, Theorem 3.30 in Öttinger]) that the Fokker–Planck equation corresponding to (2.8)–(2.9) and describing the evolution of the configuration pdf ψ is

$$\begin{aligned}\frac{\partial\psi}{\partial t} = \mathcal{L}^d_{\mathrm{FP}}(\psi) = \sum_{k=1}^{d} L^k\psi &- \frac{1}{4\lambda}\sum_{k=2}^{d}(N_x^{k-1}M_x^k + N_y^{k-1}M_y^k)\psi \\ &- \frac{1}{4\lambda}\sum_{k=1}^{d-1}(N_x^{k+1}M_x^k + N_y^{k+1}M_y^k)\psi.\end{aligned} \tag{2.10}$$

where

$$L^k = \nabla_k \cdot \left(-\nabla \boldsymbol{v} \cdot \boldsymbol{q}_k + \frac{1}{2\lambda}\boldsymbol{F}(\boldsymbol{q}_k) + \frac{1}{2\lambda}\nabla_k \right), \quad M_x^k = \frac{\partial}{\partial q_{k,x}}, \quad M_y^k = \frac{\partial}{\partial q_{k,y}},$$

$$N_x^k = F_x(\boldsymbol{q}_k) + \frac{\partial}{\partial q_{k,x}} \quad \text{and} \quad N_y^k = F_y(\boldsymbol{q}_k) + \frac{\partial}{\partial q_{k,y}}. \tag{2.11}$$

In (2) $\lambda = \zeta/4H$ is a configuration relaxation time,

$$\boldsymbol{q}_k = \begin{pmatrix} q_{k,x} \\ q_{k,y} \end{pmatrix} \quad \text{and} \quad \nabla_k = \begin{pmatrix} \frac{\partial}{\partial q_{k,x}} \\ \frac{\partial}{\partial q_{k,y}} \end{pmatrix}. \tag{2.12}$$

In the present paper the spring force law employed for each of the segments is that of the FENE model [2]

$$\boldsymbol{F}(\boldsymbol{q}) = \frac{\boldsymbol{q}}{1 - (|\boldsymbol{q}|^2)/b}, \tag{2.13}$$

where $\sqrt{b} = q_{\max}/l_0$ and $q_{\max}$ is the maximum extensibility of the segment spring.

3. A High-Order Sparse Grid Fokker–Planck-Based Method

In this section we describe how we generate numerical approximations to the leading eigenfunctions of the Fokker–Planck operator $\mathcal{L}_{\mathrm{FP}}^1$ and then employ a sparse tensor product of these eigenfunctions to represent the configuration pdf ψ for the full chain.

3.1. Construction of an eigenbasis. The connector vector $\boldsymbol{q}$ for a single FENE dumbbell ($d = 1$) lives in a disk of radius $\sqrt{b}$ and we may write the two Cartesian components (q_x, q_y) of $\boldsymbol{q}$ in terms of planar polar coordinates

$$q_x = \rho\cos\theta, \quad q_y = \rho\sin\theta, \tag{3.1}$$

with $\rho \in [0, \sqrt{b}]$ and $\theta \in [0, 2\pi]$. The first step that we take towards solving the Fokker–Planck equation (2.10) numerically is the construction of an eigenbasis consisting of solutions $\widetilde{\psi}$ to the eigenproblem[1]

$$\mathcal{L}_{\mathrm{FP}}^1 \widetilde{\psi} = \lambda_i \widetilde{\psi}. \tag{3.2}$$

We compute the leading N (say) eigenfunctions corresponding to the eigenvalues $\lambda_1, \ldots, \lambda_N$ having real parts that are smallest in absolute value: $0 \geq Re(\lambda_1) \geq Re(\lambda_2) \geq \cdots \geq Re(\lambda_N) \geq \cdots$. This is done using a Fourier–Legendre spectral Galerkin method in a manner to be described below.

We denote the even and odd[2] parts of an eigenfunction $\widetilde{\psi}$ satisfying (3.2) by $\widetilde{\psi}_0$ and $\widetilde{\psi}_1$. Then

$$\widetilde{\psi} = \widetilde{\psi}_0 + \widetilde{\psi}_1, \tag{3.3}$$

[1]The eigenvalues λ_i are not to be confused with the relaxation time λ in (2.10).

[2]A function $f(t,\rho,\theta)$ will be called even if $f(t,\rho,\theta) = f(t,\rho,\theta+\pi)$ and odd if $f(t,\rho,\theta) = -f(t,\rho,\theta+\pi)$.

where $\widetilde{\psi}_0$ and $\widetilde{\psi}_1$ are unique and defined by

$$\widetilde{\psi}_0(\rho,\theta) = \frac{\widetilde{\psi}(\rho,\theta) + \widetilde{\psi}(\rho,\theta+\pi)}{2},$$

$$\widetilde{\psi}_1(\rho,\theta) = \frac{\widetilde{\psi}(\rho,\theta) - \widetilde{\psi}(\rho,\theta+\pi)}{2}. \tag{3.4}$$

We note that since the spaces of even and odd functions are invariant under the Fokker–Planck operator $\mathcal{L}^1_{\rm FP}$ at equilibrium, both $\widetilde{\psi}_0$ and $\widetilde{\psi}_1$ are themselves eigenfunctions of this operator, belonging to the same eigenspace as $\widetilde{\psi}$.

Let us introduce the change of dependent variables $\widetilde{\psi} \to \widetilde{\alpha}$ for the eigenproblem (3.2) where

$$\widetilde{\psi}_0 = \left(1 - \frac{\rho^2}{b}\right)^s \widetilde{\alpha}_0, \quad \widetilde{\psi}_1 = \rho\left(1 - \frac{\rho^2}{b}\right)^s \widetilde{\alpha}_1, \tag{3.5}$$

$\widetilde{\alpha}_0$ and $\widetilde{\alpha}_1$ are, respectively, the even and odd parts of $\widetilde{\alpha}$, and s is a real parameter greater than unity. The motivation for the change of variables from $\widetilde{\psi}$ to $\widetilde{\alpha}$ is firstly to circumvent the numerical problems created by the unboundedness of $\boldsymbol{F}$ as $\rho^2 \to b$ in the term $\boldsymbol{F}(\boldsymbol{q})\psi$ of $\mathcal{L}^{(1)}_{\rm FP}$ (see (2.10) when $d = 1$). Secondly, the factors multiplying $\widetilde{\alpha}_0$ and $\widetilde{\alpha}_1$ in (3.5) ensure that $\widetilde{\psi} = 0$ on $\rho = \sqrt{b}$. Finally, if $\widetilde{\psi}$ is even we see from (3.5) that $\partial\widetilde{\psi}/\partial\rho = 0$ identically at $\rho = 0$ and, if odd, that $\widetilde{\psi} = 0$ at $\rho = 0$.

Using the shorthand notation

$$\psi^0_a(\rho) = \rho^a\left(1 - \frac{\rho^2}{b}\right)^s, \quad a = 0, 1, \tag{3.6}$$

we see that we may write

$$\widetilde{\psi} = \psi^0_0\widetilde{\alpha}_0 + \psi^0_1\widetilde{\alpha}_1, \tag{3.7}$$

and that therefore there exists an invertible operator $\mathcal{R}^{(1)}$ (say) such that $\widetilde{\alpha} = \mathcal{R}^{(1)}\widetilde{\psi}$ and defined for any function $\widetilde{\psi}$ of $(t, \boldsymbol{q})$ having even and odd parts $\widetilde{\psi}_0$ and $\widetilde{\psi}_1$ as

$$\mathcal{R}^{(1)}\widetilde{\psi} = (\psi^0_0)^{-1}\widetilde{\psi}_0 + (\psi^0_1)^{-1}\widetilde{\psi}_1. \tag{3.8}$$

The original eigenproblem (3.2) may now be written in terms of $\widetilde{\alpha}$ as

$$\widehat{\mathcal{L}}^1_{\rm FP}\widetilde{\alpha} = \lambda_i\widetilde{\alpha}, \tag{3.9}$$

where

$$\widehat{\mathcal{L}}^1_{\rm FP} = \mathcal{R}^{(1)}\mathcal{L}^1_{\rm FP}(\mathcal{R}^{(1)})^{-1}, \tag{3.10}$$

or, in terms of the action of $\widehat{\mathcal{L}}^1_{\rm FP}$ on the even and odd parts of $\widetilde{\alpha}$,

$$\widehat{\mathcal{L}}^1_{\rm FP}\widetilde{\alpha}_a = (\psi^0_a)^{-1}\mathcal{L}^1_{\rm FP}(\psi^0_a\widetilde{\alpha}_a), \quad a = 0, 1. \tag{3.11}$$

3.1.1. *A spectral Galerkin method.* In this paper $\widetilde{\alpha}_0$ and $\widetilde{\alpha}_1$ will be represented in terms of even and odd Fourier–Legendre series in $[-1, 1] \times [0, 2\pi]$ after performing the change of variable ρ to η, where

$$\rho^2 = b\left(\frac{1+\eta}{2}\right).$$

This yields

$$\widetilde{\alpha} = \sum_{i=0}^{1}\sum_{k=0}^{N_R}\sum_{l=i}^{2N_F} \alpha_{ikl} h_k(\eta)\Theta_{il}(\theta), \tag{3.12}$$

where

$$\Theta_{il}(\theta) = (1-i)\cos l\theta + i \sin l\theta, \tag{3.13}$$

and $\{h_k(\eta)\}_{0\le k\le N_R}$ are Lagrange interpolating polynomials based on the (N_R+1) Gauss–Legendre points in $(-1,1)$. From the representation (3.12) of $\widetilde{\alpha}$ it may be seen that the total number of Fourier modes used is $1 + 2N_F + 2N_F = 4N_F + 1$.

An algebraic system of equations corresponding to (3.9) is obtained using a Galerkin spectral method in which test functions of the form $h_m(\eta)$, $h_m(\eta)\Theta_{il}^{\text{odd}}$ or $h_m(\eta)\Theta_{il}^{\text{even}}$ $(0 \le m \le N_R, 1 \le i \le 2, 1 \le l \le N_F)$ are employed. Integrals with respect to η are evaluated using the standard Gauss quadrature rule based on the Gauss–Legendre points, and integrals with respect to θ are evaluated analytically. The eigenvalues and associated eigenvectors of the resulting $(N_R+1)(4N_F+1) \times (N_R+1)(4N_F+1)$ matrix are computed by first reducing the matrix to upper Hessenberg form and then using a hybrid double-shifted LR-QR algorithm (see Haag and Watkins [11]). All of this is done by a call to the standard Intel library routine EVLRG. We shall denote the eigenfunctions of the single segment operator $\widehat{\mathcal{L}}^1_{\text{FP}}$ that are computed in this way by $\widetilde{\varphi}_k$ $\big(k = 1, \ldots, (N_R+1)(4N_F+1)\big)$ in the sequel.

3.2. A sparse tensor product spectral method for $\psi(t, \boldsymbol{q}_1, \ldots, \boldsymbol{q}_d)$. As a generalization of the (unique) decomposition (3.3) of the function $\widetilde{\psi}$ into a sum of its even and odd parts it may be easily shown that a unique representation exists of any function ψ of the independent variables t and $\boldsymbol{q}_1, \ldots, \boldsymbol{q}_d$ in terms of 2^d new functions $\psi^{(d)}_{\boldsymbol{p}}$

$$\psi(t, \boldsymbol{q}_1, \ldots, \boldsymbol{q}_d) = \sum_{p_1,\ldots,p_d=0}^{1} \psi^{(d)}_{\boldsymbol{p}}(t, \rho_1, \theta_1, \ldots, \rho_d, \theta_d), \tag{3.14}$$

where $\boldsymbol{p} = (p_1, \ldots, p_d)$ is a parity vector, the component p_i being equal to zero if $\psi^{(d)}_{\boldsymbol{p}}$ is even in the variable $\boldsymbol{q}_i$ and p_i equalling 1 if $\psi^{(d)}_{\boldsymbol{p}}$ is odd in $\boldsymbol{q}_i$, i.e. for $k = 1, \ldots, d$

$$\psi^{(d)}_{\boldsymbol{p}}(t, \rho_1, \theta_1, \ldots, \rho_k, \theta_k + \pi, \ldots, \rho_d, \theta_d) = (-1)^{p_k}\psi^{(d)}_{\boldsymbol{p}}(t, \rho_1, \theta_1, \ldots, \rho_k, \theta_k, \ldots, \rho_d, \theta_d).$$

As in the change of variable (3.7) we now write down the solution ψ to the full Fokker–Planck equation (2.10) in terms of new functions $\alpha^{(d)}_{\boldsymbol{p}}$:

$$\begin{aligned} \psi(t, \boldsymbol{q}_1, \ldots, \boldsymbol{q}_d) &= \sum_{p_1,\ldots,p_d=0}^{1} \left(\prod_{j=1}^{d} \rho_j^{p_j} \left(1 - \frac{\rho^2}{b}\right)^s \right) \alpha^{(d)}_{\boldsymbol{p}}(t, \rho_1, \theta_1, \ldots, \rho_d, \theta_d), \\ &= \sum_{p_1,\ldots,p_d=0}^{1} \psi^0_{\boldsymbol{p}} \alpha^{(d)}_{\boldsymbol{p}}(t, \rho_1, \theta_1, \ldots, \rho_d, \theta_d) \end{aligned} \tag{3.15}$$

where

$$\psi^0_{\boldsymbol{p}}(\rho_1, \ldots, \rho_d) = \prod_{j=1}^{d} \rho_j^{p_j} \left(1 - \frac{\rho_j^2}{b}\right)^s = \prod_{j=1}^{d} \psi^0_{p_j}(\rho_j), \tag{3.16}$$

$\psi^0_{p_j}(\rho_j)$ being as defined in (3.6).

In just the same way as we introduced the change of variables $\widetilde{\alpha} = \widetilde{\alpha}_0 + \widetilde{\alpha}_1$ in Section 3.1 we now write down the new variable

$$\alpha = \sum_{p_1,\dots,p_d=0}^{1} \alpha^{(d)}_{\boldsymbol{p}}(t,\rho_1,\theta_1,\dots,\rho_d,\theta_d), \tag{3.17}$$

in terms of which we will attempt to express the Fokker–Planck equation (2.10). This is easily done by observing that there exists an invertible operator $\mathcal{R}^{(d)}$ such that

$$\alpha = \mathcal{R}^{(d)}\psi, \tag{3.18}$$

where, given any function ψ of t and $\boldsymbol{q}_1,\dots,\boldsymbol{q}_d$ written (uniquely) in the form (3.14)

$$\mathcal{R}^{(d)}\psi = \sum_{p_1,\dots,p_d=0}^{1} (\psi^0_{\boldsymbol{p}})^{-1}\psi^{(d)}_{\boldsymbol{p}}(t,\rho_1,\theta_1,\dots,\rho_d,\theta_d). \tag{3.19}$$

Equation (2.10) therefore becomes

$$\frac{\partial \alpha}{\partial t} = \widehat{\mathcal{L}}^d_{\mathrm{FP}}(\alpha) \tag{3.20}$$

where

$$\widehat{\mathcal{L}}^d_{\mathrm{FP}} = \mathcal{R}^{(d)}\mathcal{L}^d_{\mathrm{FP}}(\mathcal{R}^{(d)})^{-1}. \tag{3.21}$$

The transformed Fokker–Planck equation may be written in full as

$$\begin{aligned}\frac{\partial \alpha}{\partial t} = \widehat{\mathcal{L}}^d_{\mathrm{FP}}\alpha = \sum_{k=1}^{d}\widehat{L}^k\alpha &- \frac{1}{4\lambda}\sum_{k=2}^{d}(\widehat{N}^{k-1}_x\widehat{M}^k_x + \widehat{N}^{k-1}_y\widehat{M}^k_y)\alpha \\ &- \frac{1}{4\lambda}\sum_{k=1}^{d-1}(\widehat{N}^{k+1}_x\widehat{M}^k_x + \widehat{N}^{k+1}_y\widehat{M}^k_y)\alpha,\end{aligned} \tag{3.22}$$

where

$$\widehat{L}^k = \mathcal{R}^{(d)}L^k(\mathcal{R}^{(d)})^{-1}, \quad \widehat{N}^{k-1}_x\widehat{M}^k_x = \mathcal{R}^{(d)}N^{k-1}_x M^k_x(\mathcal{R}^{(d)})^{-1}, \text{ etc.} \tag{3.23}$$

Equation (3.22) is solved subject to an initial condition consisting of the equilibrium solution. In terms of the pdf ψ this is

$$\psi = \psi^{(\mathrm{eq})} = \left(\frac{b+2}{2\pi b}\right)^d \prod_{k=1}^{d}\left(1 - \frac{\rho_k^2}{b}\right)^{b/2}, \tag{3.24}$$

which is equivalent (see (3.17)) to

$$\alpha = \alpha^{(\mathrm{eq})} = \alpha^{(d)}_{\boldsymbol{0}} = \left(\frac{b+2}{2\pi b}\right)^d \prod_{k=1}^{d}\left(1 - \frac{\rho_k^2}{b}\right)^{b/2-s}. \tag{3.25}$$

3.2.1. *A sparse grid spectral Galerkin method.* We will now utilize the eigenfunctions $\{\widetilde{\varphi}_k\}$ described in Section 3.1 to generate a basis for the variable α (see (3.17)). Define a sequence of spaces $V_0, V_1, V_2, \dots$ such that

$$V_0 = \emptyset, \tag{3.26}$$

and

$$V_l = \mathrm{span}\{\widetilde{\varphi}_1, \dots, \widetilde{\varphi}_{N_l}\}, \tag{3.27}$$

where $N_l = \min\left(2^l - 1, (N_R + 1)(4N_F + 1)\right)$, $l = 1, \dots, L$ and L is the integer defined such that, given N_R and N_F, $2^{L-1} - 1 < (N_R + 1)(4N_F + 1) \leq 2^L - 1$.

Then the spaces $V_0, V_1, V_2, \dots$ are nested:

$$V_0 \subset V_1 \subset V_2 \subset \dots$$

and we may define difference spaces $W_1, W_2, W_3, \dots$ expressing the hierarchical excess by $W_l = V_l \setminus V_{l-1}$.

Letting $\boldsymbol{i} = (i_1, \dots, i_d)$ be a d-dimensional multi-index we introduce the product function

$$\Phi_{\boldsymbol{i}}(\eta_1, \theta_1, \dots, \eta_d, \theta_d) = \prod_{j=1}^{d} \widetilde{\varphi}_{i_j}(\eta_j, \theta_j). \tag{3.28}$$

From the manner in which the eigenproblem (3.9) is solved numerically, each eigenfunction $\widetilde{\varphi}_{i_j}$ turns out to be either odd or even (and even if they weren't we could construct such a set). We seek to approximate α at level l in one of two ways. The first is a full tensor product (FTP)

$$\alpha \approx \sum_{\boldsymbol{i}:\, 1 \leq \|\boldsymbol{i}\|_\infty \leq N_l} \alpha_{\boldsymbol{i}}(t) \Phi_{\boldsymbol{i}}(\eta_1, \theta_1, \dots, \eta_d, \theta_d) \in V_l \bigotimes \cdots \bigotimes V_l, \tag{3.29}$$

and the second a sparse tensor product (STP) [22, 28]

$$\alpha \approx \sum_{\boldsymbol{k}:\, d \leq \|\boldsymbol{k}\|_1 \leq d+l-1} \; \sum_{\boldsymbol{i} \in \mathcal{I}_{\boldsymbol{k},d}} \alpha_{\boldsymbol{i}}(t) \Phi_{\boldsymbol{i}}(\eta_1, \theta_1, \dots, \eta_d, \theta_d) \in \bigoplus_{\boldsymbol{k}:\, d \leq \|\boldsymbol{k}\|_1 \leq d+l-1} W_{k_1} \bigotimes \cdots \bigotimes W_{k_d}, \tag{3.30}$$

where the index set $\mathcal{I}_{\boldsymbol{k},d}$ in (3.30) is defined to consist of those $\boldsymbol{i}$ such that $\widetilde{\varphi}_{i_j} \in W_{k_j}$:

$$\mathcal{I}_{\boldsymbol{k},d} = \left\{\boldsymbol{i} \in \mathbb{R}^d \colon 2^{k_j - 1} \leq i_j \leq \min\left(2^{k_j} - 1, (N_R + 1)(4N_F + 1)\right)\right\}. \tag{3.31}$$

As an example of the formulae (3.29) and (3.30) we take $d = 2$ and $l = 2$. In this case, the FTP representation (3.29) becomes

$$\alpha \approx \sum_{i_1=1}^{3} \sum_{i_2=1}^{3} \alpha_{i_1,i_2}(t) \widetilde{\varphi}_{i_1} \widetilde{\varphi}_{i_2},$$

whereas (3.30) would now read

$$\alpha \approx \alpha_{1,1}\widetilde{\varphi}_1\widetilde{\varphi}_1 + \alpha_{1,2}\widetilde{\varphi}_1\widetilde{\varphi}_2 + \alpha_{2,1}\widetilde{\varphi}_2\widetilde{\varphi}_1 + \alpha_{1,3}\widetilde{\varphi}_1\widetilde{\varphi}_3 + \alpha_{3,1}\widetilde{\varphi}_3\widetilde{\varphi}_1.$$

Let us introduce a dual space V_l^* given by

$$V_l^* = \mathrm{span}\{f_1, \dots, f_{N_l}\},$$

where each function f_i $(i = 1, 2, \dots, N_l)$ in this dual basis is defined as

$$f_i(v) = \langle \widetilde{\varphi}_i^*, v \rangle_1, \quad \forall v \in V_l. \tag{3.32}$$

In (3.32) $\langle \cdot, \cdot \rangle_1$ is an inner product on V_l defined by

$$
\begin{aligned}
\langle f, g\rangle_1 &= \int_{\theta=0}^{2\pi}\int_{\rho=0}^{\sqrt{b}} f(\rho,\theta)g(\rho,\theta)\rho \, d\rho \, d\theta, \\
&= \frac{b}{4}\int_{\theta=0}^{2\pi}\int_{\eta=-1}^{1} f(\rho(\eta),\theta)g(\rho(\eta),\theta)\, d\eta \, d\theta,
\end{aligned} \tag{3.33}
$$

and the functions $\{\widetilde{\varphi}_i^*\}_{i=0}^{N_l} \in V_l$, chosen so that $f_i(\widetilde{\varphi}_j) = \delta_{ij}$. For this to be the case, we express $\widetilde{\varphi}_i^*$ as a linear combination of the basis functions $\{\widetilde{\varphi}_j\}$:

$$
\widetilde{\varphi}_i^* = \sum_{j=1}^{N_l} a_{ij}\widetilde{\varphi}_j, \tag{3.34}
$$

and see that

$$
\langle \widetilde{\varphi}_i^*, \widetilde{\varphi}_k\rangle_1 = \sum_{j=1}^{N_l} a_{ij}\langle \widetilde{\varphi}_j, \widetilde{\varphi}_k\rangle_1, \tag{3.35}
$$

so that the $N_l \times N_l$ matrix $\boldsymbol{A} = (a_{ij})$ should be chosen as the inverse of the Gram matrix $\boldsymbol{G}$ having (i, j)th component

$$
g_{ij} = \langle \widetilde{\varphi}_i, \widetilde{\varphi}_j\rangle_1. \tag{3.36}
$$

An algebraic set of discrete equations corresponding to (3.22) is found by using a Galerkin method as follows: the d-dimensional operator

$$
\frac{\partial}{\partial t} - \widehat{\mathcal{L}}_{\mathrm{FP}}^{d},
$$

is applied to either the full (3.29) or sparse (3.30) tensor product representation of α. Then the product is taken with test functions of the form

$$
\Phi_{\boldsymbol{j}}^* = \Pi_{l=1}^{d}\widetilde{\varphi}_{j_l}^*, \tag{3.37}
$$

(with $\boldsymbol{j}$ drawn from the same index set as in (3.29) or (3.30)) to get, for all suitable $\boldsymbol{j}$,

$$
\left\langle \frac{\partial \alpha}{\partial t} - \widehat{\mathcal{L}}_{\mathrm{FP}}^{d}\alpha, \Phi_{\boldsymbol{j}}^* \right\rangle_d = 0, \tag{3.38}
$$

where $\langle \cdot, \cdot \rangle_d$ generalizes the inner product introduced in (3.33) to an integral computed over the domain in $\mathbb{R}^{2d}$: $\{|\boldsymbol{q}_j| < \sqrt{b}, j = 1, \ldots, d\}$.

Writing α in its form (3.29) or (3.30), (3.38) may be expressed as

$$
\sum_{\boldsymbol{i}} \frac{d\alpha_{\boldsymbol{i}}(t)}{dt}\langle \Phi_{\boldsymbol{i}}, \Phi_{\boldsymbol{j}}^*\rangle_d - \alpha_{\boldsymbol{i}}(t)\langle \widehat{\mathcal{L}}_{\mathrm{FP}}^{d}\Phi_{\boldsymbol{i}}, \Phi_{\boldsymbol{j}}^*\rangle_d = 0, \tag{3.39}
$$

where the index set for $\boldsymbol{i}$ and $\boldsymbol{j}$ depends on whether an STP or FTP representation for α is used. Using the orthogonality properties of the bases $\{\widetilde{\varphi}_i^*\}$ and $\{\widetilde{\varphi}_j\}$ (that is, by construction, $f_i(\widetilde{\varphi}_j) = \delta_{ij}$) we then arrive at

$$\begin{aligned}
\frac{d\alpha_{\boldsymbol{j}}(t)}{dt} &- \sum_{k=1}^{d}\sum_{i_k=1}^{N_l} \alpha_{j_1,\dots,j_{k-1}i_kj_{k+1},\dots,j_d}\langle \widehat{L}^k\widetilde{\varphi}_{i_k}, \widetilde{\varphi}^*_{j_k}\rangle_1 \\
&- \frac{1}{4\lambda}\sum_{k=2}^{d}\sum_{i_{k-1}=1}^{N_l}\sum_{i_k=1}^{N_l} \alpha_{j_1,\dots,j_{k-2}i_{k-1}i_kj_{k+1},\dots,j_d} \\
&\qquad \times [\langle \widehat{N}_x^{k-1}\widetilde{\varphi}_{i_{k-1}}, \widetilde{\varphi}^*_{j_{k-1}}\rangle_1\langle \widehat{M}_x^{k}\widetilde{\varphi}_{i_k}, \widetilde{\varphi}^*_{j_k}\rangle_1 \\
&\qquad + \langle \widehat{N}_y^{k-1}\widetilde{\varphi}_{i_{k-1}}, \widetilde{\varphi}^*_{j_{k-1}}\rangle_1\langle \widehat{M}_y^{k}\widetilde{\varphi}_{i_k}, \widetilde{\varphi}^*_{j_k}\rangle_1] \\
&- \frac{1}{4\lambda}\sum_{k=1}^{d-1}\sum_{i_k=1}^{N_l}\sum_{i_{k+1}=1}^{N_l} \alpha_{j_1,\dots,j_{k-1}i_ki_{k+1}j_{k+2},\dots,j_d} \\
&\qquad \times [\langle \widehat{M}_x^{k}\widetilde{\varphi}_{i_k}, \widetilde{\varphi}^*_{j_k}\rangle_1\langle \widehat{N}_x^{k+1}\widetilde{\varphi}_{i_{k+1}}, \widetilde{\varphi}^*_{j_{k+1}}\rangle_1 \\
&\qquad + \langle \widehat{M}_y^{k}\widetilde{\varphi}_{i_k}, \widetilde{\varphi}^*_{j_k}\rangle_1\langle \widehat{N}_y^{k+1}\widetilde{\varphi}_{i_{k+1}}, \widetilde{\varphi}^*_{j_{k+1}}\rangle_1] = 0.
\end{aligned} \tag{3.40}$$

Discretization in time is effected via an implicit scheme where, denoting the vector of unknowns by $\boldsymbol{\alpha}$, the partial time derivative is approximated by a simple forward difference

$$\frac{\partial \boldsymbol{\alpha}}{\partial t} \approx \frac{\boldsymbol{\alpha}^{n+1} - \boldsymbol{\alpha}^n}{\Delta t}, \tag{3.41}$$

and $\boldsymbol{\alpha}$ is evaluated at time step $(n+1)$ in the terms in (3.22) corresponding to $\sum_k \widehat{L}^k\alpha$ and at time level n elsewhere. All of this leads to the fully discrete scheme

$$\begin{aligned}
\boldsymbol{\alpha}^{n+1} - \Delta t\sum_{k=1}^{d}\overline{L}^k\boldsymbol{\alpha}^{n+1} = \boldsymbol{\alpha}^n &- \frac{\Delta t}{4\lambda}\sum_{k=2}^{d}(\overline{N}_x^{k-1}\overline{M}_x^k + \overline{N}_y^{k-1}\overline{M}_y^k)\boldsymbol{\alpha}^n \\
&- \frac{\Delta t}{4\lambda}\sum_{k=1}^{d-1}(\overline{N}_x^{k+1}\overline{M}_x^k + \overline{N}_y^{k+1}\overline{M}_y^k)\boldsymbol{\alpha}^n,
\end{aligned} \tag{3.42}$$

where an overbar on a quantity indicates that it is a matrix. The coefficient matrix of $\boldsymbol{\alpha}^{n+1}$ in the scheme (3.42)) is approximated (to $O(\Delta t)^2$) in our code by the matrix equivalent of the operator $\bigotimes_{k=1}^{d}(\boldsymbol{I} - \Delta t\widehat{L}^k)$.

4. Stochastic Algorithm and Stress Calculator

In homogeneous polymeric flows the probability density function ψ will be independent of the centre of mass of a chain $\boldsymbol{r}_c$. We therefore neglect the stochastic equation (2.9) for the centre of mass of a polymer chain (it just moves with the local velocity) and, scaling the position vector $\boldsymbol{q}_j$ of the jth connector vector with $l_0 = \sqrt{\frac{k_BT}{H}}$, see that its equation of motion (2.8) becomes

$$d\boldsymbol{q}_j = \left[\nabla\boldsymbol{v}\cdot\boldsymbol{q}_j - \frac{1}{2\lambda}\frac{\boldsymbol{q}_j}{1-q_j^2/b}\right]dt\left[\frac{1}{4\lambda}\boldsymbol{F}(\boldsymbol{q}_{j+1}) + \frac{1}{4\lambda}\boldsymbol{F}(\boldsymbol{q}_{j-1})\right]dt + \sqrt{\frac{1}{\lambda}}d\boldsymbol{W}_j^q. \tag{4.1}$$

Equation (4.1) is solved subject to an initial condition consisting of a random sample $\{\boldsymbol{q}_1, \dots, \boldsymbol{q}_d\}$ drawn from the equilibrium distribution (3.24). Since all the equations

(4.1) ($j = 1, \ldots, d$) are coupled we advance the calculation through one time step Δt by performing one sweep from $\boldsymbol{q}_1$ to $\boldsymbol{q}_d$ of an iterative method whereby, for a given index 'j' the vectors $\{\boldsymbol{q}_1, \ldots, \boldsymbol{q}_{j-1}\}$ are assumed known at $t = (n+1)\Delta t$ and $\{\boldsymbol{q}_{j+1}, \ldots, \boldsymbol{q}_d\}$ at time $t = n\Delta t$. This leads to the scheme

$$\begin{aligned}(4.2)\qquad q_{j,x}^{n+1} + \frac{\Delta t}{2\lambda}\frac{q_{j,x}^{n+1}}{1-q^2/b} &= q_{j,x}^n + \left(\nabla \boldsymbol{v} \cdot \boldsymbol{q}_j^n\right)_x \Delta t \\ &\quad + [F_x(\boldsymbol{q}_{j+1}^n) + F_x(\boldsymbol{q}_{j-1}^{n+1})]\Delta t + \sqrt{\frac{\Delta t}{\lambda}}\widetilde{W}_{j,x} = r_{j,x}^n,\end{aligned}$$

$$\begin{aligned}(4.3)\qquad q_{j,y}^{n+1} + \frac{\Delta t}{2\lambda}\frac{q_{j,y}^{n+1}}{1-q^2/b} &= q_y^n + (\nabla \boldsymbol{v} \cdot \boldsymbol{q}_j^n)_y \Delta t \\ &\quad + [F_y(\boldsymbol{q}_{j+1}^n) + F_y(\boldsymbol{q}_{j-1}^{n+1})]\Delta t + \sqrt{\frac{\Delta t}{\lambda}}\widetilde{W}_{j,y} = r_{j,y}^n,\end{aligned}$$

where

$$q^2 = (q_{j,x}^{n+1})^2 + (q_{j,y}^{n+1})^2$$

and

$$(4.4)\qquad \widetilde{\boldsymbol{W}}_j = \frac{1}{\sqrt{\Delta t}}\left(\boldsymbol{W}_j^q(t+\Delta t) - \boldsymbol{W}_j^q(t)\right),$$

which implies that each component $\widetilde{W}_j \sim N(0,1)$. From the previous equations

$$(4.5)\qquad q_{j,x}^{n+1} = \frac{r_{j,x}^n}{1 + \Delta t/2\lambda\ 1/1 - q^2/b} \quad\text{and}\quad q_{j,y}^{n+1} = \frac{r_{j,y}^n}{1 + \Delta t/2\lambda\ 1/1 - q^2/b},$$

and therefore,

$$\boldsymbol{q}_j^{n+1}\left(1 + \frac{\Delta t}{2\lambda}\frac{1}{1-q^2/b}\right) = \boldsymbol{r}_j^n.$$

Taking absolute values in the above and setting $r = \sqrt{(r_{j,x}^n)^2 + (r_{j,y}^n)^2}$ we get the cubic equation

$$q^3 - rq^2 - bq\left(1 + \frac{\Delta t}{2\lambda}\right) + br = 0.$$

This equation has one real root q between 0 and $\sqrt{b}$. Once this has been computed $q_{j,x}$ and $q_{j,y}$ may be found from (4.5).

The elastic extra stress tensor $\boldsymbol{\tau}$ appearing in (1.1) may be written in the form

$$(4.6)\qquad \boldsymbol{\tau} = \sum_{j=1}^{d} \boldsymbol{\tau}_j,$$

where $\boldsymbol{\tau}_j$ is the contribution of the jth segment in all the polymer chains of the solution to the total elastic extra stress. In the case of a homogeneous flow the stress depends only upon time. The Kramers expression for $\boldsymbol{\tau}_j$ at time $n\Delta t$ now reads

$$(4.7)\qquad \boldsymbol{\tau}_j(n\Delta t) \approx \frac{(1-\beta)}{We}\frac{(b+4)}{b}[\boldsymbol{I} + \langle \boldsymbol{q}_j^n \otimes \boldsymbol{F}(\boldsymbol{q}_j^n)\rangle],$$

where $\langle\cdot\rangle$ denotes an average taken over all M chains in the solution and We denotes a Weissenberg number, defined as $We = \lambda\dot{\gamma}$; a dimensionless shear rate. In terms of the joint probability density function ψ, $\boldsymbol{\tau}_j(n\Delta t)$ is written as

$$(4.8)\quad \boldsymbol{\tau}_j(n\Delta t) = \frac{(1-\beta)}{We}\frac{(b+4)}{b}\left[\boldsymbol{I} + \int_{|\boldsymbol{q}_1|<\sqrt{b}} \cdots \int_{|\boldsymbol{q}_d|<\sqrt{b}} \boldsymbol{q}_j \otimes \boldsymbol{F}(\boldsymbol{q}_j) \times \psi(n\Delta t, \boldsymbol{q}_1, \ldots, \boldsymbol{q}_d) d\boldsymbol{q}_1, \ldots, d\boldsymbol{q}_d\right],$$

and the convergence of (4.7) as $M \to \infty$ to the evaluation of (4.8) with the exact ψ is assured by the strong law of large numbers.

5. Example: Steady Simple Shear Flow

In this section we present results for a homogeneous shear flow $\boldsymbol{v} = (v_x, v_y) = (y, 0)$ using both the FTP (3.29) and STP (3.30) representations for the transformed distribution function $\widetilde{\alpha}$. The stochastic method (4.2)–(4.7) was used with $M = 10000$ and 20000 chains and the tensile elastic stress τ_{xx} computed with $M = 20000$ considered to be a reference solution, against which the Fokker–Planck-based calculations of τ_{xx} could be compared. To this end, for the Weissenberg numbers considered, a time step $\Delta t = 0.01$ was used and an average value for τ_{xx} over the time interval $[7, 10]$ computed. Averages of τ_{xx} computed over later time intervals showed no significant variation and τ_{xx} was considered, as a consequence, to have reached steady state.

The CPU time in seconds on a Pentium IV machine, as well as the relative error for τ_{xx} committed by the STP and FTP methods for differing values of N_R and N_F and at two different Weissenberg numbers ($We = 1$ and 3) are shown in Figures 2 and 3. Unless otherwise stated, the level of approximation l in the STP method was taken equal to the maximum value L. At both Weissenberg numbers the STP method is considerably cheaper than an FTP method computed with the same level of discretization for the eigenbasis. At the same time, at the lower Weissenberg number, the relative error with the STP method is lower than with the corresponding FTP method and the STP method may be seen to be competitive with the stochastic approach for longer chains than its FTP counterpart. At the larger Weissenberg number, the FTP calculations are more accurate than those performed with the STP representation and the deterioration of the accuracy advantage of the STP method for small chains ($d \leq 3$) may be attributable to the use of a Galerkin formulation. FTP calculations were not considered feasible for $d > 3$.

A point of interest in Figure 3 is that by taking a finer mesh for the eigenfunction calculation ($N_R = 12, N_F = 8$) but choosing the level of approximation $l = 7 < L$ we can get a significant improvement in accuracy over the FTP results from a coarser mesh ($N_R = 5, N_F = 3$) at comparable CPU cost.

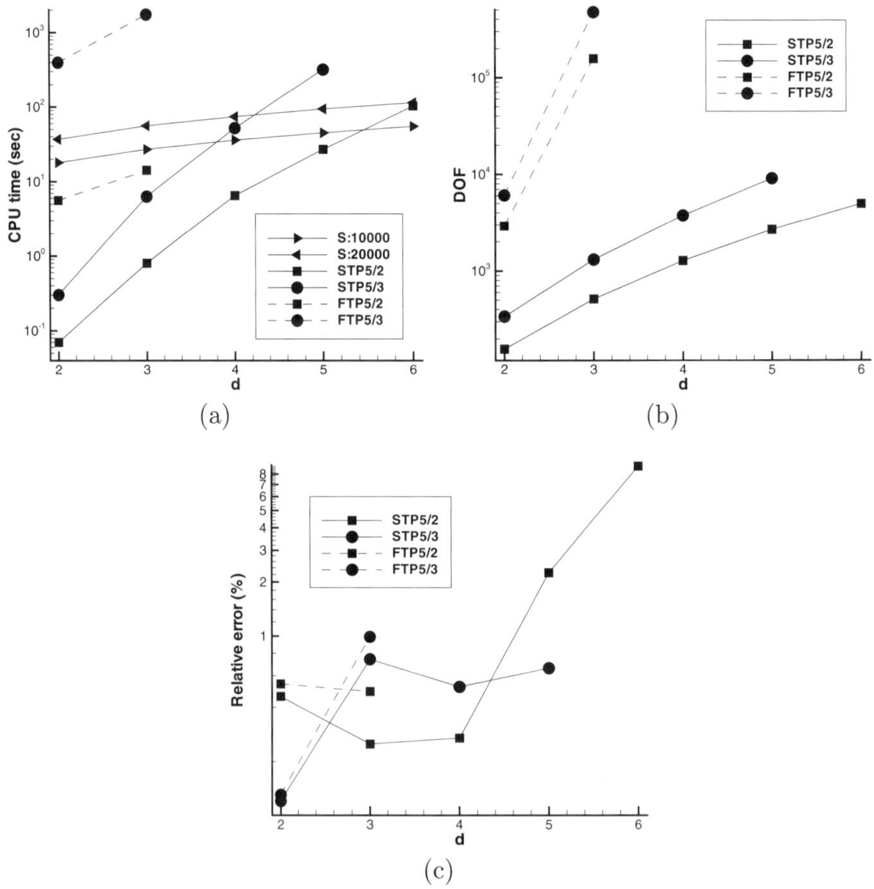

FIGURE 2. $We = 1$. (a) CPU time (s), (b) number of degrees of freedom, and (c) percentage relative error for τ_{xx} computed using (4.8). FTP i/j: full tensor product (3.29), $N_R = i, N_F = j$. STP i/j: sparse tensor product (3.30), $N_R = i, N_F = j$. Also shown are the CPU time consumed by the stochastic method (4.2)–(4.7) with $M = 10,000$ and $20,000$ realizations.

6. Conclusions

In this paper we have presented a new spectral Galerkin method for the solution of a high-dimensional Fokker–Planck equation arising in the modelling of dilute solutions of long idealized polymers. The method is inspired by ideas in the sparse grid literature and at modest levels of fluid memory and shear rate ($We \approx 1$) is demonstrably superior in accuracy and cost to a traditional full tensor product representation for the polymer configuration distribution function. The method extends the attractiveness of Fokker–Planck-based methods (as a competitive alternative to stochastic simulations) to longer polymer chains than was previously possible, although ultimately in shear flows at sufficiently high Weissenberg numbers (strong flows, highly elastic liquids) it would seem probable that

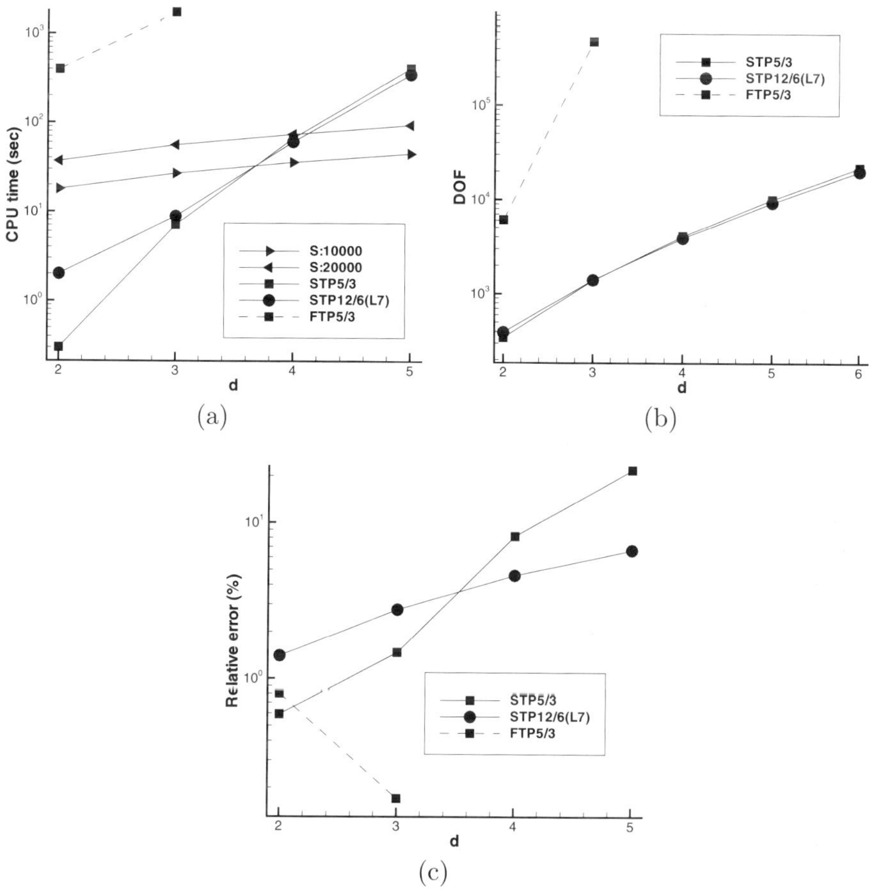

FIGURE 3. $We = 3$. (a) CPU time (s), (b) number of degrees of freedom, and (c) percentage relative error for τ_{xx} computed using (4.8). FTP 5/3: full tensor product (3.29), $N_R = 5, N_F = 3$. STP 5/3: sparse tensor product (3.30)), $N_R = 5, N_F = 3$. STP 12/8($L7$): sparse tensor product 3.30, $N_R = 12, N_F = 8, l = 7$. Also shown are the CPU time consumed by the stochastic method (4.2)–(4.7) with $M = 10,000$ and $20,000$ realizations.

well constructed stochastic methods will have the advantage in terms of cost and accuracy.

References

1. S. Achatz, *High order sparse grid methods for elliptic partial differential equations with variable coefficients*, Computing **71** (2003) 1–15.
2. R.B. Bird, C.F. Curtiss, R.C. Armstrong, and O. Hassager, *Dynamics of polymeric liquids. II. Kinetic theory*, John Wiley & Sons, New York, 1987.
3. H.-J. Bungartz and M. Griebel, *Sparse grids*, Acta Numer. **13** (2004), 147–269.
4. C. Chauvière and A. Lozinski, *Simulation of complex viscoelastic flows using Fokker–Planck equation*: 3D FENE *model*, J. non-Newtonian Fluid Mech. **122** (2004), 201–214.

5. X. Fan, *Molecular model and flow calculation*: I. *The numerical solutions to multibead-rod models in homogeneous flows*, Acta Mech. Sinica **5** (1989), 49–59.
6. ______, *Molecular models and flow calculations*: II. *Simulation of steady planar flow*, Acta Mech. Sinica **5** (1989), 216–226.
7. X.-J. Fan, *Viscosity, first normal-stress coefficient, and molecular stretching in dilute polymer solutions*, J. non-Newtonian Fluid Mech. **17** (1985), 125–144.
8. ______, *Viscosity, first normal-stress coefficients and molecular stretching in concentrated solutions and melts*, J. non-Newtonian Fluid Mech. **17** (1985), 251–265.
9. ______A numerical investigation of the hydrodynamic interaction for Hookean dumbbell suspensions in steady state shear flow, J. Chem. Phys. **85** (1985), 6237–6238.
10. M. Fixman, *Dynamics of polymer chains*, J. Chem. Phys. **42** (1965), 3831–3837.
11. J.B. Haag and D.S. Watkins, *QR–like algorithms for the nonsymmetric eigenvalue problem*, ACM Trans. Math. Software **19** (1993), 407–418.
12. M.A. Hulsen, A.P.G. van Heel, and B.H.A.A. van den Brule, *Simulation of viscoelastic flows using Brownian configuration fields*, J. non-Newtonian Fluid Mech. **70** (1997), 79–101.
13. R. Keunings, *Micro-macro methods for the multiscale simulation of viscoelastic flow using molecular models of kinetic theory*, Rheology Reviews 2004 (D.M. Binding and K. Walters, eds.), British Society of Rheology, 2004, pp. 67–98.
14. M. Laso and H.-C. Öttinger, *Calculation of viscoelastic flow using molecular models*: *the* CONNFFESSIT *approach*, J. non-Newtonian Fluid Mech. **47** (1993), 1–20.
15. A. Lozinski and C. Chauvière, *A fast solver for Fokker–Planck equation applied to viscoelastic flows calculations*: 2D FENE *model*, J. Comput. Phys. **189** (2003), 607–625.
16. A. Lozinski, C. Chauvière, J. Fang, and R.G. Owens, *Fokker–Planck simulations of fast flows of melts and concentrated polymer solutions in complex geometries*, J. Rheol. **47** (2003), 535–561.
17. R. Nayak, *Molecular simulation of liquid crystal polymer flow*: *a wavelet-finite element analysis*, Ph.D thesis, MIT, Cambridge, 1998.
18. H. Niederreiter, *Random number generation and quasi-Monte Carlo methods*, CBMS-NSF Regional Conference Series in Appl. Math., vol. 63 SIAM, Philadelphia, 1992.
19. H.C. Öttinger, *Stochastic processes in polymeric fluids*, Springer-Verlag, Berlin, Heidelberg, 1996.
20. C. Schwab and R.A. Todor, *Sparse finite elements for stochastic elliptic problems — higher order moments*, Computing **71** (2003), 43–63.
21. C. Schwab and R.A. Todor, *Sparse finite elements for elliptic problems with stochastic loading*, Numer. Math. **95** (2003), 707–734.
22. S.A. Smolyak, *Quadrature and interpolation formulas for tensor products of certain classes of functions*, Dokl. Akad. Nauk SSSR **4** (1963), 240–243.
23. J.K.C. Suen, Y.L. Joo, and R.C. Armstrong, *Molecular orientation effects in viscoelasticity*, Annu. Rev. Fluid Mech. **34** (2002), 417–444.
24. G. Venkiteswaran and M. Junk, *Quasi-Monte Carlo algorithms for diffusion equations in high dimensions*, Math. Comput. Simul. **68** (2004), 23–41.
25. ______, *A QMC approach for high dimensional Fokker–Planck equations modelling polymeric liquids*, Math. Comput. Simul. **68** (2004), 43–56.
26. T. von Petersdorff and C. Schwab, *Numerical solution of parabolic equations in high dimensions*, ESAIM: Math. Model. Numer. Anal. **38** (2004), 93–127.
27. H.R. Warner, *Kinetic theory and rheology of dilute suspensions of finitely extendible dumbbells*, Ind. Eng. Chem. Fundam. **11** (1972), 379–387.
28. C. Zenger, *Sparse grids*, Parallel Algorithms for Partial Differential Equations, Proceedings of the Sixth GAMM-Seminar Kiel (W. Hackbusch, ed.), Notes Numer. Fluid Mech., vol. 31, 1991, pp. 241–251.

CUST, GÉNIE MATHÉMATIQUE ET MODÉLISATION, UNIVERSITÉ BLAISE PASCAL II, CLERMONT-FERRAND, FRANCE.
Current address: 1, rue de la Grand Court, 10350 Saint Lupien, France.
E-mail address: pascal_delaunay@hotmail.com

CHAIRE D'ANALYSE ET SIMULATIONS NUMÉRIQUES, SECTION DE MATHÉMATIQUES, ÉCOLE POLYTECHNIQUE FÉDÉRALE DE LAUSANNE, CH–1015 LAUSANNE, SWITZERLAND.
Current address: Department of Mathematics, University of Houston, Houston, TX 77204-3008, USA.
E-mail address: alexei@math.uh.edu

DÉPARTEMENT DE MATHÉMATIQUES ET DE STATISTIQUE, UNIVERSITÉ DE MONTRÉAL, CP 6128, SUCCURSALE CENTRE-VILLE, MONTRÉAL, QC H3C 3J7, CANADA.
E-mail address: owens@dms.umontreal.ca

Centre de Recherches Mathématiques
CRM Proceedings and Lecture Notes
Volume **41**, 2007

A Partial Differential Equation for Credit Derivatives Pricing

Marcos Escobar and Luis Seco

Abstract. Partial differential equations proved to be a fundamental tool as the derivatives market developed in the seventies. As markets continue into more sophisticated territory, such as credit trading, differential equations continue to play an important role, with the added qualifier that the equations that arise are now much more complex. This paper presents a broad overview into the credit markets, and discusses a partial differential equation that arises in that context.

1. Introduction

Financial activity is an affair in risk transfer. Stocks and bonds, the financial instruments of the XIXth century, are designed to allow investors to participate in commercial enterprises; stock holders assume market risk, i.e., the risk that the firm does not meet profitability expectations; bond investors are not exposed to that market risk, and only assume default risk, i.e., the risk that the issuing entity cannot meet its financial obligations. This is also called *credit risk*, and losses can also occur without the company defaulting: a mere credit downgrade will lead to a decrease in the market value of the bond, and hence a loss, realized or not.

In the latter part of the XXth century, market risk was traded massively through the derivatives market. Investors could buy price protection related to stocks, currencies, interest rates or commodities by purchasing options or other derivatives; some are standard, others are tailor-made and labelled "over-the-counter".

But as opposed to traditional insurance premiums, financial insurance is not based merely on probabilistic considerations. To see why, imagine the following a very simple hypothetical situation.

There is an asset (a stock, a home, a currency, etc.) trading today at \$1, which can only be worth \$2 or \$0.50 next year, with equal probability; interest rates are 0%, i.e., borrowing is free. Consider also an investor which may need to buy this asset next year and is therefore concerned with increase in value; for that reason decides to buy insurance in the following form: if the asset raises to \$2, then the insurance policy will pay \$1. If the asset drops in price however, the policy pays

Research partially funded by NSERC, Mitacs and Algorithmics Inc.

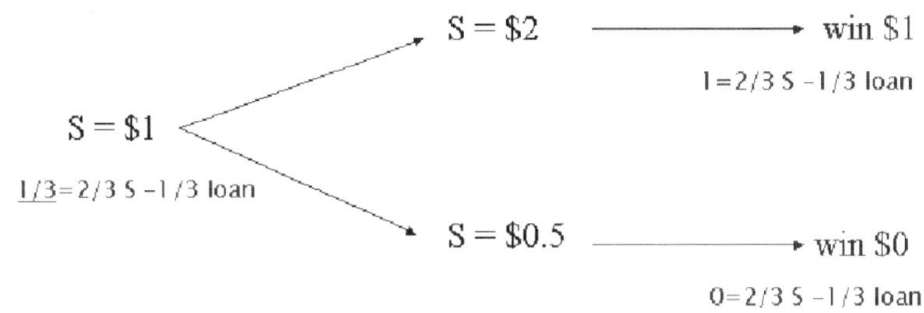

FIGURE 1. Pricing fundamentals.

nothing. One would be tempted to price this insurance policy with a premium obtained through past price movements of that asset, and it would seem that \$0.50 is the price that makes sense.

If the investor paid \$0.50, then the seller of the policy could do the following: borrows an additional \$0.10, and buys 60% of the stock. If the stock raises in value, after paying the \$1 and returning the loan they would make a profit of \$0.10. If, however, the stock drops in price, they will make a net profit of \$0.20, as the policy pays nothing and they only need to return the loan. In other words, \$0.50 is *too much*, as the issuer of the option will always make a profit: this phenomenon is called arbitrage, and it is a fundamental assumption for pricing theories that arbitrage should not exist. A simple calculation will show that the no-arbitrage price is exactly $\$\frac{1}{3}$ (see Figure 1).

This simple example (a "call option") is the basis of the no-arbitrage pricing theory ([8]), and we can quickly learn a few things from it. First, the price of a contract that depends on market moves may be replicated with buy/sell strategies, which mimic the contract pay-out but can be carried out with fixed, pre-determined costs. Second, there is a probability of events which is implied by their price, and is perhaps independent of historical events. In our example above, the implied probability of an up-move is 33%, and the probability of a down-move is 66%, because with those probabilities we can price the contract taking simple expectations. However, a more profound revision of the previous example will convince the reader with a background in diffusion processes that, if one takes the simple one-step example into a continuum of infinitesimal time/price increments, one ends with Brownian motion and associated Kolmogorov forward operator: the heat equation. One will also have a diffusion process for the asset or stock, and an associated diffusion process implied by market prices.

Black *et al* [3], derived the analog of the heat equation and Brownian motion for the case of an option with an underlying stock price that follows the Ito process given by

$$dS = \mu S\,dt + \sigma S dW^P,$$

where here S denotes the stock price, μ denotes the drift, σ is the volatility, and dW^P are infinitesimal Brownian increments. An option on a stock is as a contract that will pay a future value at expiration $f_0(S)$, which depends on the value of the underlying stock S at expiration time T. Note the similarity with our simple example above, the main difference being that in in our case now the stock trades continuously and we could therefore replicate our option by trading the stock continuously. In this case, the Black–Scholes–Merton theory shows that the price of the option contract is obtained by solving the following backward parabolic PDE (for $t < T$):

$$\begin{cases} \dfrac{\partial f}{\partial t} + \dfrac{\sigma^2}{2} S^2 \dfrac{\partial^2 f}{\partial S^2} + rS\dfrac{\partial f}{\partial S} - rf = 0, \\ f(S,T) = f_0. \end{cases}$$

At first sight, this expression has two counterintuitive features: the absence of μ and the presence of the interest rate r in the PDE. A moment's reflection however, will convince us that this is not entirely surprising: after all, in our example in Figure 1 we already saw that the price of that option was independent of the probabilities of up and down moves of the stock, and it will only depend on the cost of borrowing. This was forced on us by our no-arbitrage assumption. In more general terms, it turns out that option pricing can be established by taking expectations with respect to a "risk neutral" measure Q, which is perhaps different from the historical measure P. In our particular case, this implies that the solution to the PDE is given by

$$f(S,t) = \frac{1}{\sqrt{2\pi(T-t)}\sigma} \int_0^\infty f(u) \exp\left\{ -\frac{(\ln(u/S(t)) - (r - \sigma^2/2)(T-t))^2}{2(T-t)\sigma^2} \right\} du$$

which is easily checked. More generally, we have that

$$f(S,t) = E[f(S,T) \mid S(t)],$$
$$dS = rS\,dt + \sigma S dW^Q.$$

From this perspective, pricing becomes equivalent to finding risk neutral probabilities for a certain new measure Q (the risk neutral measure) and their pay-off expectations, and the PDE above is nothing but the Feynmann–Kac formula for this expectation.

The Black–Scholes–Merton theory also shows that one can replicate the option payoff by continuously trading the stock so that we always own $\partial_S f$ units of it.

This signified a tremendous revolution, that won Black and Scholes the Nobel prize for Economics in 1997, as it not only established a pricing mechanism for the booming options and derivative markets, but because — in highly idealized conditions, of course — it established certainty where there was risk: derivatives could be replicated by buy/sell strategies with predetermined costs.

Their discovery revolutionized market risk perspectives. But Merton, who had re-derived their pricing formalism using stochastic control theory, used this advance to start the modern theory of credit risk. His viewpoint, which we present below, was just as revolutionary.

Merton viewed a firm as shareholders and bond-holders. Bond-holders lent money to the firm, and the firm promised to pay back the loan, with interest. Shareholders own the value of the assets of the firm, minus the value of the debt (or liabilities); but firms have limited liability, which means that if the value of the assets falls below the value of the liabilities, in Merton's view, the firm defaults, shareholders owe nothing and the bondholders use the remaining value of the assets to recover a portion of their loan. In other words, the shareholders own a call option on the value of the assets of the firm, with a strike price given by the value of the liabilities at the given maturity time of the loan. The timing of his theory, which dates back to 1974, was perfect as the theory of option pricing had just been developed one year earlier, and this opened the ground for credit risk pricing and credit risk derivatives.

Strictly speaking, Merton approach assumes that the liabilities of a firm (its debt) expire at a certain time, and default could occur only at that time. Black and Cox [2] conceptually refined Merton's proposal by allowing defaults to occur at anytime within the life of the option, creating the "first passage default models". The reason for this modification is that, according to Merton's model, the firm value could dwindle to nearly nothing without triggering a default until much later; all that matters was its level at debt maturity and this is clearly not in the interest of the bond holders. Bond indenture provisions therefore often include safety covenants providing the bond investors with the right to reorganize or foreclose on the firm if the asset value hits some lower threshold for the first time. This threshold could be chosen as the firm's liabilities.

But the largest event in the credit market still had to wait until 1998, when the default of Russia and the menace of the impeachment of President Clinton over the Monica Lewinsky affair threw financial markets into disarray; the Russian default, and worries about the political stability of the United States created a *credit crunch* as bond investors fled from corporate debt for the more secure treasury bill market, introducing credit spread dislocations of historical proportions. This situation culminated with the collapse of Long Term Capital Management, a multi-billion dollar hedge fund that, anecdotally, had lured Scholes and Merton to their board of directors.

The result of these massive historical events was the explosion of the credit market. In it, financial players seek to buy and sell credit risk, either for insurance and protection in the case of default or bankruptcy of their counter-parties, or to take risk exposure which are considered either cheap or advantageous, and therefore earn above average returns. The financial instruments which are used in the credit market are numerous. In this paper, we shall focus on two closely related ones: credit default swaps (CDS), and collateralized debt obligations (CDO).

A credit default swap (CDS, see Figure 2) is a contract that provides insurance against the risk of a default by particular company (known as the reference entity). The buyer of the insurance obtains the right to sell a particular bond issued by the company for its par value when a credit event occurs. The bond is known as the

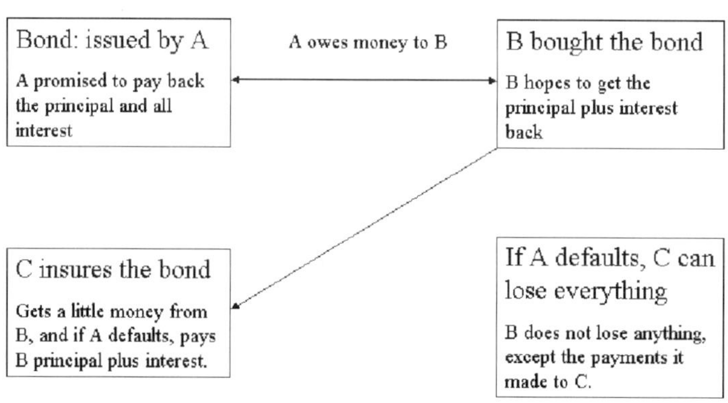

FIGURE 2. Credit derivatives.

reference obligation and the total par value of the bond that can be sold is known as the swap's notional principal. The buyer of the CDS makes periodic payments to the seller until the end of the life of the CDS or until a credit event occurs. A credit event usually requires a final accrual payment by the buyer.

A collateralized debt obligation (CDO) provides a way of creating securities with different risk characteristics from a portfolio of debt instruments. A general example would be when M types of securities are created from a portfolio of N bonds. The first tranche of securities has p_1 of the total bond principal and absorbs all credit losses from the portfolio during the life of the CDO until they have reached p_1 of the total bond principal. The second tranche has p_2 of the principal and absorbs all losses during the life of the CDO in excess of p_1 of the principal up to a maximum of $p_1 + p_2$ of the principal. The last tranche has p_M of the principal absorbs all losses in excess of $\sum_{i=1}^{M-1} p_i$ of the principal. The reason these instruments exist is that banks with large loan books, can use CDO's to effectively slice the default risk in those portfolios with credit-linked securities (the different tranches) and sell them to investors in packets which exhibit very different risk profiles: from the highly risk of the top — mezzanine — tranche (which will earn a higher fee spread), to the very secure last tranche, which will earn perhaps a minimal fee.

The building blocks of a CDO are the nth-to-default credit default swaps. An nth to default credit default swap (nCDS) is similar to a regular CDS. The buyer of protection pays a specified rate on a specified notional principal until the nth default occurs among a specified set of reference entities or until the end of the contract's life. The payments are usually made quarterly. If the nth default occurs before the contract maturity, the buyer of protection can present bonds issued by the defaulting entity to the seller of protection in exchange for the face value of the bonds. Alternatively, the contract may call for a cash payment equal to the difference between the post-default bond value and the face value. (see Figure 3).

The valuation of such structures are based on computing the probability distribution of the event "mth default". This is technically difficult because it requires one to handle the multivariate distribution of defaults, and generally most credit models fail to reliably capture multiple defaults. There are basically two procedures

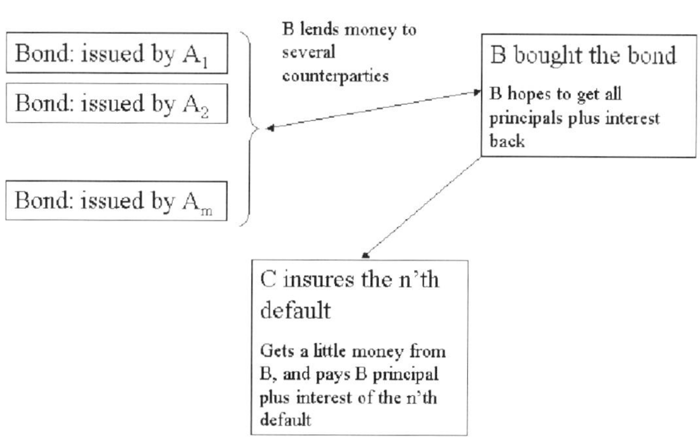

FIGURE 3. nth to default swap.

for evaluating these basket derivatives, multifactor copula models similar to that used by researchers such as [7, 13, 14] and intensity models, see Duffie-Garland [4].

In this paper we develop a partial differential equation (PDE) procedure for valuing both an nth to default CDS and a CDO. We work within the structural framework, where the default event is associated to whether the minimum value of an stochastic processes (firm's asset value) have reached a benchmark, usually the firm's liabilities (see [2, 16]). We show that the multivariate density/distribution of a vector of Brownian motions and Brownian minimums (it can be easily extended to maximums) for the case of more than two underlying components is the solution of a PDE with absorbing and boundary conditions (a Fokker–Planck equation).

The paper is organized as follows. In Section 2 we address the pricing of credit derivatives, Section 2.1 provides the set up of the economic and mathematical assumptions and the joint density/distribution functions of the minimums values of n correlated Brownian motions. Next, in Section 2.2 we do some applications, an nth to default CDS in Section 2.2.1, while a CDO is priced in 2.2.2.

2. Pricing Credit Derivatives

In a standard credit framework, there are n reference firms, $C_1, \ldots, C_n$. The protection buyer A and the protection seller B. A pays a regular fee (initial fee) to B until a pre-specified event occur or maturity. B agrees on paying the total default. One of the method for pricing credit derivatives is the structural approach.

In the structural approach, one makes explicit assumptions about the dynamics of a firm's assets, its capital structure, as well as its debt and shareholders. The philosophy of the structural approach, which goes back to [3, 16], is to consider corporate liabilities as contingent claims on the assets of the firm. Here the firm's market value (the total market value of the firm's assets, denoted V_T) is the fundamental state variable. The most popular structure in practice is based on Merton's [16] framework, where default can only occur at maturity (whenever $V_T < D$). Recognizing that a firm may default well before the maturity of the debt, one may alternatively assume that the firm goes bankrupt when the value of its assets falls below some lower threshold (that is, when $\min_{0\leq s\leq t} V(s) < D$, see [2].

The most prominent and challenging feature of a credit instruments is the dependency structure of defaults. This is why most recent efforts have been directed to the capture of these dependency structures. Structural framework's handling of this problem rest on the finding of the multivariate distribution of first-passage-time. This distribution can be inferred from the multivariate distribution of a vector of Brownian minimums (see [5, 21]).

We know, based on the Black–Scholes framework, that the pricing of derivatives can be reduced to finding the expected value of the derivative's payoff under a "free-arbitrage" equivalent measure.

2.1. Assumptions and results.

We first present the very basic assumptions required in our work.

- Interest rate, r is constant.
- The value of the assets, $V_i(t)$, follows an Ito process with constant drift r and volatility $\sigma_i^2(t)$ (an adapted process such that $\int_0^t (\sigma_i(s))^2\, ds < \infty$ a.s.) under the risk neutral measure:

$$dV_i = rV_i\, dt + \sigma_i(t)V_i dW_i(t).$$

- Firm i defaults as soon as its asset value V_i reaches the liabilities, denoted as D_i. This is the definition of default within the structural framework (see [2, 5, 16]).

Let us take $X(t) = \ln V(t)$ as the n-dimensional Brownian motion vector with drift $\mu = (\mu_1, \dots, \mu_n)$ (where $\mu_i(t) = r - \frac{1}{2}\sigma_i^2(t)$) and covariance matrix $\Sigma = (\sigma_{i,j}(t)) = (\rho_{ij}\sigma_i(t)\sigma_j(t))$,

$$X(t) = \mu(t) \cdot t + \operatorname{diag}(\Sigma(t)) \cdot W(t), \quad t \geq 0, \tag{2.1}$$

where $W(t)$ is a correlated Brownian vector such that $E[dW_i dW_j] = \rho_{ij}\, dt \ \forall i \neq j = 1, \dots, n$.

The two most important variables in our framework are the running minimum of the process $X(t)$ and the time of default t_i, they are defined as follows:

The running minimum is defined as:

$$\underline{X}_i(t) = \min_{0 \leq s \leq t} X_i(s).$$

While the time of default for firm i is defined using the running minimum as:

$$t_i = \inf\{t \mid \underline{V}_i(t) \leq D_i\} = \inf\{t \mid \underline{X}_i(t) \leq \ln D_i\}. \tag{2.2}$$

When handling several companies simultaneously, those previous definitions are not enough. The time of default of a given company is a marginal piece of information which fails to hint at other companies' movements. Therefore it is useful to create variables aimed at describing joint defaults scenarios. In this regard the variables τ_j: time of j-defaults is defined next:

The time of j-defaults is defined as: $\tau_j = t_{i_j}$ where $t_{i_1} \leq \cdots \leq t_{i_n}$.

Having defined all these key variables, we next show that their probability distribution can be obtained from the joint probabilities of running minimums. The reason why we emphasize this relationship is because the joint probability of running minimums (as well as running maximum in this regard) is known. It is actually the solution of a partial differential equation.

Denote $\pi_t(1)$ for the probability of exactly one default before t. Then

$$\begin{aligned}\pi_t(1) &= P(\tau_1 \leq t, \tau_2 > t) \\ &= \sum_{i=1}^{n} P(\underline{V}_1(t) > D_1, \dots, \underline{V}_i(t) \leq D_i, \dots, \underline{V}_n(t) > D_n).\end{aligned} \tag{2.3}$$

In general, the probability of exactly j defaults before t could be found as:

$$\pi_t(j) = P(\tau_j \leq t, \tau_{j+1} > t) \tag{2.4}$$

$$\begin{aligned} = \sum_{i_1 \neq i_j = 1} P(\underline{V}_1(t) > D_1, \dots, \underline{V}_{i_1}(t) \leq D_{i_1}, \dots, \underline{V}_{i_j}(t) \\ \leq D_{i_j}, \dots, \underline{V}_n(t) > D_n),\end{aligned} \tag{2.5}$$

where the summation is taken over the $k!/(j!(k-j)!)$ different ways in which the $\underline{V}_i$ can be chosen.

The probability of at least one default before t, $P(1,t)$, would then be:

$$\begin{aligned}P(1,t) &= P(\tau_1 \leq t) \\ &= [1 - P(\underline{V}_1(t) > D_1, \dots, \underline{V}_i(t) > D_i, \dots, \underline{V}_n(t) > D_n)].\end{aligned} \tag{2.6}$$

Similarly, the probability of at least j defaults before t would be:

$$P(j,t) = P(\tau_j \le t) = \sum_{k=j}^{N} \pi_t(k). \tag{2.7}$$

Let us also denote $f_j(.)$ as the density of τ_j. Then

$$f_j(t) = \frac{\partial P(j,t)}{\partial t}. \tag{2.8}$$

Notice that in order to compute those previously defined probabilities, we need the multivariate distribution of the vector of minimums of X. This is tackled in the next result.

Theorem 2.1. *Let us denote:*

$$\begin{aligned} P(X_1(t) \in dx_1, \dots, X_n(t) \in dx_n, \underline{X}_1(t) > m_1, \dots, \underline{X}_n(t) > m_n) \\ = p(x_1, \dots, x_n, t, m_1, \dots, m_n, \mu, \Sigma)\, dx_1 \dots dx_n, \end{aligned} \tag{2.9}$$

where $m_i = \ln(D_i)$, $p(x_1, \dots, x_n, t, m_1, \dots, m_n, \mu, \Sigma)$ *is the joint distribution/density function of the minimums and endpoints of* $X(t)$, $x_i > m_i$, $m_i \le 0$; $\forall i$. *Then* p *satisfies the following partial differential equation (Fokker–Planck equation):*

$$\begin{cases} \dfrac{\partial p}{\partial t} = -\sum_{i=1}^{n} \mu_i(t) \cdot \dfrac{\partial p}{\partial x_i} + \sum_{i,j=1}^{n} \dfrac{\sigma_{ij}(t)}{2} \cdot \dfrac{\partial^2 p}{\partial x_i \partial x_j}, \\ p(x, t=0) = \prod_{i=1}^{n} \delta(x_i), \\ p(x_1, \dots, x_i = m_i, \dots, x_n, t) = 0, \quad i = 1, \dots, n, \\ x_i > m_i, \ m_i \le 0, \quad i = 1, \dots, n. \end{cases}$$

PROOF. What follows is a sketch of the proof of Theorem 2.1. Suppose that $X(t)$ follows the process described in (2.1), consider any twice differential function $p(X,t)$ that is a conditional expectation of some function of X at a later date T, $g(X(T))$:

$$p(X,t) = E\big[g\big(X(T)\big) \mid X(t) = x\big]. \tag{2.10}$$

By Ito's lemma, we can compute the stochastic process for f. Moreover, note that $p(X,T) = g\big(X(T)\big)$, therefore f is a martingale. This implies that the drift of the process for p has to be zero, leading to the equation:

$$\frac{\partial p}{\partial t} = -\sum_{i=1}^{n} \mu_i(t) \cdot \frac{\partial p}{\partial x_i} + \sum_{i,j=1}^{n} \frac{\sigma_{ij}(t)}{2} \cdot \frac{\partial^2 p}{\partial x_i \partial x_j}$$

with boundary condition $p(X,T) = g\big(X(T)\big)$.

Selecting a convenient function g is key for getting a meaning for the function p, for example if $g(v, \sigma_i) = \exp\{i\theta_1 v_1 + i\theta_2 v_2\}$ then the solution is the characteristic function for X. In our case we take $g(X) = \delta(X - X_0)$ therefore the solution $p(X,t)$ is the conditional probability density at time t that $X(T) = X_0$.

To see the absorbing boundary condition notice that we ask for the probability density $p(x_1, \dots, x_n, t)$ for the stochastic vector $X(t)$, starting at $t = 0$ with $X(0) = 0$ to reach $x = (x_1, \dots, x_n)$ at time t. We do the following in order to make the path of $X_i(t)$ always greater than m_i, ($\underline{X}_i(t) > m_i$, $\forall i$): if the stochastic component X_i reaches m_i for the first time we no longer count these vector realizations (this would assure $\underline{X}_i(t) > m_i$). Therefore $p(x, t, \mu, \Sigma)$ must be zero for $X_i \le m_i$, $\forall i$. The prescription that the processes are no longer counted if they passed a boundary is substituted by an absorbing n-dimensional wall ($p(x_1, \dots, x_i = m_i, \dots, x_n, t) = 0$, $i = 1, \dots, n$) [1]. For the domain $m \leq X$, $p(x, t, m, \mu, \Sigma)$ must satisfy the previous Fokker–Planck equation [2] (see [18]). □

Remark 2.11. Note that integrating p with respect to x leads to the distribution function of the minimum:

$$(2.12)\quad P(\underline{X}_1(t) > m_1, \dots, \underline{X}_n(t) > m_n) = \iint p(x_1, \dots, x_n, t, m_1, \dots, m_n, \mu, \Sigma)\, dx_1, \dots, dx_n,$$

which is the key element for computing the quantities $\pi_t(j)$, $P(j,t)$ and $f_j(t)$ on (2.5), (2.7) and (2.8) respectively.

In a one-dimensional setting, assuming constant drift and volatility $P(X_1(t) \in dx_1, \underline{X}_1(t) > m_1) = p(x_1, t, m_1, \mu_1, \sigma_1^2)\, dx_1$, the PDE becomes:

$$\begin{cases} \dfrac{\partial p}{\partial t} = -\mu_1 \cdot \dfrac{\partial p}{\partial x_1} + \dfrac{\sigma_1^2}{2} \cdot \dfrac{\partial^2 p}{\partial x_1^2}, \\ p(x_1, t = 0) = \delta(x_1), \\ p(x_1 = m_1, t) = 0, \\ x_1 > m_1, \ m_1 \leq 0. \end{cases}$$

A solution to this PDE can be obtained combining transformations and the reflection principle. First removing the drift term by a transformation $q(x_1, t) = p(x_1, t)e^{ax_1 + bt}$ for convenient values of a and b. Notice that $q(x_1, t)$ can be seen as the solution to $q(x_1, t)\, dx_1 = P(X_1(t) \in dx_1, \underline{X}_1(t) > m_1)$ where $X_1(t)$ is a plain Brownian process without drift. Then we find $q(x_1, t)$ using the reflection principle.

The reflection principle is a symmetry principle that states that $P(W_1(t) < w_1, \underline{W}_1(t) < m_1) = P(W_1(t) < 2m_1 - w_1)$, or,

$$P(W_1(t) < w_1, \underline{W}_1(t) > m_1) = P(W_1(t) < w_1) - P(W_1(t) < 2m_1 - w_1).$$

It then follows that:

$$q(x_1, t) = \frac{\partial P(X_1(t) < x_1, \underline{X}_1(t) > m_1)}{\partial x_1}.$$

Substituting $q(x_1, t)$ back into the equation $p(x_1, t) = q(x_1, t)e^{-ax_1 - bt}$ leads to:

$$p(x_1, t, m_1, \mu, \sigma^2) = \frac{1}{\sigma_1 t}\phi\left(\frac{x_1 - \mu_1 t}{\sigma_1 \sqrt{t}}\right)\left(1 - \exp\left\{\frac{-4m_1^2 + 4m_1 x_1}{2\sigma_1^2 t}\right\}\right).$$

Notice that integrating with respect to x_1 leads to the distribution function of the minimum, which is called the inverse Gaussian distribution

$$(2.13)\qquad p(\underline{X}_1(t) \geq m_1) = \Phi\left(\frac{m_1 - \mu_1 t}{\sigma_1 \sqrt{t}}\right) - \exp\left\{\frac{2m_1\mu_1}{\sigma_1^2}\right\}\Phi\left(\frac{-m_1 - \mu_1 t}{\sigma_1 \sqrt{t}}\right)$$

(see for example [5]).

[1] For a given t, p gives the probability that the path lies within the domain, so not only $x_i(t) \geq m_i$ but also $\underline{X}_i(t) \geq m_i$.

[2] It describes the time evolution of the probability density function of position and velocity of a particle, it is also known as forward Kolmogorov equation.

He *et al* [6] provided the joint density for the case of two Brownian processes $(X_1, \underline{X}_1, X_2, \underline{X}_2)$ assuming constant drift and volatility. This can also be seen as a particular case of this framework, Remark 2.11 taking $k = 2$, $n = 2$. They pointed out that the joint distribution/density $p(x_1, x_2, t, m_1, m_2, \mu, \Sigma)$ for $x_1 \geq m_1$, $x_2 \geq m_2$ where $m_i \leq 0$ must satisfy the following Partial differential equation:

$$\begin{cases} \dfrac{\partial p}{\partial t} = -\sum_{i=1}^{2} \mu_i \cdot \dfrac{\partial p}{\partial x_i} + \sum_{i,j=1}^{2} \dfrac{\sigma_{ij}}{2} \cdot \dfrac{\partial^2 p}{\partial x_i \partial x_j}, \\ p(x, t = 0) = \prod_{i=1}^{2} \delta(x_i), \\ p(x_i = m_i, t) = 0, \quad i = 1, 2, \\ x_i > m_i, \ m_i \leq 0, \quad i = 1, 2. \end{cases}$$

The solution of this PDE is:

$$p(x_1, x_2, t, m_1, m_2, \mu, \Sigma) = \frac{e^{a_1 \cdot x_1 + a_2 \cdot x_2 + b \cdot t}}{\sigma_1 \sigma_2 \sqrt{1 - \rho^2}} \cdot h(x_1, x_2, t, m_1, m_2, \mu, \Sigma), \tag{2.14}$$

where: a and b are the solution of the following system of equations:

$$0 = \mu + \sigma_{i,j} \cdot a_j, \tag{2.15}$$

$$b = \sum_{i,j=1}^{2} \sigma_{ij} \cdot a_i \cdot a_j + \sum_{i=1}^{2} \mu_i \cdot a_i, \tag{2.16}$$

while $h =:$

$$\frac{2}{\beta t} \sum_{l \in \mathbf{N}} e^{-(r^2 + r_0^2)/2t} \cdot \sin \frac{l \cdot \pi \cdot \theta_0}{\beta} \cdot \sin \frac{l \cdot \pi \cdot \theta}{\beta} \cdot I_{(l\pi/\beta)} \left(\frac{r \cdot r_0}{t} \right). \tag{2.17}$$

Here:

$$z_1 = \frac{1}{\sqrt{1 - \rho^2}} \left(\frac{x_1 - m_1}{\sigma_1} - \rho \frac{x_2 - m_2}{\sigma_2} \right), \quad z_2 = \left(\frac{x_2 - m_2}{\sigma_2} \right),$$

$$\tan \theta = \frac{z_2}{z_1}, \ \theta \in [0, \pi], \quad \tan \beta = -\frac{\sqrt{1 - \rho^2}}{\rho}, \ \beta \in [0, \pi],$$

$$z_{10} = \frac{1}{\sqrt{1 - \rho^2}} \left(\frac{-m_1}{\sigma_1} + \rho \frac{m_2}{\sigma_2} \right), \quad z_{20} = -\frac{m_2}{\sigma_2},$$

$$r = \sqrt{z_1 + z_2}, \quad r_0 = \sqrt{z_{10} + z_{20}},$$

$$\tan \theta_0 = \frac{z_{20}}{z_{10}}, \ \theta_0 \in [0, \beta], \quad \rho = \frac{\sigma_{i,j}}{\sigma_i \sigma_j}.$$

$I_n(x)$ is the Bessel function of the second kind, fundamental solution of the Bessel equation:

$$x^2 \frac{d^2 Y}{dx^2} + x \frac{dY}{dx} + (x^2 - n^2) Y = 0.$$

Integrating with respect to (x_1, x_2) leads to the joint distribution function of the minimums.

The main reason why this PDE could be solved, in closed form, for dimension two is the existence of an orthogonal coordinate system (polar coordinates) that allows for the removal of the mixing derivative term in the PDE equation while keeping a separable PDE equation with orthogonal boundary conditions. The procedure for finding the solution was based on three steps, first removing the drift term (using a similar transformation to that in the uni-dimensional case), then a change to polar coordinates removes the mixing derivative while keeping the boundary conditions orthogonal. The remaining PDE after those transformation is a Laplace equation which was solved explicitly using separation of variables.

A closed form solution for the n-dimensional case, assuming constant drift and volatility, has not been found yet. The most important reason for this failure could be traced to the non existence of an orthogonal coordinate system (within systems of degree two) capable of removing the mixing derivative while keeping the boundary conditions orthogonal. For example in the three dimensional case, it is known that there are 11 orthogonal coordinates systems in which the Laplace equation can be solved by separation of variables, but none of them fulfills both objectives: removal of mixing derivatives while orthogonalizing the boundary conditions (see Moon and Spencer 1988).

Remark 2.18. This joint distribution/density can be manipulated in order to obtain the joint distribution/density of a subset of endpoints and minimums; for example $P(X_1(t) \in dx_1, \ldots, X_n(t) \in dx_n, \underline{X}_1(t) \geq m_1, \ldots, \underline{X}_k(t) \geq m_k)$, $P(X_{k+1}(t) \in dx_{k+1}, \ldots, X_n(t) \in dx_n, \underline{X}_1(t) \in dx_1, \ldots, \underline{X}_k(t) \in dx_k)$ by integration.

Remark 2.19. The previously defined quantities, $\pi_t(j)$ and $P(j,t)$ $\forall j$, can be obtained by using Theorem 2.1. Specifically we can imply the joint density of $g(t, m_1, \ldots, m_n)$ of the multivariate distribution of $(\underline{V}_1, \ldots, \underline{V}_n)$.

2.2. Applications.

2.2.1. *nth to default* CDS. As explained previously, an nth to default credit default swap (CDS) is similar to a regular CDS.

The buyer of protection:

- **Pays** a specified rate on a specified notional principal until the nth default occurs among a specified set of N reference entities or until the end of the contract's life. The payments are usually made quarterly (we assume only an initial payment).
- **Receives** If the nth default occurs before the contract maturity, the buyer of protection can present bonds issued by the defaulting entity to the seller of protection in exchange for the face value of the bonds.

Assumptions. (1) Principals associated with all the underlying entities are the same, $L = 1$ and the expected recovery rates are the same, $R(\tau_j)$.

(2) In the event of jth defaults, the seller pays the notional principal times $\left(1 - R(\tau_j)\right)$.

(3) For the sake of simplicity we assume only one payment (from the buyer) at the beginning of the contract. (Usually the buyer of protection makes quarterly payments in arrears at a specified rate until the jth default occurs. Is is also common to define it so that there is a payoff for the first j defaults (rather than just for the j default); also, sometimes the rate of payment reduces as defaults occur.)

Proposition 2.1. *The present value* (t) *of a* j*th to default* CDS *can be computed, based on Black–Scholes arguments, as follows:*

$$E_t^Q\left[\left(1 - R(\tau_j)\right) \cdot e^{r(\tau_j - t)} \cdot 1_{\{\tau_j < T\}}\right] = \int_t^T \left(1 - R(s)\right) \cdot e^{r(s-t)} \cdot f_j(s)\, ds. \tag{2.20}$$

Notice that the density $f_j(t)$ can be computed from the solution to the Fokker–Planck equation (see Theorem 2.1, Eq. (2.8) and Remark 2.11).

2.2.2. *Collateralized debt obligation.* We keep the same assumptions as in a CDS, i.e. the expected recovery rates (R) and principals (L) associated with all the underlying names are the same.

A CDO is a way of creating M securities with different risk characteristics from a portfolio of N debt instruments (i.e. defaultable bonds). A quite general example would be as follow:

Tranche 1: Absorbs all credit losses from the portfolio during the life of the CDO until they have reached $p_1\%$ of the total bond principal.

Tranche 2: It has $p_2\%$ of the principal and absorbs all losses in excess of $p_1\%$ of the principal up to a maximum of $q_2\% = (p_1+p_2)\%$ of the principal.

Generally, Tranche i is responsible for defaults between $q_i\%$ and $q_{i+1}\%$. (Same as losses cause of the same principal L).

The M tranche has $1 - \sum_{i=1}^{M-1} p_i\%$ of the principal and absorbs all losses in excess of $q_M\% = \sum_{i=1}^{M-1} p_i\%$ of the principal. Notice $\sum_{i=1}^{M} p_i\% = 1$. $(q_i+p_i = q_{i+1})$.

Each Tranche has specific yields (r_i), which represent the rates of interest paid to tranche holders. These rates are paid on the balance of the principal remaining in the tranche after losses have been paid. The principal to which the promised payments are applied declines as defaults occur. Tranche 1 is sometimes referred to as toxic waste while Tranche N by contrast is usually given an Aaa rating.

Notice that the present value of the expected cost of defaults for this tranche is the sum of the cost of defaults for nth to default CDS for values of n between $q_i\%$ and $q_{i+1}\%$. Suppose that there is a promised percentage payment of r_i at time τ. In our case the payment is:

(1) $p_i\% \cdot L \cdot r_i$ with probability $1 - P(q_i, \tau)$.
(2) $(p_i\% - 1\%) \cdot L \cdot r_i$ with probability $\pi_\tau(q_i)$.
(3) $(p_i\% - 2\%) \cdot L \cdot r_i$ with probability $\pi_\tau(q_i+1)$.
(4) So on.

Proposition 2.2. *The expected payment, payoff, for Tranche i, would be:*

$$p_i \cdot r_i \cdot L \cdot [1 - P(q_i, \tau)] + \sum_{j=1}^{p_i-1} [(p_i - j) \cdot r_i \cdot L \cdot \pi_\tau(q_i + j - 1)]. \tag{2.21}$$

The present value (t) of tranche i can be computed, based on a Black–Scholes argument as:

$$\begin{aligned} &\int_t^T e^{-r\cdot(s-t)} \cdot (p_i \cdot r_i \cdot L)\, ds - \int_t^T e^{-r\cdot(s-t)} \cdot (p_i \cdot r_i \cdot L) \cdot f_{q_i}(s)\, ds \\ &\qquad + \sum_{j=1}^{p_i-1} \left\{ \int_t^T e^{-r\cdot(s-t)} \cdot [(p_i - j) \cdot r_i \cdot L] \cdot f_{q_i+j-1}(s)\, ds \right\}. \end{aligned} \tag{2.22}$$

Proposition 2.3. *The distribution of losses $Lo(t)$ for tranche i can be obtained by noticing the following relation:*

$$P(Lo(t) = L \cdot q_i + s) = \frac{\pi_t(q_i + s)}{\sum_{l=0} \pi_t(q_i + l)} \tag{2.23}$$

where π_t were defined in (2.3), (2.4), (2.5), (2.6), (2.7), (2.8).

Similarly to pricing CDSs, the price of tranches are explicit functions of joint probabilities of defaults, therefore the densities $f_i(t)$ are the key element in those equation. This density can be computed from the solution to the Fokker–Planck equation (see Theorem 2.1, Eq. (2.8) and Remark 2.11.

3. Conclusion

A partial differential equations with absorbing boundary conditions is shown to provide the joint distribution of minimums and endpoints of a vector of correlated diffusion processes. The implications of this PDE for pricing derivatives is presented along the lines of [3, 16] structural models. The closed form solution for this joint distribution in the cases of one and two-dimensional cases is described.

References

1. G.B. Arfken and H.J. Weber, *Mathematical methods for physicists*, 4th ed., Academic Press, Inc., San Diego, CA, 1995.
2. F. Black and J.C. Cox, *Valuing corporate securities: some effects of bond indenture provisions*, AFAJ **31** (1976), 351–367.
3. F. Black and M.S. Scholes, *The pricing of options and corporate liabilities*, JPE **81** (1973), 81–98.
4. D. Duffie and N. Garland, *Risk and valuation of* CDO, Financial Analysts J. **57** (2001), no. 1, 41–59.
5. K. Giesecke, *Default and information*, 2003 (working paper).
6. H. He, W. Keirstead, and J. Rebholz, *Double lookbacks*, J. Math. Finance **8** (1998), 201–228.
7. J. Hull and A. White, *Valuation of a* CDO *and nth to default* CDS *without Monte Carlo simulation*, J. Derivatives **12** (2004), no. 2, 8–23.
8. ———, *Options, futures and other derivatives*, 5th ed., Prentice Hall, 2003.
9. J. Jarrow and S. Turnbull, *The intersection of market and credit risk*, J. Banking and Finance,**24** (2000), no. 1-2, 271–299.
10. H. Johnson and R. Stultz, *The pricing of options with default risk*, J. Finance **42** (1987), 267–280.
11. I. Karatzas and S. Shreve, *Brownian Motion and stochastic calculus*, Grad. Texts in Math., vol. 113. Springer-Verlag, New York, 1988.
12. J.P. Laurent, *Comparing copula models for the pricing of basket credit derivatives and* CDO*'s*, 2004 (working paper).
13. J.P. Laurent and J. Gregory, *Basket default swaps,* CDO*s and factor copulas*, J. of Risk **7** (2005), no. 4, 103–122.
14. D.X. Li, *On default correlation: a copula function approach*, 1999; SSRN: `http://ssrn.com/abstract=187289`.
15. D. Lund and B. Oksendal, eds., *Stochastic models and option values: applications to resources, environment, and investment problems*, New York, NY, North-Holland, 1991.
16. R. Merton, *On the pricing of corporate debt: the risk structure of interest rates*, J. Finance **29** (1974), 449–470.
17. P. Moon and D.E. Spencer, *Eleven coordinate systems*, Field Theory Handbook, Including Coordinate Systems, Differential Equations, and Their Solutions, 2nd ed. New York, Springer-Verlag, New York, 1988, pp. 1–48.
18. H. Risken, *The Fokker–Planck equation.* 2nd ed., Springer-Verlag, New York, 1989.
19. P. Schonbucher, *Credit derivatives pricing models*, John Wiley & Sons Inc., 2003.
20. P. Wilmott, *Derivatives. The theory and practices of financial engineering*, John Wiley & Sons Inc., 1988.
21. Ch. Zhou, *An analysis of default correlations and multiple defaults*, The Review of Financial Studies **14** (2001), no. 2, 555–576.

Department of Mathematics, Ryerson University, 350 Victoria Street Toronto, Ontario, Canada, M5B 2K3.

E-mail address: escobar@ryerson.ca

Department of Mathematics, University of Toronto, Toronto, Ontario, Canada M5S 2E4.

E-mail address: seco@math.toronto.edu

Centre de Recherches Mathématiques
CRM Proceedings and Lecture Notes
Volume **41**, 2007

A Short Review on Computational Issues Arising in Relativistic Atomic and Molecular Physics

Maria J. Esteban

Abstract. This paper is a short review of computational issues arising when trying to compute electronic wave functions in atoms or molecules containing heavy nuclei. In this case, relativistic effects can play an important role and *ad hoc* models have to be used. The basic operator of this theories is the Dirac operator, which unlike the Schrödinger operator, is unbounded both from below. This feature creates difficulties both at the theoretical and the computational levels. In this review we discuss some of these issues. A much detailed review about this topic can be found in [7].

1. Introduction

Quantum Mechanics and Quantum Chemistry are large field of investigation for the improvement of numerical and computational methods: the models involved are quite complex and so the computations are slow, heavy and often not accurate enough. Computing the electronic structure of an atom can be heavy, depending on the number of electrons involved, but the problems are still not too big. In the case of a large molecule, the number of electrons can be very large and moreover, there are nuclear singularities at different locations. But all these complications become much more delicate and heavy when one deals with relativistic computations. The relativistic effects are not negligible as soon as heavy nuclei are present in the atom or the molecule, and taking then into account often means replacing the Schrödinger operator with the Dirac operator. This operator is unbounded from below and electronic energies are to be found in the gap of its continuous spectrum. Minimization procedures leading to electronic energies in the nonrelativistic case should now be replaced by min-max procedures and this is quite difficult to implement at the numerical level. So, other alternatives have to be found to deal with the unboundedness of the operator.

2000 *Mathematics Subject Classification.* Primary: 47A75, 81V45, 81V55; Secondary: 65N25, 35Q40, 35Q75, 47A11, 47N50, 49R50.

MJE acknowledges partial support of European Programs HPRN-CT # 2002-00277 & 00282.

The author would like to thank the referee for very useful and interesting comments and new references.

This is the final form of the paper.

The basic theory in relativistic quantum mechanics is QED (Quantum Electrodynamics) but this is too complicated for computations, except in very simple situations. So, for an N-electron system the most commonly used models used in numerical computations are based on the Hamiltonian

$$H = -\sum_{i=1}^{N} \left(D_{c,x_i} + V(x_i)\right) + \frac{1}{2}\sum_{\substack{i,j=1\\ i\neq j}}^{N} \frac{1}{|x_i - x_j|}, \tag{1.1}$$

where D_{c,x_i} is the Dirac operator acting on the x_i variable:

$$D_c = -ic\hbar\boldsymbol{\alpha}\cdot\nabla + mc^2\beta = -ic\hbar\sum_{k=1}^{3}\boldsymbol{\alpha}_k\partial_k + mc^2\boldsymbol{\beta}. \tag{1.2}$$

The usual representation of the matrices α_1, α_2, α_3 and β in 2×2 blocks is given by

$$\beta = \begin{pmatrix} I_2 & 0 \\ 0 & -I_2 \end{pmatrix}, \quad \alpha_k = \begin{pmatrix} 0 & \sigma_k \\ \sigma_k & 0 \end{pmatrix}, \qquad (k=1,2,3),$$

where the Pauli matrices σ_k's are defined as

$$\sigma_1 = \begin{pmatrix} 0 & 1 \\ 1 & 0 \end{pmatrix}, \quad \sigma_2 = \begin{pmatrix} 0 & -i \\ i & 0 \end{pmatrix}, \quad \sigma_3 = \begin{pmatrix} 1 & 0 \\ 0 & -1 \end{pmatrix}.$$

On the other hand, V is the nuclear potential:

$$V(x) = \sum_{j=1}^{M} \frac{Z_j}{|x - X_j|}. \tag{1.3}$$

M being the number of nuclei and Z_j, the charge of the jth nuclei, which is supposed to be fixed at the point X_j. Note that this model supposes that the nucleii are point-like. One can also consider extended nucleus models in which the support of the nuclear distribution of charge contains an open set. Actually, modern atomic and molecular relativistic calculations often use finite nuclear models, see for instance reference [51].

As far as this review is concerned, the main properties satisfied by the operator D_c are:

$$D_c^2 = -c^2\hbar^2\Delta + m^2c^4, \tag{1.4}$$

and the fact that the spectrum of D_c is unbounded both from above and below and has a gap between $-mc^2$ and mc^2:

$$\sigma(D_c) = (-\infty, -mc^2] \cup [mc^2, +\infty). \tag{1.5}$$

Note that the operator H acts of functions $\psi\colon \mathbb{R}^3 \to \mathbb{C}^4$, the so-called 4-spinors. Let us write 4-spinors in the form $\psi = \binom{\varphi}{\chi}$, $\varphi, \chi\colon \mathbb{R}^3 \to \mathbb{C}^2$. In what follows we will often refer to the φ (resp. χ) as the upper or larger (resp. lower or smaller) spinor.

In the choice of the above Hamiltonian there are a number of simplifications which have been made: the nuclei are supposed to be point-like and all electrostatic interactions (electron-electron and nuclei-electrons) are described as Coulomb interactions. Moreover, this Hamiltonian is not Lorentz invariant as any correct relativistic theory should be. Nevertheless this model seems to describe quite accurately the electrodynamics of an atom or a molecule. Note also that appart from the above questions, there are other issues related to the use of this Hamiltonian. The first one is that the whole real line is in its spectrum, and so there cannot be stable bound states, since all possible eigenvalues would be embedded in the continuum. The second issue is related to the so-called *variational collapse*. This means that computations can be easily lead to very negative values for the energies, since D_c is unbounded from below. Finally, the appearance of *spurious eigenvalues* in the implementation of many available numerical methods is also of concern.

The operator H acts on functions which are defined in $\mathbb{R}^{3N}$ and with values in $\mathbb{C}^4$. So, as soon as N is not too small, computations are almost impossible to be performed. One of the solutions found in Physics and Chemistry to deal with this difficulty is to consider functions in a smaller space. More concretely, since the Pauli Principle forces electronic wavefunctions to be antisymmetric, Swirles [48] proposed to look for wave functions which are stationary with respect to the Hamiltonian H in the space of the so-called Slater determinants (note that the first to have proposed such a procedure in the non relativistic context was Hartree). With this procedure, called the Dirac–Fock approximation, one computes an approximation of the eigenvalues of H by restricting the quadratic form $\Psi \mapsto \langle \Psi, H\Psi \rangle$ to the class of the functions Ψ which are a simple Slater determinant:

$$\Psi = \psi_1 \wedge \cdots \wedge \psi_N \tag{1.6}$$

where $(\psi_1, \ldots, \psi_N)$ is an orthonormal system of $L^2(\mathbb{R}^3, \mathbb{C}^4)$, $\int_{\mathbb{R}^3} (\psi_i, \psi_j)_{\mathbb{C}^2} = \delta_{ij}$. The precise meaning of (1.6) is:

$$\Psi_k(x_1, \ldots, x_N) = \frac{1}{\sqrt{N!}} \det\left(\psi_{i,k}(x_j)\right), \quad k = 1, \ldots, 4, \; i, j = 1, \ldots, N.$$

Since the set of all the Ψ's having the form (1.6) is not a vector subspace of $\bigwedge_{i=1}^N L^2(\mathbb{R}^3 \times \{\pm\}, \mathbb{C}^4)$, one then obtains an energy functional which is nonlinear in terms of $\psi_1, \ldots, \psi_N$. The associated Euler–Lagrange equations form a system of N coupled nonlinear PDEs, the so-called DF (Dirac–Fock) equations, which can be written as follows

$$D_{c,\psi}\psi_k = \varepsilon_k \psi_k, \quad k = 1, \ldots, N, \tag{1.7}$$

where $\psi = (\psi_1, \ldots, \psi_N)$ satisfies $\int_{\mathbb{R}^3} \left(\psi_i(x), \psi_j(x)\right) dx = \delta_{ij}$, i.e.

$$\mathrm{Gram}(\psi) = \mathbb{1}, \tag{1.8}$$

and

$$D_{c,\psi} = D_c + (\rho_\psi - \nu) * \frac{1}{|x|} - \frac{\gamma_\psi(x,y)}{|x-y|}, \tag{1.9}$$

$$\begin{aligned} \gamma_\psi(x,y) &= \sum_{\ell=1}^N \psi_\ell(x) \otimes \psi_\ell(y)^*, \\ \rho_\psi(x) &= \mathrm{tr}_{\mathbb{C}^4}\left(\gamma_\psi(x,x)\right) = \sum_{\ell=1}^N \left(\psi_\ell(x), \psi_\ell(x)\right). \end{aligned} \tag{1.10}$$

Equations (1.7) are the Euler–Lagrange equations of the Dirac–Fock functional

$$\begin{aligned} \mathcal{E}_{\mathrm{DF}}^{\nu,c}(\psi) = \sum_{\ell=1}^N (\psi_\ell, D_c\psi_\ell)_{L^2} - \sum_{\ell=1}^N \left(\psi_\ell, \left(\nu * \frac{1}{|x|}\right)\psi_\ell\right)_{L^2} \\ + \frac{1}{2} \iint_{\mathbb{R}^3 \times \mathbb{R}^3} \frac{\rho_\psi(x)\rho_\psi(y) - \mathrm{tr}(\gamma_\psi(x,y)\gamma_\psi(y,x))}{|x-y|}\, dx\, dy \end{aligned} \tag{1.11}$$

on the manifold

$$\mathcal{M}_{\rm DF} := \left\{\psi = (\psi_1, \ldots, \psi_N) \in \left(H^{\frac{1}{2}}(\mathbb{R}^3, \mathbb{C}^4)\right)^N, \mathrm{Gram}(\psi) = \mathbb{1}\right\}.$$

Using rather complicated variational methods, the existence of solutions to (1.7) when $Z := \sum_{j=1}^{M} Z_j > N-1$ has been proved in [18,44]. But in this topic the theoretical issues of existence are not, for the moment, very linked to the computational issues, except in the case $N = 1$, where interesting computational methods have been found recently by using the variational structure highlighted by the theoretical existence results (see Section 3).

The Dirac–Fock equations seem to give an accurate description of the electronic structure in closed-shell[1] atoms, but neither in open-shell atoms, nor in molecules. For this, the MCDF (multi-configuration Dirac–Fock) equations are more appropriate: here the idea is not to work in the set $\mathcal{M}_{\rm DF}$ of the Slater determinants, but rather consider well-chosen linear combinations of Slater determinants. The idea here is the following: in order to construct a basis of Slater determinants, one chooses as basis functions for the ψ_i's the eigenfunctions of the operator $D_c + V$. These eigenfunctions correspond to different eigenvalues and have attached to them some numbers, denoted by Π, J, M, corresponding to parity and angular momentum. Multiconfiguration methods consider linear combinations of all Slater determinants which have the same numbers Π, J and M.

In the coming sections we present some of the methods used to compute approximate solutions for the above models. Section 2 deals with the general case $M \geq 1$, $N \geq 1$, while Section 3 is devoted to the monoelectronic case: $M \geq 1$, but $N = 1$.

2. Computations for Atoms and Molecules: The Multi-Electronic Case

Form the computational viewpoint, the difference between the DF and the MCDF equations is that the latter are more complicated and cannot be diagonalized as the former ones. But regarding the numerical work needed to solve them, they are quite similar in structure and properties. So, in this section we will deal with both of them at the same time.

Up to now there are two basic methods to solve numerically the MCDF equations: either by using analytical or numerical basic sets and then perform a Galerkin approximation, or find directly a numerical solution of the equations by discretizing them. It seems that the direct numerical resolution is better suited for small systems, but for cases with millions of determinants (N very large) analytical basis set techniques seem better suited.

The methods based on discretization of the equations are mostly predictor-corrector or finite difference methods, but recently there have been also some works based on finite element methods. The main idea is to solve the highly nonlinear equations by an iteration procedure (a fixed-point argument): at every step of the iteration, the eigenvalues correspond to the Hamiltonian constructed with the eigenfunctions obtained at the previous step. Much of the work found in the Physics and Chemistry literature is concerned with the calculation of the coefficients of the different terms, that is with the algebraic structure of the system of unknowns. This reduces the number of terms to be computed numerically to the strict minimum,

[1]Closed shells are the shells filled with the maximum number of electrons as allowed by the Pauli principle (see [50])

which has a consequence the reduction of the numerical errors. Good references for this are [6, 21, 23].

In order to deal with the singularities created by those of the nuclear potential and to get better accuracy for the eigenfunctions in the region where they are singular, the equations are often discretized after a logarithmic change of variables has been performed around the singularities. In the diatomic case, the elliptic hyperbolic coordinates are often used to reduce the complexity and better discretize the problem around the two singularities. In more complex molecules little can be done to reduce the numerical work.

Good general references about the numerical methods presented above, their specificities, and some comparison between them can be found in [7, Chapters 3 and 4]. In particular, one can see the reasons why predictor-corrector methods can have some interest to treat part of the system, but cannot avoid some problems that finite difference schemes can. Some algorithms are mixed predictor-corrector/finite differences.

Some important difficulties appearing in many of the existing algorithms are: keeping the continuity of some of the solution's components, speeding the sometimes very slow convergence rate, avoiding (or dealing with) spurious roots and most of all, not falling into the negative continuum (the so-called variational collapse that "haunts" all people making relativistic computations). One of the solutions proposed by some authors [31] is to consider *ad hoc* projector operators to avoid too negative energies. Let us discuss this issue a little bit in detail, since it is of major importance in relativistic computations (and absent in nonrelativistic ones). Avoiding negative energies is often interpreted as avoiding positronic states. When trying to obtain a Hamiltonian from QED one sees that electronic states appear as eigenfunctions of a Hamiltonian which contains the Dirac operator but "sandwiched" between two projector operators, which project onto the positive energy space. These operators are *self-consistent* (depend on the solutions) and so they are not known *a priori*. Taking the projectors associated to the free Dirac operator (i.e. considering the so-called Brown–Ravenhall operator) does not seem a good idea, since the real projectors depend strongly on the external potential. Constructing the projection operator using the solutions of the free-particle Dirac equation introduces errors, as pointed out in [29]. Many propositions have been made and we will discuss them more in detail in the next section. Mittleman [43] suggested using continuously updated projection operators in SCF-calculations. In the finite-basis (algebraic) approximation such projection operators are implicitly used by selecting for the next iteration the lowest *positive-energy* solutions. Correct convergence is then straightforwardly obtained by combining such vector selection with kinetic balance. In [31] it has been proposed to use positive projection operators corresponding to finite basis composed of the eigenfunctions of a simpler problem, but many options are possible, like for instance choosing well the boundary conditions in the discretized problem, which is in fact a similar projection procedure and works well. On the other hand, when using basis sets techniques, it is clearly much easier to "project out" the very negative energies, since the choice of the basis (the behaviour of the basis elements at infinity for instance) determines the space of the final solutions, and therefore, their behavior also. A final (very extended and used) technique to avoid the variational collapse, is to use the so-called *kinetic-balance* principle (see [37, 47]), which forces the spinor lower component to have a specific form (as a function of the upper spinor). This is somehow equivalent to using a Brown–Ravenhall operator, since the link between the upper and the lower spinors are the ones satisfied by (generalized) eigenfunctions of the free Dirac operator. When solving the Dirac equation or mean-field generalisations of it for atoms by finite-difference methods, spurious solutions are avoided by imposing the correct boundary conditions near the origin and near infinity in the radial equations. In molecular calculations using basis set expansions the correct coupling of the large and small components are obtained by employing the kinetic balance relation (see references above). In the next section we explain in detail a recent method allowing to avoid the variational collapse when $N = 1$ hoping that this sheds some light on the general problem.

Let us add that negative eigenvalues do not necessarily correspond to positronic states, and that one can have negative electronic states in molecules. So, all the efforts to avoid these negative eigenvalues have to be weighed carefully, determining in which context the measures taken are needed and/or useful.

Regarding molecular computations, most of the existing methods are based on the use of basis sets (a Galerkin-like method), the basis sets being analytical or numerical. The former allow to perform many computations exactly or independently before launching the final numerical computation (for a good description see [46] and more in particular the article of K. Faegri and K.G. Dyall in [46]). Their use and choice relies on a sound chemical knowledge about the solutions' properties. Note that it is quite usual for molecular calculations to avoid the variational collapse by defining the small component in terms of the large one [37], by using for instance the kinetic-balance principle, which is not very accurate but helps to avoid variational collapse.

A good discussion about the appearance of spurious solutions in the molecular computations and also about the possible reasons for poor convergence (or lack of convergence in some cases) is found in [16, 45].

The question of how to choose a good basis set has been addressed in many papers, for instance in [12, 13, 20, 22, 32, 37].

Finally note that there exist some packages to perform full relativistic atomic and molecular computations, as GRASP (`http://web.am.qub.ac.uk/DARC/grasp.html`), BERTHA (see [24]), DIRAC (`http://dirac.chem.sdu.dk/`) and of course, the first one to have existed, due to Desclaux [6]. A more recent, but quite complete code is the UTCHEM code (`http://utchem.qcl.t.u-tokyo.ac.jp/`).

3. Computations in the Case of a Single Electron

The case of a single electron seems quite simple at first sight, since the problem is now to find eigenvalues and eigenfunctions of operators $D_c + V$ where V is an external electrostatic potential, which is in general nonpositive: in atomic and molecular physics, V stands in general for a nuclear potential, and it is in general nonpositive, i.e. *attractive*. We are interested in eigenvalues not lying in the continuum spectrum, since they are supposed to represent stable bound state energies. If the Dirac operator were bounded from below this question would be quite simple to deal with, both analytically and numerically, since the ground state energy, $\lambda_1(V)$, if it existed, would be the minimum of the Rayleigh quotient

$$R_V(\psi) := \frac{(\psi, (D_c + V)\psi)}{(\psi, \psi)}. \tag{3.1}$$

The other eigenvalues above $\lambda_1(V)$ would be obtained as minimizers of the Rayleigh quotient on the orthogonal space of the previously found eigenfunctions.

But in the case of the Dirac operator, its unboundedness from below forces us to look for eigenvalues in gaps of the continuum (essential) spectrum, and therefore, no eigenvalues can be found by a global minimization of R_V since the infimum of R_V is equal to $-\infty$.

When the potential V is not too singular (and we are going to assume this always from now on), the essential spectrum of the operator $D_c + V$ is, as for the free operator D_c, the set $(-\infty, -mc^2] \cup [mc^2, +\infty)$. So, in a time independent model eigenvalues corresponding to stable bound states of an electron in the electrostatic field corresponding to the potential V have to be in the gap $(-mc^2, mc^2)$ and formally correspond to min-max values of the Rayleigh quotient R_V (note that in a time dependent model, eigenvalues can enter the gap by its both ends). This problem is not easy to solve, and different ways have been explored so far. Indeed, even if this case seems not very interesting because *few* atoms and molecules have a single electron, the linear Dirac equation appears also when dealing with the multielectronic case and solving the corresponding equations by a iterative procedure (see Section 2). Hence, this problem has attracted a lot of attention both in the Physics and in the Chemistry literature, as a first (and often intermediate) step to deal with real and complex atoms and molecules.

As in the multielectron case, various alternative approaches have been used: variational methods, basis set techniques or derivation of effective bounded-from-below operators which in some sense approach well the unbounded Dirac operator. From the mathematical viewpoint little had been done until 10 years ago when rigorous variational (min-max) approaches were proposed to characterize the eigenvalues [8,9,17,25,26]. Soon after, efficient and robust algorithms based on some of the rigorous variational approaches were proposed [10,11]. In the remainder of this section we describe some of the existing approaches to compute the eigenvalues of operators of the form $D_c + V$.

At this point let us stress that in order to apply almost all that is going to be said below, we need to ask that the potential V is not too singular, which means that it can have only a finite number of point singularities, and that around any of those singular points, $\bar{x}$, it has to behave less singularly than the Coulomb potential $-c/|x - \bar{x}|$. More precisely, we need that for all $\bar{x} \in \mathbb{R}^3$,

$$\limsup_{x \to \bar{x}} |x - \bar{x}| V(x) < c. \tag{3.2}$$

This condition is there to ensure that the operator $D_c + V$ has a well-defined self-adjoint extension. Moreover, without (3.2), the operator D_c would not be able to control the potential V, what is necessary to define the min-max characterizations of the eigenvalues of $D_c + V$ (it is interesting to see the relation between this and a Hardy-like inequality that was proved in [9]).

Coming back to the methods proposed to compute the eigenvalues, let us start their description by the ones consisting in a discretization of the problem in a finite basis set of dimension n. Once the basis is chosen, the problem consists, of course, in finding the eigenvalues of the corresponding $n \times n$ matrix. The choice of the basis is crucial, since at a discrete level, if the basis set is just any, we will find eigenvalues corresponding to the discretization of the continuum, that is, above mc^2 and below $-mc^2$ and also eigenvalues in the gap $(-mc^2, mc^2)$. Only the latter are of interest to us. The efficiency of the method depends on the good choice of the basis set, and this efficiency is related to the behavior at infinity of the basis set functions and also to the sparseness properties of the discrete matrix. A difficulty that has been often encountered is the appearance of spurious eigenvalues, that is eigenvalues of the discretized $n \times n$ matrix which do not approach eigenvalues of the operator $D_c + V$ (see for instance [12]). W. Kutzelnigg has written two excellent reviews [37, 38] about the use of basis sets in Dirac calculations, where many relevant references can be found. About this approach let us say that in the choice of the basis sets different strategies have been taken: either look for efficiency of the numerical method by choosing for instance B-splines [32], or choose analytical basis sets which help to compute some of the integrals almost explicitly and which represent well the expected behavior of the eigenfunctions at infinity [12,13]. About how to choose a good basis sets, see [20, 22, 37].

The second approach consists in searching for operators which approach the Dirac operator $D_c + V$ in some sense and which are bounded below. The real reason why this can work is that at the nonrelativistic limit (for c going to $+\infty$), the spectrum of the operator D_c+V, shifted by $-mc^2$, is known to converge towards the spectrum of the Schrödinger operator $-\hbar^2\Delta/2+V$, operator which is bounded from below if V satisfies the conditions described above. A long series of papers propose to approach $D_c + V$, or operators which are spectrally equivalent to it, by operators which equal to $-\hbar^2\Delta/2+V$ plus terms which are "small" when c is large. The first series of works considers unitary transformations W which do not change the spectrum, and writes down an expansion of $W^\dagger(D_c + V)W$ in powers of c^{-2}:

$$W^\dagger(D_c + V)W = H_0 + \frac{H_1}{c^2} + \frac{H_2}{c^4} + \cdots + \frac{H_k}{c^{2k}} + \cdots, \tag{3.3}$$

and cut it at a finite power $k \geq 1$. There are many ways to choose W, the different strategies trying to optimize some given criterium. In many of the proposed methods, $H_0 = -\hbar^2\Delta/2 + V + mc^2$, and the operators $H_1, H_2, \ldots$ are easy to compute formally. Some of the approaches are aimed at diagonalizing the operator $D_c + V$. The first attempt in this direction was the so-called Foldy–Wouthuysen transformation [19] which is not known in closed form, but which can be again expanded in powers of c^{-2}. A very good discussion of approaches made in this direction can be found in [1]. Many other proposals have followed in order to avoid some of the problems encountered by the Foldy-Wouthuysen transformation. Some of the above expansions present convergence problems near the singularities of the potential, and to remedy this, Chang *et al* [3], proposed a new method to write the expansion, called ZORA, convergent this time: the expansion is now made in powers of $(2mc^2 - V)^{-1}$ (see also [14, 15]). A similar idea appears in the work of Heully *et al* [28, 29] and this method has been studied in detail by van Lenthe *et al* [41]. A good review about this kind of methods has been written by Kutzelnigg [39] and more recent proposals for exact 2-component relativistic Hamiltonians can be found in [40].

Before we come to the third kind of strategy to find and compute the eigenvalues of Dirac operators, the variational one, let us remark that something that has been noticed by many people is that the eigenvalue problem

$$(D_c + V)\psi = \lambda\psi, \tag{3.4}$$

is equivalent to an eigenvalue problem for the upper spinor φ, where $\psi = \binom{\varphi}{\chi}$, $\varphi, \chi . \mathbb{R}^3 \to \mathbb{C}^4$. Indeed (3.4) is equivalent to the system

$$
\begin{cases} R\chi = (\lambda - mc^2 - V)\varphi, \\ R\varphi = (\lambda + mc^2 - V)\chi, \end{cases} \tag{3.5}
$$

with $R = -ic\hbar(\vec{\sigma}.\vec{\nabla}) = -ic\hbar \sum_{k=1}^{3} \sigma_k \partial/\partial x_k$ where σ_1, σ_2, σ_3 are the Pauli matrices defined in Section 1. Then, if ψ is an eigenfunction of H associated with the eigenvalue λ, and if the function $\lambda + mc^2 - V$ is never equal to 0, we have:

$$
\chi = \frac{R\varphi}{\lambda + mc^2 - V}, \tag{3.6}
$$

and

$$
R\left(\frac{R\varphi}{\lambda + mc^2 - V}\right) + (mc^2 + V)\varphi = \lambda\varphi. \tag{3.7}
$$

The above equation can be seen as an eigenvalue problem for the eigenvalue-dependent operator $H(\lambda) := R(\lambda + mc^2 - V)^{-1}R + mc^2 + V$ which acts on 2-component functions, unlike the Dirac operator $D_c + V$ which acts on 4-component functions. And solutions of (3.7) are eigenfunctions corresponding to the zero eigenvalue for the operator $H(\lambda) - \lambda$. To those knowing about pencil problems, this is one of them.

For many years the eigenvalue dependence of H has been an obstacle in the way of finding eigenvalues of Dirac operators. As we will see below, one can keep it and still derive from it a very good method to find and compute the Dirac eigenvalues. But this will be the case only for 1-electron systems!

The third way to find and compute the eigenvalues of operator $D_c + V$ is based on variational ideas. As we said above, minimizing the Rayleigh quotient is out of question, since the operator is not bounded from below. A possibility is to minimize it in a set of functions in which it is bounded from below, that is, find a subspace which is large enough to contain, or almost contain, all the eigenfunctions, but not too large to allow the Rayleigh quotient to take values below $-mc^2$. This is tricky, specially because the space certainly depends on V. Other ways have been explored : to look for eigenvalues of $(D_c + V)^2$ [2, 52] or of $(D_c + V)^{-1}$ [30]. The first idea is good if the potential V does not support any negative eigenvalue, which is not always the case. Moreover, practically speaking, the integrals entering into the computation become much more difficult to compute. The second one needs that there are no eigenvalues of $D_c + V$ around 0. And both would had difficulties to distinguish electronic from positronic states.

A last idea is to use the fact that the eigenvalues are actually stationary values (min-maxes) for the Rayleigh quotient, and under the appropriate assumptions on V, they correspond rigorously to min-max critical values for R_V [8, 9, 17, 25, 26]. Computationally speaking, Talman [49] and Datta *et al* [5] seem to have been the first ones to have the idea of computing the ground-state energy for $D_c + V$ by taking a min-max of R_V by first maximizing in the lower componer of the spinors ψ and then minimizing in the upper component. The work in [9, 25] has showed that Talman's and Datta *et al's* idea can be made rigorous and that it actually leads to very efficient algorithms to compute the eigenvalues of $D_c + V$. Indeed, as shown in [9], for a giver upper spinor φ: $\mathbb{R}^3 \to \mathbb{C}^4$, the maximum of R_V over all possible lower spinors χ, $\lambda(\varphi)$, satisfies a scalar equation :

$$\lambda(\varphi)\int_{\mathbb{R}^3}|\varphi|^2\,dx = \int_{\mathbb{R}^3}\left(\frac{mc^2\hbar^2|(\sigma\cdot\nabla)\varphi|^2}{mc^2 - V + \lambda(\varphi)} + (mc^2+V)|\varphi|^2\right)dx. \tag{3.8}$$

The ground state energy is now the infimum of a function, $\lambda(\varphi)$ over all φ's, it is so the solution of a minimization problem which is well posed, since the function $\lambda(\varphi)$ is bounded from below. Discretizing this minimization procedure is now easy and it has been made in [10, 11]. Later on Kolb *et al* have used finite element and LCAO (linear combination of atomic orbitals) methods to solve the same problem [34–36, 55]. In our opinion Talman's paper [49] has played a key role in realizing the min-max nature of the Dirac eigenvalue problem.

This method works perfectly, without fear of variational collapse or other diseases present in most computations in relativistic quantum mechanics. But only for potentials which are attractive (this is not a drawback in applications, since nuclear potentials are attractive). The main drawback is that it is not clear how to export the underlying ideas to the multielectronic case, except if one considers DFT (density functional theory) or OEP (optimized effective potential) models, in which all the potentials are local, even those modelling inter-electronic interaction.

Note that this method is directly related to the scalar nonlinear problem (3.7), that is, to the search of λ's such that 0 is an eigenvalue of $H(\lambda)-\lambda$. This is the eigenvalue-dependent operator that so many people have tried to avoid. In this method, we do not avoid it, on the contrary we fully use it to obtain an exact minimization characterization for the ground state and good and efficient minimization algorithms for its computation. Actually, we discovered recently that this method is equivalent to a method used by van Lenthe *et al* [42] (see also [27]) to solve (3.7) in an iterative way. In this paper they give computational results showing that the method is quite effective, even if not much better than ZORA except for deep core electrons. Some final references for eigenvalue-dependent 2-component Hamiltonian are due to Cowan and Griffin [4] in the framework of the Hartree–Fock model and to Wood and Boring [54] in a DFT framework. A recent version was proposed by Malkin and co-workers [33].

References

1. M. Barysz and A. Sadlej, *Two-component methods of relativistic quantum chemistry: from the Douglas–Kroll approximation to the exact two-component formalism*, J. Mol. Struct. (Theochem) **573** (2001), 181–200.
2. W.E. Bayliss and S.J. Peel, *Stable variational calculations with the Dirac Hamiltonian*, Phys. Rev. A **28** (1983), no. 4, 2552–2554.
3. Ch. Chang, J.-P. Pelissier, and Ph. Durand, *Regular two-component Pauli-like effective Hamiltonians in Dirac theory*, Phys. Scripta **34** (1986), 394–404.
4. R.D. Cowan and D.C. Griffin, *Approximate relativistic corrections to atomic radial wave functions*, J. Opt. Soc. Am. **66** (1976), no. 10, 1010.
5. S.N. Datta and G. Deviah, *The minimax technique in relativistic Hartree–Fock calculations*, Pramana **30** (1988), no. 5, 393–416.
6. J.P. Desclaux, *A relativistic multiconfiguration Dirac–Fock package*, Methods and Techniques in Computational Chemistry (E. Clementi, ed.) (STEF, Cagliari, 1993), METTEC-94, vol. A, chap. 5, PubliCEA, Saclay, France, 1993, pp. 253–274.
7. J.P. Desclaux, J. Dolbeault, M.J. Esteban, P. Indelicato, and E. Séré, *Computational approaches of relativistic models in quantum chemistry*, Handbook Numer. Anal., vol. 10, North-Holland, Amsterdam, 2003, pp. 453–483.
8. J. Dolbeault, M.J. Esteban, and E. Séré, *Variational characterization for eigenvalues of Dirac operators*, Calc. Var. Partial Differential Equations **10** (2000), 321–347.

9. ———, *On the eigenvalues of operators with gaps. Application to Dirac operators*, J. Funct. Anal. **174** (2000), 208–226.
10. J. Dolbeault, M.J. Esteban, and E. Séré, *A variational method for relativistic computations in atomic and molecular physics*, Int. J. Quantum. Chemistry **93** (2003), 149–155.
11. J. Dolbeault, M.J. Esteban, E. Séré, and M. Vanbreugel, *Minimization methods for the one-particle Dirac equation*, Phys. Rev. Lett. **85** (2000), no. 19, 4020–4023.
12. G.W.F. Drake and S.P. Goldman, *Application of discrete-basis-set methods to the Dirac equation*, Phys. Rev. A (3) **23** (1981), 2093–2098.
13. G.W.F. Drake and S.P. Goldman, *Relativistic Sturmian and finite basis set methods in atomic physics*, Advances in Atomic and Molecular Physics (Sir D. Bates and B. Bederson, eds.), vol. 25, Academic Press, New York, 1988, pp. 393–416
14. Ph. Durand, *Transformation du Hamiltonien de Dirac en Hamiltoniens variationnels de type Pauli. Application à des atomes hydrogenoïdes*, C. R. Acad. Sc. Paris Sér. II **303** (1986), no. 2, 119–124.
15. Ph. Durand and J.-P. Malrieu, *Effective Hamiltonians and pseudo-potentials as tools for rigorous modelling*, Ab initio Methods in Quantum Chemistry I (K.P. Lawley, ed.), J. Wiley & Sons, PUBLISHER ADDRESS, 1987.
16. K.G. Dyall, I.P. Grant, and S. Wilson, *Matrix representation of operator products*, J. Phys. B **17** (1984), 493–503.
17. M.J. Esteban and E. Séré, *Existence and multiplicity of solutions for linear and nonlinear Dirac problems*, Partial Differential Equations and Their Applications (P.C. Greiner, V. Ivrii, L.A. Seco and C. Sulem, eds.) (Toronto, 1995), CRM Proc. Lecture Notes, vol. 12, Amer. Math. Soc, Providence, RI, 1997, pp. 107–118.
18. ———, *Solutions for the Dirac–Fock equations for atoms and molecules*, Comm. Math. Phys. **203** (1999), 499–530.
19. L.L. Foldy and S.A. Wouthuysen, *On the Dirac theory of spin $\frac{1}{2}$ particles and its non relativistic limit*, Phys. Rev. (2) **78** (1950), 29–36.
20. I.P. Grant, *Conditions for convergence of variational solutions of Dirac's equation in a finite basis*, Phys. Rev. A **25** (1982), no. 2, 1230–1232.
21. ———, *Relativistic atomic structure calculations*, Meth. Comp. Chem. **2** (1987), 132.
22. ———, *Notes on basis sets for relativistic atomic structure and* QED, AIP Conference Proc. (P.J. Mohr, W.R. Johnson, and J. Sucher, eds.), vol. 189, Amer. Inst. Phys., New York, 1989, pp. 235–0253.
23. I.P. Grant and H.M. Quiney, *Foundation of the relativistic theory of atomic and molecular structure*, Adv. Atom. Mol. Phys. **23** (1988), 3786.
24. I.P. Grant, H.M. Quiney, and H. Skaane, BERTHA — *4-component relativistic molecular quantum mechanics*, Supercomputers, Collision Processes and Applications (K.L. Bell, K.A. Berrington, D.S.F. Crothers and K.T. Taylor, eds.), Plenum, New York, 1999, pp. 213–224.
25. M. Griesemer, R.T. Lewis, and H. Siedentop, *A minimax principle in spectral gaps: Dirac operators with Coulomb potentials*, Doc. Math. **4** (1999), 275–283 (electronic).
26. M. Griesemer and H. Siedentop, *A minimax principle for the eigenvalues in spectral gaps*, J. London Math. Soc. (2) **60** (1999), no. 2, 490–500.
27. D. Hegarty, *On the Dirac equation in the algebraic approximation*, Theor. Chim. Acta **70** (1986), 351–356.
28. J.L. Heully, I. Lindgren, E. Lindroth, S. Lundqvist, and A.M. Mårtensson-Pendrill, *Diagonalisation of the Dirac Hamiltonian as a basis for a relativistic many-body procedure*, J. Phys. B **19** (1986), 2799–2815.
29. ———, *Comment on relativistic wave equations and negative-energy states*, Phys. Rev. A **33** (1986), 4426–4429.
30. R.N. Hill and C. Krauthauser, *A solution to the problem of variational collapse for the one-particle Dirac equation*, Phys. Rev. Lett. **72** (1994), no. 14, 2151–2154.
31. P. Indelicato and J.-P. Desclaux, *Projection operators in the multiconfiguration Dirac–Fock method*, Phys. Scr. **T46** (1993), 110–114.
32. W.R. Johnson, S. Blundell, and J. Sapirstein, *Finite basis sets for Dirac equation constructed from B splines*, Phys. Rev. A **37** (1988), no. 2, 307–315.

33. S. Komorský, M. Repiský, O.L. Malkina, V.G. Malkin, I. Malkin, and M. Kaupp, *Resolution of identity Dirac–Kohn–Sham method using the large component only: calculations of g-tensor and hyperfine tensor*, J. Chem. Phys. **124** (2006), 084108-1–084108-8.
34. O. Kullie and D. Kolb, *High accuracy Dirac-finite-element* (FEM) *calculations for H_2^+ and* Th_2^{179+}, Eur. Phys. J. D **17** (2001), no. 2, 167–173.
35. ______, *Dirac finite element method calculations for* Th_2^{179+}, J. Phys. B **36** (2003), 4361–4366.
36. O. Kullie, D. Kolb, and A. Rutkowski, *Two-spinor fully relativistic finite-element* (FEM) *solution of the two-center Coulomb problem*, Chem. Phys. Letters **383** (2004), 215–221.
37. W. Kutzelnigg, *Basis set expansion of the Dirac operator without variational collapse*, Int. J. Quant. Chem. **25** (1984), 107–129.
38. ______, *Relativistic one-electron Hamiltonians "for electrons only" and the variational treatment of the Dirac equation*, Chemical Physics **225** (1997), 203–222.
39. ______, *Effective Hamiltonians for degenerate and quasidegenerate direct perturbation theory of relativistic effects*, J. Chem. Phys. **110** (1999), no. 17, 8283–8294.
40. W. Kutzelnigg and W. Liu, *Quasirelativistic theory equivalent to fully relativistic theory*, J. Chem. Phys. **123** (2005), 241102-1–241102-4.
41. E. van Lenthe, E.J. Baerends, and J.G. Snijders, *Relativistic regular two-component Hamiltonians*, J. Chem. Phys. **99** (1993), 4597–4610.
42. ______, *Solving the Dirac equation, using the large component only, in a Dirac-type Slater orbital basis set*, Chem. Phys. Lett. **236** (1995), 235–241.
43. M.H. Mittleman, *Theory of relativistic effects on atoms: configuration-space Hamiltonian*, Phys. Rev. A **24** (1981), 1167–1175.
44. E. Paturel, *Solutions of the Dirac equations without projector*, Ann. Henri Poincaré **1** (2000), 1123–1157.
45. W.H.E. Schwarz and H. Wallmeier, *Basis set expansions of relativistic molecular wave equations*, Mol. Phys. **46** (1982), no. 5, 1045–1061.
46. P. Schwerdtfeger (ed.), *Relativistic electronic structure theory. I. Fundamental aspects*, Elsevier, Amsterdam, 2002.
47. R.E. Stanton and S. Havriliak, *Kinetic balance: a partial solution to the problem of variational safety in Dirac calculations*, J. Chem. Phys. **81** (1984), 1910–1918.
48. B. Swirles, *The relativistic self-consistent field*, Proc. Roy. Soc. A **152** (1935), 625–649.
49. J.D. Talman, *Minimax principle for the Dirac equation* Phys. Rev. Lett. **57** (1986), no. 9, 1091–1094.
50. B. Thaller, *The Dirac equation*, Springer-Verlag, Berlin, 1992.
51. L. Visscher and K.G. Dyall, *Dirac–Fock atomic electronic structure calculations using different nuclear charge distributions*, At. Data Nucl. Data Tables **67** (1997), 207.
52. H. Wallmeier and W. Kutzelnigg, *Use of the squared Dirac operator in variational relativistic calculations*, Chem. Phys. Lett. **78** (1981), no. 2, 341–346.
53. S. Wilson, *On the use of many-body perturbation theory and quantum electrodynamics in molecular electronic structure theory*, J. Mol. Struct. (Theochem) **547** (2001), 279–291.
54. J.H. Wood and A.M. Boring, *Improved Pauli Hamiltonian for local-potential problems*, Phys. Rev. B **18** (1978), 2701–2711.
55. H. Zhang, O. Kullie, and D. Kolb, *Minimax* LCAO *approach to the relativistic two-centre Coulomb problem and its finite element* (FEM) *spectrum*, J. Phys. B **37** (2004), no. 4, 905–916.

CEREMADE (UMR CNRS 7534), Université Paris-Dauphine, 75775 Paris Cedex 16, France.

E-mail address: `esteban@ceremade.dauphine.fr`

Centre de Recherches Mathématiques
CRM Proceedings and Lecture Notes
Volume **41**, 2007

Model Hamiltonians in Density Functional Theory

Paola Gori-Giorgi, Julien Toulouse, and Andreas Savin

ABSTRACT. The formalism of Kohn and Sham uses a specific (model) Hamiltonian which highly simplifies the many-electron problem to that of noninteracting fermions. The theorem of Hohenberg and Kohn tells us that, for a given ground state density, this Hamiltonian is unique. In principle, this density can be chosen as that of the real, interacting system. To obtain the energy, or other properties of the real system, approximations are needed. Working with non interacting fermions is an important simplification, but it may be easier to produce approximations with different choices of the model Hamiltonian. The feature that the exact density is (ideally) reproduced can be kept in the newly defined fictitious systems. Using model Hamiltonians having the same form as the physical one, that is, being built of one- and two-body operators, allows to approach the physical Hamiltonian arbitrarily close, and thus a systematic reduction of the approximations.

1. Introduction

1.1. General context. The knowledge from first principles of the electronic structure of atoms, molecules and solids is contained in the N-electron Schrödinger equation that, within the Born–Oppenheimer approximation (i.e., at fixed nuclei positions $\boldsymbol{R}_\alpha$), reads

$$\widehat{H}(\mathbf{r}_1,\ldots,\mathbf{r}_N)\Psi(\mathbf{r}_1\sigma_1,\ldots,\mathbf{r}_N\sigma_N) = E\Psi(\mathbf{r}_1\sigma_1,\ldots,\mathbf{r}_N\sigma_N), \tag{1.1}$$

with

$$\widehat{H} = \widehat{T} + \widehat{W}_{ee} + \widehat{V}_{ne}, \tag{1.2}$$

$$\widehat{T} = -\frac{1}{2}\sum_{i=1}^{N}\nabla^2_{\mathbf{r}_i}, \tag{1.3}$$

$$\widehat{W}_{ee} = \frac{1}{2}\sum_{i\neq j}^{N}\frac{1}{|\mathbf{r}_i-\mathbf{r}_j|} \equiv \frac{1}{2}\sum_{i\neq j}^{N} w_{ee}(|\mathbf{r}_i-\mathbf{r}_j|), \tag{1.4}$$

$$\widehat{V}_{ne} = -\sum_{i=1}^{N}\left(\sum_{\alpha=1}^{M}\frac{Z_\alpha}{|\boldsymbol{R}_\alpha-\mathbf{r}_i|}\right) \equiv \sum_{i=1}^{N} v_{ne}(\mathbf{r}_i), \tag{1.5}$$

2000 *Mathematics Subject Classification.* Primary: 81-08; Secondary: 81Q05.
This is the final form of the paper.

where v_{ne} is the external potential due to the M nuclei of charges Z_α at positions $\boldsymbol{R}_\alpha$, and Hartree atomic units, $\hbar = 1$ (reduced Planck's constant) $m = 1$ (electron mass), $a_0 = 1$ (Bohr radius), $e = 1$ (electron charge), have been used. Since electrons obey the Fermi–Dirac statistics, Ψ must be antisymmetric under particle exchange, $\Psi(\mathbf{r}_1\sigma_1, \ldots, \mathbf{r}_i\sigma_i, \ldots, \mathbf{r}_j\sigma_j, \ldots, \mathbf{r}_N\sigma_N) = -\Psi(\mathbf{r}_1\sigma_1, \ldots, \mathbf{r}_j\sigma_j, \ldots, \mathbf{r}_i\sigma_i, \ldots, \mathbf{r}_N\sigma_N)$, where $\mathbf{r}$ denotes the three-dimensional electronic position and σ the spin degree of freedom ($\uparrow$ or $\downarrow$). In what follows we will be only concerned with the search for the ground-state energy, i.e., the lowest eigenvalue E_0 of (1.1).

The methods that both chemists and physicists have developed to find approximate solutions of (1.1) can be roughly divided in two large groups: wave-function methods (traditional quantum chemistry methods [28], quantum Monte Carlo [7]) and density methods (density functional theory [18], and density matrix functional theory [4], that is somehow in between the two groups). Simplistically, wave-function methods start from an approximation for the antisymmetric, normalized, N-electron wave-function, Ψ_{approx}, and take advantage of the variational principle,

$$E_0 \leq E_0^{\text{approx}} = \min_{\Psi_{\text{approx}}} \langle \Psi_{\text{approx}} | \widehat{H} | \Psi_{\text{approx}} \rangle. \tag{1.6}$$

Since Ψ_{approx} is a $3N$-dimensional object, wave-function methods are in general computationally expensive when the number of particles increases.

Density Functional Theory (DFT) uses the electron density $n(\mathbf{r})$ as a basic variable,

$$n(\mathbf{r}) = N \sum_{\sigma_1, \ldots, \sigma_N} \int |\Psi(\mathbf{r}\sigma_1, \mathbf{r}_2\sigma_2, \ldots, \mathbf{r}_N\sigma_N)|^2 \, d\mathbf{r}_2, \ldots, d\mathbf{r}_N, \tag{1.7}$$

a much simpler quantity to handle, resulting in a low computational cost that allows to reach system sizes much larger than those accessible to wave-function methods. At given number of electrons N, the Hohenberg and Kohn theorem [17] tells us that the ground-state density $n(\mathbf{r})$ of (1.7) completely determines (except for an additive constant) the external potential $v_{ne}(\mathbf{r})$ of (1.5) (for the sake of simplicity we only consider physical Hamiltonians with a nondegenerate ground state). Since the kinetic energy operator of (1.3) and the electron-electron interaction of (1.4) are the same for all systems, a *universal* functional F of the density $n(\mathbf{r})$ can be defined as [22]

$$F[n; \widehat{W}_{ee}, \widehat{T}] = \min_{\Psi \to n} \langle \Psi | \widehat{T} + \widehat{W}_{ee} | \Psi \rangle, \tag{1.8}$$

where, to keep the connection with what we will do in the next sections, we have explicitly shown the dependence of F on the electronic interaction $\widehat{W}_{ee}$ and on the kinetic energy operator $\widehat{T}$. The minimum search in (1.8) is performed over all antisymmetric wave-functions Ψ that yield the density $n(\mathbf{r})$ (by definition $n(\mathbf{r})$ also gives, by integration, the number of electrons N). The universal functional F can be also defined as a Legendre transform [23]

$$F[n; \widehat{W}_{ee}, \widehat{T}] = \sup_{v} \left\{ \min_{\Psi} \langle \Psi | \widehat{T} + \widehat{W}_{ee} + \widehat{V} | \Psi \rangle - \int n(\mathbf{r}) v(\mathbf{r}) \, d\mathbf{r} \right\}, \tag{1.9}$$

where, as in the rest of this paper, $\widehat{V}$ denotes a local one-body operator of the form of (1.5) with $v_{ne}(\mathbf{r})$ replaced by $v(\mathbf{r})$. If the exact form of the functional F was known, the variational principle would tell us that

$$E_0 = \min_n \left\{ F[n; \widehat{W}_{ee}, \widehat{T}] + \int n(\mathbf{r}) v_{ne}(\mathbf{r})\, d\mathbf{r} \right\}. \tag{1.10}$$

In practice, because we rely on approximations for F, the energy estimated by carrying out the minimization in (1.10) can be lower than the exact E_0.

1.2. Kohn–Sham density functional theory. The Kohn–Sham [19] approach to DFT introduces another density functional $T_s[n]$,

$$T_s[n] \equiv F[n; 0, \widehat{T}] = \sup_v \left\{ \min_\Phi \langle \Phi | \widehat{T} + \widehat{V} | \Phi \rangle - \int n(\mathbf{r}) v(\mathbf{r})\, d\mathbf{r} \right\}, \tag{1.11}$$

where, as in the rest of this work, Φ always denotes the wave-function of a spin-$\frac{1}{2}$ fermionic system with zero electron-electron interaction, i.e., in the majority of cases, a single Slater determinant (the fundamental idea of Kohn–Sham has been to introduce the fermionic statistic in the construction of $T_s[n]$). Since dealing with noninteracting particles is computationally simple, Kohn and Sham [19] proposed to search for that particular noninteracting system which has the same ground-state density of the physical one. This defines a model system which is usually called Kohn–Sham (KS) system. The difference between $F[n; \widehat{W}_{ee}, \widehat{T}]$ and $T_s[n]$ defines the Hartree-exchange-correlation functional $E_{\mathrm{Hxc}}[n]$,

$$E_{\mathrm{Hxc}}[n] = F[n; \widehat{W}_{ee}, \widehat{T}] - T_s[n] \tag{1.12}$$

that needs to be approximated. The functional derivative, $\delta E_{\mathrm{Hxc}}[n]/\delta n(\mathbf{r}) = v_{\mathrm{Hxc}}(\mathbf{r})$, determines the one-body KS potential $v_{\mathrm{KS}}(\mathbf{r}) = v_{ne}(\mathbf{r}) + v_{\mathrm{Hxc}}(\mathbf{r})$ that forces the N noninteracting electrons to have the same density of the physical system. The KS-DFT relies thus on the assumption that, given a physical ground-state density $n(\mathbf{r})$, it is possible to find a noninteracting system which has the *same ground-state density*.

From the functional $E_{\mathrm{Hxc}}[n]$ the classical electrostatic Hartree term is usually extracted,

$$E_{\mathrm{H}}[n] = \frac{1}{2} \int d\mathbf{r} \int d\mathbf{r}' \frac{n(\mathbf{r}) n(\mathbf{r}')}{|\mathbf{r} - \mathbf{r}'|}, \tag{1.13}$$

and the remaining, unknown, part is called exchange-correlation energy,

$$E_{\mathrm{xc}}[n] = F[n; \widehat{W}_{ee}, \widehat{T}] - T_s[n] - E_{\mathrm{H}}[n]. \tag{1.14}$$

The Hartree functional $E_{\mathrm{H}}[n]$ describes the electrons as if they were classical charge distributions. It is a simple functional of the density, and yields an important part of the energy of a many-electron system, but is nonzero also for a one-electron density. This "self-interaction part" of $E_{\mathrm{H}}[n]$ is cancelled by the exact $E_{\mathrm{xc}}[n]$, but this does not occur for most of the current approximate functionals, which suffer of the so called "self-interaction error".

The success of KS-DFT is mostly due to the fact that simple approximations (local-density approximation and generalized gradient corrections) for $E_{\mathrm{xc}}[n]$ and its functional derivative provide practical estimates of thermodynamical, structural and spectroscopic properties of atoms, molecules and solids. However, with the current approximations, KS-DFT is still lacking in several aspects, in particular it fails to handle near-degeneracy correlation effects (rearrangement of electrons within partially filled shells) and to recover long-range van der Waals interaction energies. The inaccuracy of KS-DFT stems from our lack of knowledge of $E_{\rm xc}[n]$, and much effort is put nowadays in finding new approximations to this term [24]. A trend in the current research is to construct *implicit* functionals of the density: in particular, the exchange-correlation functional is divided into exact exchange

$$E_x[n] = \langle\Phi|\widehat{W}_{ee}|\Phi\rangle - E_{\rm H}[n], \tag{1.15}$$

and the remaining correlation energy $E_c[n] = E_{\rm xc}[n] - E_x[n]$. The exact exchange cancels the self-interaction error of $E_{\rm H}[n]$, and is an implicit functional of $n(\mathbf{r})$ through the Slater determinant Φ. The corresponding Kohn–Sham potential must be determined via the optimized effective potential method (OEP) [20, 37],

$$E_0 = \inf_v\{\langle\Phi_v|\widehat{T} + \widehat{W}_{ee} + \widehat{V}_{ne}|\Phi_v\rangle + E_c[n_{\Phi_v}]\}, \tag{1.16}$$

where Φ_v is the ground state of the noninteracting Hamiltonian, $\widehat{T} + \widehat{V}$. The construction of a correlation energy functional $E_c[n]$ to be used with exact exchange in (1.16) is still an open problem.

In this work, we review some basic ideas, results, and open questions of a different approach: instead of trying to approximate the KS $E_{\rm xc}[n]$, we change the model system defined by (1.11).

1.3. Adiabatic connection formula. Before discussing the choice of different model Hamiltonians, we report some equations that will be used in the next sections. An exact formula for the functional $E_{\rm Hxc}[n]$ can be obtained via the adiabatic connection formalism [30, 36]: by varying a real parameter λ, the interaction $w^\lambda(r_{12})$ between the electrons (we have defined $r_{12} = |\mathbf{r}_1 - \mathbf{r}_2|$ to denote pairwise interactions that only depend on the electron-electron distance) is switched on continuously from zero to $1/r_{12}$, while the density is kept fixed by an external one-body potential $\widehat{V}^\lambda$. Each Hamiltonian $\widehat{H}^\lambda$ along this adiabatic connection has a ground-state wavefunction Ψ^λ that yields, by construction, the same density $n(\mathbf{r})$ for each λ. If $w^{\lambda=0} = 0$ and $w^{\lambda=a} = 1/r_{12}$, the KS Hartree-exchange-correlation energy is given by [30, 36]

$$E_{\rm Hxc}[n] = \int_0^a d\lambda \int_0^\infty dr_{12} 4\pi r_{12}^2 f^\lambda(r_{12}) \frac{\partial w^\lambda(r_{12})}{\partial\lambda}, \tag{1.17}$$

where the spherically and system-averaged pair density (APD) $f^\lambda(r_{12})$ is obtained by integrating $|\Psi^\lambda|^2$ over all variables but $r_{12} = |\mathbf{r}_2 - \mathbf{r}_1|$,

$$f^\lambda(r_{12}) = \frac{N(N-1)}{2} \sum_{\sigma_1,\ldots,\sigma_N} \int |\Psi^\lambda(\mathbf{r}_{12}, \boldsymbol{R}, \mathbf{r}_3, \ldots, \mathbf{r}_N)|^2 \frac{d\Omega_{\mathbf{r}_{12}}}{4\pi}\, d\boldsymbol{R}\, d\mathbf{r}_3, \ldots, d\mathbf{r}_N. \tag{1.18}$$

where $\boldsymbol{R} = (\mathbf{r}_1 + \mathbf{r}_2)/2$.

2. Changing the Model Hamiltonian

2.1. General considerations. Equation (1.11) shows that the KS approach to DFT introduces a model Hamiltonian in which $\widehat{W}_{ee}$ is set equal to zero and the one-body potential, $v_{\rm KS}(\mathbf{r})$, is different from $v_{ne}(\mathbf{r})$ (and is obtained by imposing the condition that the ground-state electron density of the model Hamiltonian be the same of the physical system). Solving the noninteracting KS Hamiltonian instead of the fully interacting Hamiltonian of (1.1) is obviously very practical. The exact

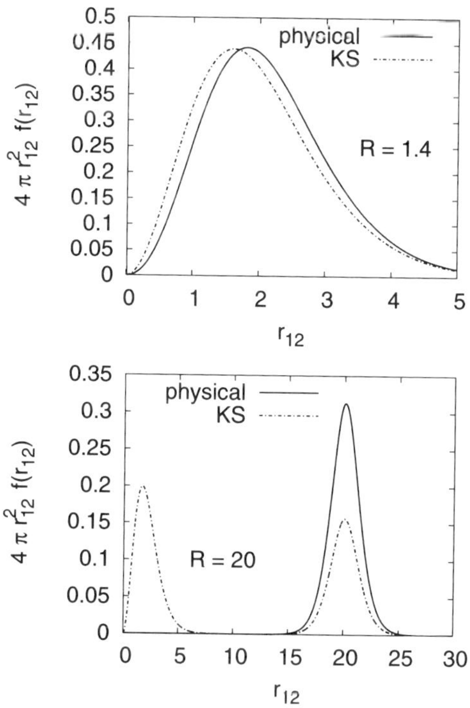

FIGURE 1. The spherically and system-averaged pair density $f^{\lambda}(r_{12})$ of (1.18) for the H_2 molecule in the KS system ($\lambda = 0$) and in the physical system (Coulombic interaction). The internuclear equilibrium distance $R = 1.4$ and the extreme stretched molecule at $R = 20$ are considered.

ground-state energy of the physical system can, in principle, be obtained from the KS Slater determinant Φ via the functional $E_{\text{xc}}[n]$. However, there are cases in which the restriction that the model system be noninteracting makes the search for an approximate $E_{\text{xc}}[n]$ seem like a daunting task. Consider the simple example of the H_2 molecule (two electrons in the field of two nuclei of unitary charge spaced by a distance R). In Figure 1 we report the quantity $4\pi r_{12}^2 f^{\lambda}(r_{12})$ that enters in (1.17) for the KS system and for the physical system ($\lambda = 0$ and $\lambda = a$, respectively) at two internuclear distances R. As shown by (1.17), the change in the APD $f^{\lambda}(r_{12})$ when we switch from the KS system to the physical (fully interacting) system determines the functional $E_{\text{Hxc}}[n]$. If this change is not drastic, we can expect that universal approximations for $E_{\text{Hxc}}[n]$ can work relatively well. This is, e.g., the case of the H_2 molecule at the equilibrium distance $R = 1.4$ a.u., reported in Figure 1. But when we stretch the molecule (e.g., in the extreme case $R = 20$ considered in the same figure), we see that the two APD are completely different, which means that the term $E_{\text{Hxc}}[n]$ becomes very important: it has to correct the very different nature of the KS wavefunction with respect to the physical one. This effect is completely system-dependent and it is thus very difficult to include in a *universal* functional of the density.

Cases like this appear when we have near-degenerate levels in the Kohn–Sham system (in the case of the H_2 molecule described above, the energies of the two KS states σ_g and σ_u become closer and closer as R increases). The basic idea reviewed in this paper is to remove the constraint that the model system be noninteracting in order to keep a reasonable resemblance between the model wavefuntion and the real one.

2.2. Using a modified electron-electron interaction. Obviously, the model system must still be less expensive to solve than the physical one. A possible approach is to define a density functional for a "partial" electron-electron interaction $w(r_{12}) \geq 0$, smaller than the Coulomb interaction $\forall r_{12}$, $w(r_{12}) \leq 1/r_{12}$, but different from zero (we restrict our choice to pairwise interactions that only depend on the electron-electron distance). Following the idea of the adiabatic connection of Section 1.3, we can make our partial interaction depend on a real parameter μ in such a way that $w^\mu(r_{12}) \to 0$ when $\mu \to 0$ and $w^\mu(r_{12}) \to 1/r_{12}$ when μ tends to some positive value a, and define

$$F^\mu[n] \equiv F[n; \widehat{W}^\mu, \widehat{T}] = \sup_v \left\{ \min_\Psi \langle \Psi | \widehat{T} + \widehat{W}^\mu + \widehat{V} | \Psi \rangle - \int n(\mathbf{r}) v(\mathbf{r}) \, d\mathbf{r} \right\}. \tag{2.1}$$

With this definition, we see that the density functional $F^\mu[n]$ switches from the Kohn–Sham one of (1.11) to the physical one of (1.9),

$$F^{\mu\to 0}[n] = F[n; 0, \widehat{T}] = T_s[n], \tag{2.2}$$

$$F^{\mu\to a}[n] = F[n; \widehat{W}_{ee}, \widehat{T}]. \tag{2.3}$$

In analogy with the KS approach, we can ask that the model system with interaction $w^\mu(r_{12})$ have the same density of the physical one. This fixes the external one-body potential (that we call $v^\mu(\mathbf{r})$) in our model Hamiltonian $\widehat{H}^\mu$. The ground-state wavefunction of our model system is denoted Ψ^μ: it is a multideterminantal wavefunction that must be computed with one of the standard methods of quantum chemistry (configuration interaction, coupled cluster, multi-configurational self-consistent field,...). In general, if μ is not too large (i.e., if w^μ is still much smaller than the full Coulomb interaction), few determinants describe Ψ^μ quite accurately, so that the computational cost can be kept low. The difference $\overline{F}^\mu[n]$ between the physical functional $F[n; \widehat{W}_{ee}, \widehat{T}]$ and the partially-interacting $F^\mu[n]$,

$$\overline{F}^\mu[n] = F[n; \widehat{W}_{ee}, \widehat{T}] - F[n; \widehat{W}^\mu, \widehat{T}], \tag{2.4}$$

is what we need to approximate, together with its functional derivative that determines $v_{ne}(\mathbf{r}) - v^\mu(\mathbf{r})$.

How to choose $w^\mu(r_{12})$? Two points are important to determine the partial interaction: (i) that we can design reasonable approximations for the corresponding $\overline{F}^\mu[n]$, and (ii) that we can solve adequately the Hamiltonian $\widehat{H}^\mu$ of the model system. A convenient choice seems to be a long-ranged interaction, i.e., a $w^\mu(r_{12})$ that behaves as $1/r_{12}$ for large r_{12}, but that is softer than $1/r_{12}$ for small r_{12}. The reasons for this choice are (i) short-range correlation effects seem to be more transferable from one system to another [2], and they should thus be more easily described in terms of an approximate *universal* density functional; (ii) the traditional wavefunction methods of quantum chemistry can reasonably describe Ψ^μ at a lower cost than the fully interacting system, because of the smaller interaction and because of the absence of the electron-electron cusp. For practical reasons (analytic matrix elements for both Gaussians and plane-waves, i.e. the most commonly used basis sets in quantum chemistry and solid state physics, respectively), a common choice is

$$w^{\mu}(r_{12}) = \frac{\text{erf}(\mu r_{12})}{r_{12}}, \tag{2.5}$$

where $\text{erf}(x)$ is the error function (and we thus have $a = \infty$). Other possibilities for $w^{\mu}(r_{12})$ have been also explored: they are all, like (2.5), arbitrary, and an open question is wether there is a way to determine $w^{\mu}(r_{12})$ according to some optimal criteria. The approach resulting from the choice of (2.5) is the one that we describe more in details in the next Section 3. Before doing that, for completeness we also briefly introduce a different choice of the model Hamiltonian.

2.3. Adding a nonlocal one-body operator. Another way to modify the model system is by adding a nonlocal one-body operator, $\widehat{O}_{\text{NL}}$, multiplied by a real, positive constant g [15, 29],

$$F^{g}[n] \equiv F[n; \widehat{T} + g\widehat{O}_{\text{NL}}, \widehat{W}_{ee}]. \tag{2.6}$$

By setting $\widehat{O}_{\text{NL}}$ equal to the projector (here reported for closed-shell systems) onto the virtual Kohn–Sham orbitals $\phi_i(\mathbf{r})$,

$$\widehat{O}_{\text{NL}}(\mathbf{r}, \mathbf{r}') = \delta(\mathbf{r} - \mathbf{r}') - \sum_{i=1}^{N/2} \phi_i(\mathbf{r})\phi_i^*(\mathbf{r}'), \tag{2.7}$$

the functional $F^{g}[n]$ of (2.6) switches from the physical one at $g = 0$ to the Kohn–Sham one of (1.11) in the limit $g \to \infty$ (since $\langle \Psi | \widehat{O}_{\text{NL}} | \Psi \rangle$ is always ≥ 0, as $g \to \infty$ the model system minimizes its energy by making $\langle \Psi | \widehat{O}_{\text{NL}} | \Psi \rangle = 0$, that is, by occupying only the KS orbitals). This choice of the model system is not further detailed here; the interested reader may find additional information in the literature [6, 15, 29].

3. Multideterminantal DFT from a Long-Range-Only Interaction

3.1. The functionals. Following the same steps of Section 1.3, we can write an exact formula for the functional $\overline{F}^{\mu}[n]$ of (2.4) in terms of the APD $f^{\mu}(r_{12})$. It is convenient to use for the adiabatic connection the same partial interaction $w^{\mu}(r_{12})$ of (2.5) chosen to determine the model system,

$$\overline{F}^{\mu}[n] = \int_{\mu}^{\infty} d\mu' \int_{0}^{\infty} dr_{12} 4\pi r_{12}^2 f^{\mu'}(r_{12}) \frac{\partial w^{\mu'}(r_{12})}{\partial \mu'}. \tag{3.1}$$

This formula is identical to (2.4), except from the fact that the integration over the coupling constant starts from the positive value μ instead than zero.

As in KS-DFT, the functional $\overline{F}^{\mu}[n]$ can be divided into an Hartree term,

$$\overline{E}_{\text{H}}^{\mu}[n] = \frac{1}{2} \int d\mathbf{r} \int d\mathbf{r}' n(\mathbf{r}) n(\mathbf{r}') \left[\frac{1}{|\mathbf{r} - \mathbf{r}'|} - w^{\mu}(|\mathbf{r} - \mathbf{r}'|) \right], \tag{3.2}$$

and an exchange-correlation term

$$\overline{E}_{\text{xc}}^{\mu}[n] = \overline{F}^{\mu}[n] - \overline{E}_{\text{H}}^{\mu}[n] \tag{3.3}$$

that needs to be approximated. The functional $\overline{E}^{\mu}_{\rm xc}[n]$ can, in turn, be divided into exchange and correlation in two different ways. We can, in fact, define an exchange functional by using the Kohn–Sham determinant Φ,

$$\overline{E}^{\mu}_{\rm x}[n] = \langle \Phi | \widehat{W}_{ee} - \widehat{W}^{\mu} | \Phi \rangle - \overline{E}^{\mu}_{\rm H}[n], \tag{3.4}$$

and then define the usual correlation energy functional $\overline{E}^{\mu}_{c}[n]$,

$$\overline{E}^{\mu}_{c}[n] = \overline{E}^{\mu}_{\rm xc}[n] - \overline{E}^{\mu}_{\rm x}[n], \tag{3.5}$$

but we can also define a multideterminantal (md) exchange functional [34] by using the wavefunction Ψ^{μ},

$$\overline{E}^{\mu}_{\rm x,md}[n] = \langle \Psi^{\mu} | \widehat{W}_{ee} - \widehat{W}^{\mu} | \Psi^{\mu} \rangle - \overline{E}^{\mu}_{\rm H}[n], \tag{3.6}$$

and then a corresponding correlation energy,

$$\overline{E}^{\mu}_{\rm c,md}[n] = \overline{E}^{\mu}_{\rm xc}[n] - \overline{E}^{\mu}_{\rm x,md}[n]. \tag{3.7}$$

These two ways of splitting exchange and correlation play a role only if we implement an optimized effective potential-like scheme, in analogy with what is usually done for the exact-exchange KS-DFT [20, 37] [see (1.16)]. In the case of the multideterminantal exchange of (3.6) the problem can be reformulated as [34]

$$E_0 = \inf_{v^{\mu}} \{ \langle \Psi^{\mu}_{v^{\mu}} | \widehat{T} + \widehat{W}_{ee} + \widehat{V}_{ne} | \Psi^{\mu}_{v^{\mu}} \rangle + \overline{E}^{\mu}_{\rm c,md}[n_{\Psi^{\mu}_{v^{\mu}}}] \}, \tag{3.8}$$

where $\Psi^{\mu}_{v^{\mu}}$ is obtained by solving the Schrödinger equation corresponding to the Hamiltonian $\widehat{H}^{\mu} = \widehat{T} + \widehat{W}^{\mu} + \widehat{V}^{\mu}$. This equation is the generalization of (1.16) to the case of the multideterminantal model system Ψ^{μ}.

3.2. Exact properties of the functionals. From (3.1) it is possible to derive exact properties of the functionals of Eqs. (3.4)–(3.7) in the $\mu \to 0$ limit [25, 32] and in the $\mu \to \infty$ limit [14, 32].

In the first case, all the functionals $\overline{E}^{\mu}_{x}$, $\overline{E}^{\mu}_{c}$ [Eqs. (3.4) and (3.5)] and $\overline{E}^{\mu}_{\rm x,md}$, $\overline{E}^{\mu}_{\rm c,md}$ [Eqs. (3.6) and (3.7)] tend to the KS functionals of Section 1.2. This limit can be studied with perturbation theory: one finds that the way in which the functionals approach the KS ones depends on whether the system is confined [32] (atoms, molecules) or extended [25].

The large-μ limit is the most interesting, since, as shown by (3.1), it always lies in the range of μ-values for which we want to construct approximations. In this limit, all the functionals of Eqs. (3.4)–(3.7) vanish and, since $\partial w^{\mu}(r_{12})/\partial\mu = 2/\sqrt{\pi} e^{-\mu^2 r_{12}^2}$, equation (3.1) shows that their large-μ behavior is determined by the short-range part (small r_{12}) of the spherically and system-averaged pair density $f^{\mu}(r_{12})$. It is possible to show [14] that, when $\mu \to \infty$, the small-r_{12} part of $f^{\mu}(r_{12})$ is dominated by

$$f^{\mu}(r_{12}) = f(0) \left[1 + 2 r_{12} p_1(\mu r_{12}) + \frac{2}{\sqrt{\pi}\mu} \right], \tag{3.9}$$

where $f(0)$ is the "on-top value" (zero electron-electron distance) of the physical (i.e., corresponding to the Coulomb interaction) $f(r_{12})$, and the function $p_1(y)$ is equal to [14]

$$p_1(y) = \frac{e^{-y^2} - 2}{2\sqrt{\pi} y} + \left(\frac{1}{2} + \frac{1}{4y^2} \right) \mathrm{erf}(y). \tag{3.10}$$

Inserting (3.9) (and, for the case of $E_x^\mu[n]$, the KS $f^{\mu=0}(r_{12})$, corresponding to the KS determinant Φ) into (3.1) we find, for unpolarized systems [8, 14, 32]

$$(3.11)\quad \overline{E}_x^{\mu\to\infty}[n] = -\frac{\pi}{4\mu^2}\int d\mathbf{r}\, n(\mathbf{r})^2 + O\left(\frac{1}{\mu^4}\right),$$

$$(3.12)\quad \overline{E}_c^{\mu\to\infty}[n] = \frac{\pi}{\mu^2}\left[f(0) - \frac{1}{4}\int d\mathbf{r}\, n(\mathbf{r})^2\right] + f(0)\frac{4\sqrt{2\pi}}{3\mu^3} + O\left(\frac{1}{\mu^4}\right),$$

$$(3.13)\quad \overline{E}_{\text{x,md}}^{\mu\to\infty}[n] = \frac{\pi}{\mu^2}\left[f(0) - \frac{1}{2}\int d\mathbf{r}, n(\mathbf{r})^2\right] + f(0)\frac{4\sqrt{\pi}(2\sqrt{2}-1)}{3\mu^3} + O\left(\frac{1}{\mu^4}\right),$$

$$(3.14)\quad \overline{E}_{\text{c,md}}^{\mu\to\infty}[n] = -f(0)\frac{4\sqrt{\pi}(\sqrt{2}-1)}{3\mu^3} + O\left(\frac{1}{\mu^4}\right).$$

These equations tell us that (i) if we use the definition of exchange with the KS determinant, the functional $\overline{E}_x^\mu[n]$ is, for large μ, exactly described by a local functional of the density; (ii) all the other three functionals involve the physical "on-top" $f(0)$, which is usually not available (its knowledge would require a very accurate calculation for the physical system!). However, it is possible to construct approximations for $f(0)$. For instance, for many systems the estimate of $f(0)$ from the local density approximation (i.e., by transfer from the uniform electron gas model) is rather good [2]. Another possibility is to use (3.9) to estimate $f(0)$ from $f^\mu(0)$. This estimate has the advantage of being without self-interaction error [14].

3.3. Building approximate functionals. Following the same historical path of KS-DFT, the simplest approximation one can think of is the local density (LDA),

$$(3.15)\qquad \overline{E}_{\text{xc}}^\mu[n] = \int d\mathbf{r}\, n(\mathbf{r})\big[\varepsilon_{\text{xc}}(n(\mathbf{r})) - \varepsilon_{\text{xc}}^\mu(n(\mathbf{r}))\big],$$

where $\varepsilon_{\text{xc}}(n)$ and $\varepsilon_{\text{xc}}^\mu(n)$ are the exchange-correlation energy per particle of an electron gas of uniform density n with interaction $1/r_{12}$ [3] and $w^\mu(r_{12})$ [25, 35], respectively.

Generalized-gradient approximations (GGA) for $\overline{E}_{\text{xc}}^\mu$ have been also constructed [9, 10, 16, 33] along similar lines of KS-DFT [26]. In this context, we only mention a conceptually simple approximation based on the exact properties of the previous Section 3.2. It consists [33] in a rational interpolation between a given GGA KS functional at $\mu = 0$, and zero at $\mu \to \infty$. The rational interpolation is constrained to recover the exact large-μ behavior of Eqs. (3.11)–(3.12), using the LDA approximation for the physical on-top $f(0)$.

All these ways of constructing approximate functionals are based on the assumption of transferability of short-range exchange and correlation effects from one system (the uniform electron gas) to the others. Another strategy that came out recently, and that seems very well suited to describe short-range correlation effects, consists in generating realistic APD $f^{\mu'}(r_{12})$ along the adiabatic connection [with $\mu' \geq \mu$, to be inserted in (3.1)] by solving simple "radial" (unidimensional) equations [12, 13] for a set of "effective geminals" $\psi_i^{\mu'}(r_{12})$,

$$(3.16)\qquad \begin{gathered}\left[-\frac{1}{r_{12}}\frac{d^2}{dr_{12}^2}r_{12} + \frac{\ell(\ell+1)}{r_{12}^2} + w_{\text{eff}}^{\mu'}(r_{12})\right]\psi_i^{\mu'}(r_{12}) = \varepsilon_i^{\mu'}\psi_i^{\mu'}(r_{12}) \\ \sum_i \vartheta_i|\psi_i^{\mu'}(r_{12})|^2 = f^{\mu'}(r_{12}).\end{gathered}$$

The effective interaction $w^{\mu'}_{\text{eff}}(r_{12})$ is (in principle) the Lagrange multiplier for the exact $f^{\mu'}(r_{12})$ (i.e., corresponding to the wavefunction $\Psi^{\mu'}$ along the adiabatic connection; in practice $w^{\mu'}_{\text{eff}}(r_{12})$ is approximated). In these equations the rule for the occupancy ϑ_i of the effective geminals is chosen, for spin compensated systems, to be Slater-determinant-like: occupancy 1 for even ℓ (singlet symmetry), occupancy 3 for odd ℓ (triplet symmetry), up to $N(N-1)/2$ occupied geminals. This rule has been applied to solve the effective Eqs. (3.16) in the uniform electron gas at $\mu = \infty$ (Coulombic interaction), with rather accurate results [5, 11] when combined with simple approximations for the effective interaction potential $w_{\text{eff}}(r_{12})$.

3.4. The calculation of Ψ^μ: an illustrative example. As said in the previous paragraphs, the model wavefunction Ψ^μ is, in most cases, computed with one of the traditional wavefunction methods of quantum chemistry. We report here an illustrative example of such calculation, and we then give in the next Section 4 an overview of recent results obtained using different methods for computing Ψ^μ, combined with different approximations for the functionals.

Since the only purpose of this paragraph is to investigate the wavefunction part of the multideterminantal DFT, we only consider two very small systems, the He and the Be atoms, for which it has been possible to construct very accurate potentials v^μ [32]. In this way (i) we have essentially no approximation on the functional part and we can focus our attention on the effect of approximations on the calculation of Ψ^μ, and (ii) we can produce accurate benchmark results to test our calculations. Of course, this is not the general procedure in which we are finally interested. The general procedure rather consists in using a given approximation (see Section 3.3) for the functional $\overline{E}^\mu_{\text{xc}}[n]$ and the corresponding potential, and combining it with a wavefunction method to calculate Ψ^μ, to finally obtain the total ground state energy of the physical Hamiltonian. This procedure has been followed with very promising results in the works reviewed in the next Section 4.

Here, in order to illustrate the efficiency of approximate wavefunction methods to treat Ψ^μ, we thus proceed as follows [32]. We first construct, for each μ, the model Hamiltonian $\widehat{H}^\mu$ using an accurate potential v^μ [32] and compute accurately its ground-state energy, $E^\mu = \langle\Psi^\mu|\widehat{H}^\mu|\Psi^\mu\rangle$, at the multi-reference configuration interaction with singles and doubles (MRCISD) level. We then compute various approximate ground-state energies, $E^\mu_S = \langle\Psi^\mu_S|\widehat{H}^\mu|\Psi^\mu_S\rangle$, by using approximate configuration interaction (CI) type wave functions Ψ^μ_S expanded into linear combinations of all the few Slater determinants generated from small orbital spaces S. The orbitals used are the natural orbitals of the Coulombic system calculated at the MRCISD level. The accuracy of the approximation for Ψ^μ_S can be assessed by looking at the difference between E^μ_S and E^μ

$$\Delta E^\mu_S = E^\mu_S - E^\mu. \tag{3.17}$$

The differences ΔE^μ_S are plotted as a function of μ in Figure 2 for the He atom with the orbital spaces $S = 1s$, $S = 1s2s$ and $S = 1s2s2p$. One sees that, in the Coulombic limit, $\mu \to \infty$, the reduction of the orbital space leads to important errors in the energy. When μ is decreased, i.e. when the interaction is reduced, the errors due to limited orbital spaces get smaller and smaller. For instance, at $\mu = 1$, using only the single-determinant wave function Ψ^μ_{1s}, leads to an error ΔE^μ_{1s} of less than 0.005 Hartree.

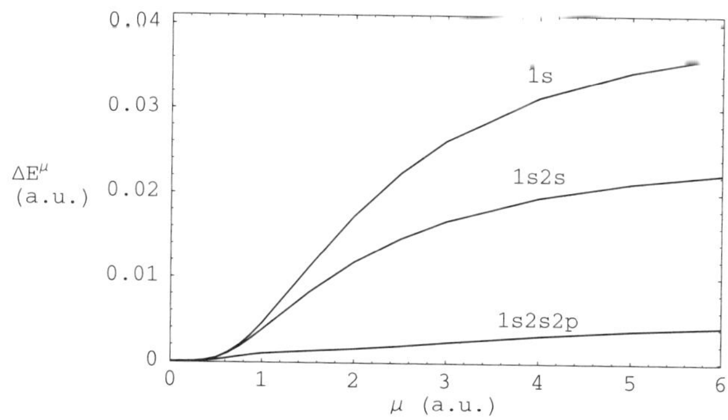

FIGURE 2. Ground-state energy differences $\Delta E_S^\mu = \langle \Psi_S^\mu | \widehat{H}^\mu | \Psi_S^\mu \rangle - \langle \Psi^\mu | \widehat{H}^\mu | \Psi^\mu \rangle$ where Ψ^μ is an accurate wave function and Ψ_S^μ are approximate wave functions generated from small orbital spaces $S = 1s$, $S = 1s2s$ and $S = 1s2s2p$, as a function of μ for the He atom.

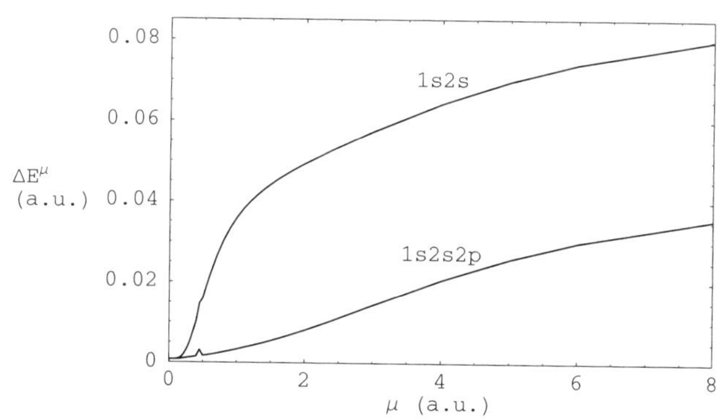

FIGURE 3. Ground-state energy differences $\Delta E_S^\mu = \langle \Psi_S^\mu | \widehat{H}^\mu | \Psi_S^\mu \rangle - \langle \Psi^\mu | \widehat{H}^\mu | \Psi^\mu \rangle$ where Ψ^μ is an accurate wave function and Ψ_S^μ are approximate wave functions generated from small orbital spaces $S = 1s2s$ and $S = 1s2s2p$, as a function of μ for the Be atom.

The case of the Be atom with the orbital spaces $S = 1s2s$ and $S = 1s2s2p$ is reported in Figure 3. Because of the near-degeneracy of the $2s$ and $2p$ levels, the inclusion of $2p$ configurations in the wave function is important, quite independently of the electron-electron interaction. Indeed, the difference $E_{1s2s}^\mu - E_{1s2s2p}^\mu$ remains large for almost all μ's. On the contrary, the error of the calculation where the $2p$ orbitals are included, ΔE_{1s2s2p}^μ, quickly falls off when μ is decreased. Again, for $\mu = 1$ for instance, the error ΔE_{1s2s2p}^μ given by the few-determinant CI-type wave function Ψ_{1s2s2p}^μ is less than 0.005 Hartree.

Therefore, the modification of the interaction enables to increase the accuracy of CI-type wave function calculations, or equivalently for a fixed target accuracy, decrease the effort of the calculation by reducing the orbital space. The crucial point for this effect to appear seems to be the reduction of the electron-electron interaction compared to the Coulomb interaction rather than the long-range character of the modified interaction.

4. Some Recent Results: A Short Overview

The idea of using the partially interacting model system of (2.1) has been explored with different techniques to compute the wavefunction Ψ^{μ}, and using different approximations for the functionals of Section 3.1. We review very briefly some of the corresponding results.

The LDA functional for $\overline{E}^{\mu}_{\rm xc}[n]$ has been combined with configuration interaction (CI) to handle the wavefunction Ψ^{μ} in [21, 27]; the method has been applied to atoms and small molecules. Again using LDA for the short-range functional, in [1] it has been shown that long-range van der Waals forces in rare gas dimers can be well described by using second-order perturbation theory for Ψ^{μ}.

A short-range GGA functional has been combined with the coupled-cluster (CC) method to describe Ψ^{μ}, with very good results for small molecules, both for the closed and the open shell cases [9, 10]: in particular, the results from multideterminantal DFT for small and medium basis-set sizes are better of both the pure DFT result and the pure CC result. In [31], a different GGA functional [16, 32] has been used, and Ψ^{μ} has been determined by multi-configuration self consistent field (MC-SCF). The corresponding application to systems in which near-degeneracy effects play a major role is promising, although self-interaction errors in the functionals are still a problem in some cases [31].

Realistic $f^{\mu'}(r_{12})$ from Eqs. (3.16) have been generated for two-electron atoms in [12]: with a very simple approximation for $w^{\mu'}_{\rm eff}(r_{12})$ (inspired to the one used for the electron gas at $\mu = \infty$), extremely accurate $\overline{E}^{\mu}_{\rm xc}[n]$ have been obtained. This new strategy of producing functionals is still at a very early stage, and several steps are needed for its development. In particular, we need better approximations for $w^{\mu'}_{\rm eff}(r_{12})$, and the implementation of an efficient algorithm to couple the radial equations (3.16) with the method used to solve the model Hamiltonian $\widehat{H}^{\mu}$.

In conclusions, changing the model Hamiltonian in DFT calculations seems to be a promising alternative to the construction of better exchange-correlation functionals for standard Kohn–Sham DFT. Of course, the method has been under development only in the last decade, and it still needs further improvement and investigation in many aspects.

References

1. J.G. Angyan, I.C. Gerber, A. Savin, and J. Toulouse, *van der Waals forces in density functional theory: perturbational long-range electron interaction corrections*, Phys. Rev. A **72** (2005), no. 012510, 1–9.
2. K. Burke, J.P. Perdew, and M. Ernzerhof, *Why semilocal functionals work: accuracy of the on-top pair density and importance of system averaging*, J. Chem. Phys. **109** (1998), 3760–3771.
3. D. M. Ceperley and B.J. Alder, *Ground state of the electron gas by a stochastic method*, Phys. Rev. Lett. **45** (1980), 566–569.

4. A.J. Coleman and V.I. Yukalov, *Reduced density matrices: Coulson's challenge*, Springer, Berlin, 2000.
5. B. Davoudi, M. Polini, R. Asgari, and M.P Tosi, *Self consistent Overhauser model for the pair distribution function of an electron gas in dimensionalities $D = 3$ and $D = 2$*, Phys. Rev. B **66** (2002), no. 075110, 1–8.
6. G.E. Engel, and W.E. Pickett, *Investigation of density functionals to predict both ground-state properties and band structures*, Phys. Rev. B **54** (1996), 8420–8429.
7. W.M.C. Foulkes, L. Mitas, R.J. Needs, and G. Rajagopal, *Quantum Monte Carlo simulations of solids*, Rev. Modern Physics **73** (2001), 33–83.
8. P.M.W. Gill and R.D. Adamson, *A family of attenuated Coulomb operators*, Chem. Phys. Lett. **261** (1996), 105–110.
9. E. Goll, H.-J. Werner, and H. Stoll, *A short-range gradient-corrected density functional in long-range coupled-cluster calculations for rare gas dimers*, Phys. Chem. Chem. Phys. **7** (2005), 3917–3923.
10. E. Goll, H.-J. Werner, H. Stoll, T. Leininger, P. Gori-Giorgi, and A. Savin, *A short-range gradient-corrected spin density functional in combination with long-range coupled-cluster methods: application to alkali-metal rare-gas dimer*, Chem. Phys. **329** (2006), 276–282.
11. P. Gori-Giorgi and J.P. Perdew, *Short-range correlation in the uniform electron gas: extended Overhauser model*, Phys. Rev. B **64** (2001), no. 155102, 1–8.
12. P. Gori-Giorgi and A. Savin, *Simple model for the spherically- and system-averaged pair density: results for two-electron atoms*, Phys. Rev. A **71** (2005), no. 032153, 1–9.
13. ———, *System-adapted correlation energy density functionals from effective pair interactions*, Phil. Mag. **86** (2006), 2643–2659.
14. ———, *Properties of short-range and long-range correlation energy density functionals from electron-electron coalescence*, Phys. Rev. A **73** (2006), no. 032506, 1–9.
15. C. Gutlé, *Espace orbitalaires et théorie de la fonctionnelle de la densité*, Ph.D thesis, Université Pierre et Marie Curie, 2003; http://www.lct.jussieu.fr/DFT/ (in french).
16. J. Heyd, G.E. Scuseria, and M. Ernzerhof *Hybrid functionals based on a screened Coulomb potential*, J. Chem. Phys. **118** (2003), 8207–8215.
17. P. Hohenberg and W. Kohn, *Inhomogeneous electron gas*, Phys. Rev. B **136** (1964), 864–871.
18. W. Kohn, *Nobel lecture: electronic structure of matter — wave functions and density functionals*, Rev. Modern Physics **71** (1999), 1253–1266.
19. W. Kohn and L.J. Sham, *Self-Consistent equations including exchange and correlation effects*, Phys. Rev. A **140** (1965) , 1133–1138.
20. S. Kümmel and J.P. Perdew, *Optimized effective potential made simple: orbital functionals, orbital shifts, and the exact Kohn–Sham exchange potential*, Phys. Rev. B **68** (2003), no. 035103, 1–15; and references therein.
21. T. Leininger, H. Stoll, H.-J. Werner, and A. Savin, *Combining long-range configuration interaction with short-range density functionals*, Chem. Phys. Lett. **275** (1997), 151–160.
22. M. Levy, *Universal variational functionals of electron densities, 1st-order reduced density-matrices, and natural spin-orbitals and solution of the v-representability problem*, Proc. Nat. Acad. Sci. U.S.A. (1979) **76**, 6062–6065.
23. E.H. Lieb, *Density functionals for Coulomb-systems*, Int. J. Quant. Chem. **24** (1983), 243–277.
24. A.E. Mattsson, *In pursuit of the "divine" functional*, Science **298** (2002), 759–760.
25. S. Paziani, S. Moroni, P. Gori-Giorgi, and G.B. Bachelet, *Local-spin-density functional for multideterminant density functional theory*, Phys. Rev. B **73** (2006), no. 155111, 1–9.
26. J.P. Perdew, K. Burke, and M. Ernzerhof, *Generalized gradient approximation made simple*, Phys. Rev. Lett. **77** (1996), 3865–3868.
27. R. Pollet, A. Savin, T. Leininger, and H. Stoll, *Combining multideterminantal wave functions with density functionals to handle near-degeneracy in atoms and molecules*, J. Chem. Phys. **116** (2002), 1250–1258.
28. J.A. Pople, *Nobel lecture: quantum chemical models*, Rev. Modern Physics **71** (1999), 1267–1274.
29. J. Rey and A. Savin, *Virtual space level shifting and correlation energies*, Int. J. Quant. Chem. **69** (1998), 581–590.
30. A. Savin, F. Colonna, and R. Pollet, *Adiabatic connection approach to density functional theory of electronic systems*, Int. J. Quant. Chem. **94** (2003), 166–190.

31. J. Toulouse, *Extension multidéterminantale de la méthode de Kohn–Sham en théorie de la fonctionnelle de la densité par décomposition de l'interaction électronique en contributions de longue portée et de courte portée*, Ph.D thesis, Université Pierre et Marie Curie, 2005 ; `http://www.lct.jussieu.fr/pagesperso/toulouse/` (in french).
32. J. Toulouse, F. Colonna, and A. Savin, *Long-range/short-range separation of the electron-electron interaction in density functional theory*, Phys. Rev. A **70** (2004), no. 062505, 1–16.
33. ———, *Short-range exchange and correlation energy density functionals: beyond the local density approximation*, J. Chem. Phys. **122** (2005), no. 014110, 1–10.
34. J. Toulouse, P. Gori-Giorgi, and A. Savin, *A short-range correlation energy density functional with multi-determinantal reference*, Theor. Chem. Acc. **114** (2005), 305–308.
35. J. Toulouse, A. Savin, and H.-J. Flad, *Short-range exchange-correlation energy of a uniform electron gas with modified electron-electron interaction*, Int. J. Quantum Chem. **100** (2004), 1047–1056.
36. W. Yang, *Generalized adiabatic connection in density functional theory*, J. Chem. Phys. **109** (1998), 10107–10110.
37. W. Yang and Q. Wu, *Direct Method for Optimized Effective Potentials in Density-Functional Theory*, Phys. Rev. Lett. **89** (2002), no. 143002, 1–4; and references therein.

Laboratoire de Chimie Théorique, CNRS, Université Pierre et Marie Curie, 4 Place Jussieu, F-75252 Paris, France.
E-mail address: `gori@lct.jussieu.fr`

Cornell Theory Center, Frank H.T. Rhodes Hall, Cornell University, Ithaca, New York, 14853-3801, USA.
E-mail address: `toulouse@cornell.edu`

Laboratoire de Chimie Théorique, CNRS, Université Pierre et Marie Curie, 4 Place Jussieu, F-75252 Paris, France
E-mail address: `savin@lct.jussieu.fr`

Centre de Recherches Mathématiques
CRM Proceedings and Lecture Notes
Volume **41**, 2007

Simulation of Quantum-Classical Dynamics by Surface-Hopping Trajectories

Hyojoon Kim and Raymond Kapral

ABSTRACT. Methods for simulating the dynamics of composite systems, where part of the system is treated quantum mechanically and its environment is treated classically, are discussed. Such quantum-classical systems arise in many physical contexts where certain degrees of freedom have an essential quantum character while the other degrees of freedom can be treated classically. The dynamics of these composite systems is governed by a quantum-classical Liouville equation, which is a high-dimensional matrix partial differential equation. Its solution can be obtained using hybrid Monte Carlo-Molecular Dynamics methods in terms of an ensemble of surface-hoping trajectory segments interspersed with quantum transitions. The method is illustrated by computing the rate of a nonadiabatic chemical reaction in which a two-level system is coupled to a bistable oscillator which is, in turn, coupled to a bath of harmonic oscillators.

1. Introduction

Classical molecular dynamics simulation has proven to be an extremely valuable tool for obtaining insight into physical, chemical, and biological phenomena. The greatest virtue of molecular dynamics is that the coupled Newtonian equations of motion for a large number of degrees of freedom can be solved by trajectory calculations using well-established algorithms [1, 2].

A fully quantum mechanical version of molecular dynamics for systems containing many degrees of freedom is not computationally feasible; however, in many applications, the quantum character of only small number of degrees of freedom is crucial while the dynamics of the remaining degrees of freedom can be approximated accurately by classical mechanics. A decomposition of this type is especially appropriate for a subsystem composed of light particles, such as electrons or protons, interacting with a solvent of massive molecules. This fact has motivated the construction of a quantum-classical Liouville equation description of the dynamics of such composite systems. The quantum-classical Liouville equation, when expressed in a suitable basis of quantum states, is a high-dimensional matrix partial differential equation [5, 6, 18, 19, 28].

2000 *Mathematics Subject Classification.* 60H15.

This work was supported in part by a grant from the Natural Sciences and Engineering Research Council of Canada.

Quantum (and classical) transport coefficients can be expressed in terms of time integrals of flux-flux correlation functions. Their evaluation involves sampling from suitable equilibrium initial conditions and time evolution of observables. We have shown that it is possible to compute such correlation functions by propagating observables with quantum-classical Liouville dynamics, while retaining the full quantum-mechanical equilibrium structure of system [3, 4, 8–10, 23]. In this way one can estimate quantum transport properties for a large many-body system. The calculation of the quantum equilibrium structure of a system, although a difficult problem, is far more tractable than that of the carrying out quantum time evolution. One difficulty that appears in the calculation of the quantum equilibrium structure is the fact that the statistical weight for each trajectory can be a complex number and the existence of rapidly varying phases often make sampling of initial states a challenging problem. Effective sampling algorithms are required to tackle quantum-classical systems with a large number of degrees of freedom [8, 15].

In this chapter we describe how it is possible to solve the quantum-classical Liouville equation in terms of an ensemble of classical trajectory segments interspersed by quantum transitions. The techniques we describe here are applicable to complex many-body systems and have been used to study a model proton transfer reaction in a polar molecular solvent [3]. Here, in order to illustrate the method in more a transparent context, we show how the reaction rate can be computed for a two-level system, coupled to a bistable oscillator which is, in turn, coupled to a bath of harmonic oscillators. The model captures many of the essential features of real proton transfer processes in the condensed phase, especially the coupling of the quantum protonic degree of freedom to the classical solvent polarization coordinate which, in turn, is coupled to other bath degrees of freedom.

2. Reaction Rate Constant Expressions

While the computational scheme we describe can be used to obtain a variety of transport properties, we limit our discussion here to the computation of the rate of a chemical reaction $A \rightleftharpoons B$ occurring in a many-body environment. For a quantum mechanical system in thermal equilibrium the time dependent reaction rate coefficient $k_{\mathrm{AB}}(t)$ for this reaction may be determined from the time integral of a flux-flux correlation function [11, 27],

$$k_{\mathrm{AB}}(t) = \int_0^t dt' \langle \hat{\jmath}_A; \hat{\jmath}s_B(t') \rangle = \langle \dot{\widehat{N}}_A; \widehat{N}_B(t) \rangle. \tag{2.1}$$

where $\hat{\jmath}_A = \dot{\widehat{N}}_A = (i/\hbar)[\widehat{H}, \widehat{N}_A]$ is the flux of $\widehat{N}_A$, with an analogous expression for $\hat{\jmath}_B$, $[\cdot,\cdot]$ is the commutator and the angular brackets $\langle \hat{A}; \widehat{B} \rangle = 1/\beta \int_0^\beta d\lambda \langle e^{\lambda \widehat{H}} \hat{A} e^{-\lambda \widehat{H}} \widehat{B} \rangle$ denote a Kubo transformed correlation function, with $\beta = (k_B T)^{-1}$. The equilibrium quantum canonical average is $\langle \cdots \rangle = Z_Q^{-1} \,\mathrm{Tr} \cdots e^{-\beta \widehat{H}}$ where Z_Q is the partition function. In the reaction rate expression, $\widehat{N}_{A(B)}$ is the operator that characterizes species $A(B)$. The reaction rate can be found from the plateau value of $k_{\mathrm{AB}}(t)$.

Introducing Wigner transforms that double the phase-space dimension, the time dependent reaction rate coefficient can be written in the form [8, 9, 23]

$$k_{\mathrm{AB}}(t) = \int d\mathcal{X}\, N_{BW}(\mathcal{X}, t) \overline{W}_A(\mathcal{X}, 0), \tag{2.2}$$

where the phase space variable $\mathcal{X} \equiv (\mathcal{R}, \mathcal{P})$. From this expression we see that the calculation of the reaction rate involves two aspects: how to carry out the time evolution of the observable and how to sample initial equilibrium distribution. The time evolution of the Wigner transform of the operator $\widehat{N}_B$ can be written formally as

$$\frac{d}{dt} N_{BW}(\mathcal{X}, t) = iL_W(\mathcal{X}) N_{BW}(\mathcal{X}, t), \tag{2.3}$$

where iL_W is the quantum Liouville operator in Wigner-transformed form [8, 9]. The quantum equilibrium structure is described by the spectral density function W_A which is given by

$$\overline{W}_A(\mathcal{X}, 0) \equiv \frac{1}{\beta} \int_0^\beta d\lambda \int d\mathcal{X}' \big(iL_W(\mathcal{X}') A_W(\mathcal{X}')\big) W(\mathcal{X}', \mathcal{X}, i\hbar\lambda), \tag{2.4}$$

where

$$W(\mathcal{X}', \mathcal{X}, t) = \frac{1}{(2\pi\hbar)^{2\nu} Z_Q} \int d\mathcal{Z}\, d\mathcal{Z}' e^{-i/\hbar(\mathcal{P}\cdot\mathcal{Z} + \mathcal{P}'\cdot\mathcal{Z}')} \times \left\langle \mathcal{R} + \frac{\mathcal{Z}}{2} \right| e^{-\beta\widehat{H} - i/\hbar\widehat{H}t} \left| \mathcal{R}' - \frac{\mathcal{Z}'}{2} \right\rangle \left\langle \mathcal{R}' + \frac{\mathcal{Z}'}{2} \right| e^{i/\hbar\widehat{H}t} \left| \mathcal{R} - \frac{\mathcal{Z}}{2} \right\rangle. \tag{2.5}$$

and ν denotes the coordinate space dimension. Both the quantum dynamics and quantum equilibrium structure are difficult to compute for general systems with a large number of degrees of freedom. Below, we describe algorithms that can be used to approximately compute such quantum rate coefficient correlation functions.

3. Trajectory Description of Dynamics

3.1. Quantum-classical evolution. Since it is not computationally feasible to carry out full quantum time evolution of large many-body systems, we consider situations where the quantum system can be partitioned into two subsystems: $\mathcal{S}$ the subsystem in whose quantum dynamics we are directly interested and its environment $\mathcal{E}$. We suppose that the dynamics of the environment, in the absence of coupling to $\mathcal{S}$, can be accurately approximated by classical equations of motion. The phase space set of coordinates $\mathcal{X}$ can be partitioned into two sets corresponding to the $\mathcal{S}$ and $\mathcal{E}$ subsystems as $\mathcal{X} = (x, X)$ with $x = (r, p)$ and $X = (R, P)$. In order to obtain a representation in terms of surface-hopping dynamics, it is convenient to write the subsystem operators in the adiabatic basis. The adiabatic basis states are defined as follows: The partial Wigner transform of the Hamiltonian is $\widehat{H}_W = P^2/2M + \hat{p}^2/2m + \widehat{V}_W(\hat{q}, R) \equiv P^2/2M + \hat{h}_W(R)$, where $\hat{h}_W(R)$ is the Hamiltonian for the subsystem $\mathcal{S}$ in the presence of fixed particles of the bath. The adiabatic eigenstates are the solutions of the eigenvalue problem, $\hat{h}_W(R)|\alpha; R >= E_\alpha(R)|\alpha; R >$ and the eigenvalues determine the Born–Oppenheimer surfaces that depend on the bath configuration. Changes of the adiabatic wavefunctions with respect to the bath coordinates lead to the breakdown of the Born–Oppenheimer approximation and give rise to transitions between different adiabatic surfaces. We term such transitions nonadiabatic effects. Using the adiabatic basis we can write

$$A_W(\mathcal{X}) = \sum_{\alpha\alpha'} \int dz\, e^{i/\hbar p\cdot z} \left\langle r - \frac{z}{2} \right| \alpha; R \Big\rangle A_W^{\alpha\alpha'}(X) \Big\langle \alpha'; R \left| r + \frac{z}{2} \right\rangle, \tag{3.1}$$

where $A_W^{\alpha\alpha'}(X) = \langle\alpha; R|\hat{A}_W(X)|\alpha'; R\rangle$. Equation (2.2) can be rewritten with the operators expressed in the adiabatic basis as

$$k_{\rm AB}(t) = \sum_{\alpha\alpha'} \int dX\, N_{BW}^{\alpha\alpha'}(X,t)\overline{W}_A^{\alpha'\alpha}(X,0). \tag{3.2}$$

This expression for the quantum mechanical time dependent rate constant is exact and still intractable. We approximate it by replacing the full quantum evolution of the observable by quantum-classical Liouville evolution. In the quantum-classical limit, $N_{BW}^{\alpha'\alpha}(X,t)$ satisfies the following evolution equation, similar to (2.3) [5]:

$$\frac{d}{dt}N_{BW}^{\alpha\alpha'}(X,t) = \sum_{\beta\beta'} i\mathcal{L}_{\alpha\alpha',\beta\beta'}(X)N_{BW}^{\beta\beta'}(X,t). \tag{3.3}$$

The quantum-classical Liouville operator in the adiabatic basis $i\mathcal{L}$ is given by [5]

$$\begin{aligned} i\mathcal{L}_{\alpha\alpha',\beta\beta'}(X) &= [i\omega_{\alpha\alpha'}(R) + iL_{\alpha\alpha'}(X)]\delta_{\alpha\beta}\delta_{\alpha'\beta'} - J_{\alpha\alpha',\beta\beta'}(X) \\ &\equiv i\mathcal{L}^0_{\alpha\alpha'}(X)\delta_{\alpha\beta}\delta_{\alpha'\beta'} - J_{\alpha\alpha',\beta\beta'}(X), \end{aligned} \tag{3.4}$$

where the classical evolution operator is defined as

$$iL_{\alpha\alpha'} = \frac{P}{M}\frac{\partial}{\partial R} + \frac{1}{2}\left[F_W^{\alpha}(R) + F_W^{\alpha'}(R)\right]\frac{\partial}{\partial P}, \tag{3.5}$$

with

$$\begin{aligned} J_{\alpha\alpha',\beta\beta'}(X) = &-\frac{P}{M}d_{\alpha\beta}\left[1 + \frac{1}{2}S_{\alpha\beta}(R)\frac{\partial}{\partial P}\right]\delta_{\alpha'\beta'} \\ &- \frac{P}{M}d^*_{\alpha'\beta'}\left[1 + \frac{1}{2}S^*_{\alpha'\beta'}(R)\frac{\partial}{\partial P}\right]\delta_{\alpha\beta}. \end{aligned} \tag{3.6}$$

Here the frequency $\omega_{\alpha\alpha'}(R) \equiv [E_\alpha(R) - E_{\alpha'}(R)]/\hbar$, the Hellmann–Feynman force $F_W^\alpha = -\langle\alpha; R|\partial\widehat{V}_W(\hat{q}, R)/\partial\hat{R}|\alpha; R\,\rangle$, the nonadiabatic coupling matrix element is $d_{\alpha\beta} = \langle\alpha; R|\nabla_R|\beta; R\rangle$, and $S_{\alpha\beta} = (E_\alpha - E_\beta)d_{\alpha\beta}[(P/M)d_{\alpha\beta}]^{-1}$ [5]. This dynamics is exact for any multi-level system bilinearly coupled to a harmonic bath and will provide a good approximation to the dynamics for general baths and couplings for systems where the bath degrees of freedom are massive compared to those of the quantum subsystem.

The formal solution of (3.3) is given by

$$N_{BW}^{\alpha\alpha'}(X,t) = \sum_{\beta\beta'}(e^{i\mathcal{L}t})_{\alpha\alpha',\beta\beta'}N_{BW}^{\beta\beta'}(X,t). \tag{3.7}$$

The propagator in each time interval satisfies the following Dyson equation [13]

$$\begin{aligned} &(e^{-i\mathcal{L}t})_{\alpha\alpha',\beta\beta'} \\ &\quad = e^{-i\mathcal{L}^0_{\alpha\alpha'}t}\delta_{\alpha\beta}\delta_{\alpha'\beta'} + \sum_{\nu\nu'}\int_0^t dt' e^{-i\mathcal{L}^0_{\alpha\alpha'}(t-t')}J_{\alpha\alpha',\nu\nu'}(e^{-i\mathcal{L}t})_{\nu\nu',\beta\beta'}, \end{aligned} \tag{3.8}$$

where

$$\begin{aligned} e^{-i\mathcal{L}^0_{\alpha\alpha'}(t-t')} &= e^{-i\int_t^{t'} d\tau\, \omega_{\alpha\alpha'}(R_{\alpha\alpha',\tau})}e^{-iL_{\alpha\alpha'}(t-t')} \\ &\equiv \mathcal{W}_{\alpha\alpha'}(t,t')e^{-iL_{\alpha\alpha'}(t-t')}\,, \end{aligned} \tag{3.9}$$

The terms involving the bath momentum derivative in J are difficult to evaluate exactly. In order to approximate their action on any function of the momenta, we introduce the momentum-jump approximation. In this approximation, J takes the form of a momentum translation operator whose effect on a function of the momentum is to shift the momentum by a certain value. Using the operator identity, $(S_{\alpha\beta}/2)\cdot\partial/\partial P = \Delta E_{\alpha\beta} M\partial/(\partial(P\cdot\hat{d}_{\alpha\beta})^2)$, the action of the operator on any function $f(P)$ of the momentum may be written approximately as [5, 24]

$$
\begin{aligned}
(3.10)\quad \left(1+\Delta E_{\alpha\beta}M\frac{\partial}{\partial(P\cdot\hat{d}_{\alpha\beta})^2}\right)f(P) &\approx e^{\Delta E_{\alpha\beta}M\partial/\partial(P\cdot\hat{d}_{\alpha\beta})^2}f(P)\\
&= f(\hat{d}^{\perp}_{\alpha\beta}(P\cdot\hat{d}^{\perp}_{\alpha\beta})+\hat{d}_{\alpha\beta}\mathrm{sgn}(P\cdot\hat{d}_{\alpha\beta})\sqrt{(P\cdot\hat{d}_{\alpha\beta})^2+\Delta E_{\alpha\beta}M})\\
&= f(P+\hat{d}_{\alpha\beta}\Delta P),
\end{aligned}
$$

where $\Delta P = -(\hat{d}_{\alpha\beta}\cdot P)+\mathrm{sgn}(P\cdot\hat{d}_{\alpha\beta})\sqrt{(P\cdot\hat{d}_{\alpha\beta})^2+\Delta E_{\alpha\beta}M}$. In the second approximate equality on the right hand side of this equation we have replaced the sum by an exponential. The momentum vector may be written in terms of its components along $\hat{d}_{\alpha\beta}$ and a perpendicular vector $\hat{d}^{\perp}_{\alpha\beta}$, $P=\hat{d}_{\alpha\beta}(\hat{d}_{\alpha\beta}\cdot P)+\hat{d}^{\perp}_{\alpha\beta}(\hat{d}^{\perp}_{\alpha\beta}\cdot P)$. In the last line, we used the fact that the exponential operator is a translation operator in the variable $(P\cdot\hat{d}_{\alpha\beta})^2$. Note that if there is insufficient energy in the bath to accommodate a quantum transition, the argument in the square root factor in the momentum shift will be negative. This will place a restriction on allowed transitions. Using the Dyson representation of the quantum-classical propagator in conjunction with the momentum jump approximation, Eq. (3.7) can be written as an ensemble of trajectories whose form is shown schematically in Figure 1. In this figure diagonal matrix elements of the species operator, $N^{\alpha\alpha}_{BW}$, evolve according to Newton's equations of motion on the α adiabatic surface. A quantum transition

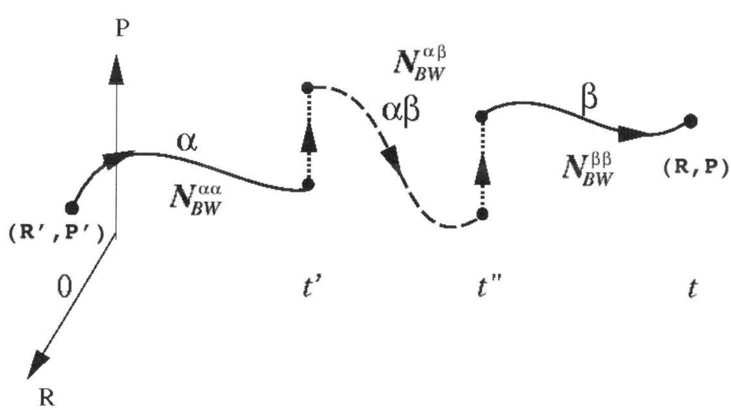

FIGURE 1. Schematic picture of a trajectory that enters into the computation of the time evolution of operator. A diagonal matrix element of the species operator $N^{\alpha\alpha}_{BW}$ evolves classically on the α adiabatic surface. The first quantum transition makes an off-diagonal matrix element of the operator $N^{\alpha\beta}_{BW}$ evolve on the mean of the α and β adiabatic surfaces with a phase factor. And the second quantum transition yields a diagonal matrix element $N^{\beta\beta}_{BW}$ on the β adiabatic surface.

gives rise to an off-diagonal matrix element of the operator, $N_{BW}^{\alpha\beta}$, and this transition is accompanied by a jump in the momentum of environment. This transition places the system in a coherent mixture of states α and β. The system evolves classically on the mean of the α and β adiabatic surfaces and a phase factor enters in the expression for the observable. A second quantum transition destroys the coherent state and yields a diagonal matrix element, $N_{BW}^{\beta\beta}$, again accompanied by a change in the momentum of the bath. Subsequent evolution takes place on the β adiabatic surface.

A similar interpretation can be given to the evolution of the density matrix. In this context we remark that the time evolution of the off-diagonal density matrix elements, integrated over bath degrees of freedom to give the quantum subsystem density matrix, is customarily used to define the decoherence time. In the above picture of the dynamics, we see from our trajectory simulation scheme that coherence is created and destroyed in passages between the diagonal and off-diagonal states. In this way, decoherence is automatically incorporated in our formalism and appears from the average over the ensemble of trajectories. In our model simulations, the average time spent in the coherent state is small compared to the typical time scale of the chemical rate process.

3.2. Simulation algorithm. The quantum-classical Liouville operator $\hat{\mathcal{L}}$ is time independent and the evolution operator can be simply decomposed into a composition of evolution operators in time segments of arbitrary length. Therefore, the evolution of a dynamical variable can be obtained by the successive application of evolution operators in small time intervals. Dividing the time interval t into N segments of lengths Δt_j such that the jth segment $\Delta t_j = t_j - t_{j-1}$, we obtain

$$(3.11)\qquad (e^{i\hat{\mathcal{L}}t})_{\alpha_0\alpha_0',\alpha_N\alpha_N'} = \sum_{(\alpha_1\alpha_1')\cdots(\alpha_{N-1}\alpha_{N-1}')} \prod_{j=1}^{N} (e^{i\hat{\mathcal{L}}(t_j-t_{j-1})})_{\alpha_{j-1}\alpha_{j-1}',\alpha_j\alpha_j'},$$

where $(\alpha_0\alpha_0') \equiv (\alpha\alpha')$. If Δt is sufficiently small, one can make a one-point approximation to the time integral in (3.8) as

$$\begin{aligned}(3.12)\quad (e^{i\hat{\mathcal{L}}(t_j-t_{j-1})})_{\alpha_{j-1}\alpha_{j-1}',\alpha_j\alpha_j'} &\approx s e^{i\mathcal{L}^0_{\alpha_{j-1}\alpha_{j-1}'}\Delta t}(\delta_{\alpha_j\alpha_j',\alpha_{j-1}\alpha_{j-1}'} - \Delta t J_{\alpha_{j-1}\alpha_{j-1}',\alpha_j\alpha_j'}) \\ &= \mathcal{W}_{\alpha_{j-1}\alpha_{j-1}'}(t_{j-1},t_j) e^{iL_{\alpha_{j-1}\alpha_{j-1}'}\Delta t} \\ &\quad \times (\delta_{\alpha_j\alpha_j',\alpha_{j-1}\alpha_{j-1}'} - \Delta t J_{\alpha_{j-1}\alpha_{j-1}',\alpha_j\alpha_j'}),\end{aligned}$$

where the phase factor $\mathcal{W}_{\alpha_{j-1}\alpha_{j-1}'}(t_{j-1},t_j) = e^{i\omega_{\alpha_{j-1}\alpha_{j-1}'}(t_j-t_{j-1})}$ associated with time segment (t_j, t_{j-1}) arises from the action of $\exp\left(i\mathcal{L}^0_{\alpha_{j-1}\alpha_{j-1}'}(t_j - t_{j-1})\right)$ on any phase space function. At the end of each segment, the system may either stay on the same energy surface or make a transition to a new state. The above expression for the propagator converges to the exact Dyson integral equation form in the limit when the number of time slices tends to infinity, the time slice interval tends to zero and the product of these two quantities is the finite total time over which the dynamics is followed [13].

The sequential short-time algorithm is easily implemented. The total time of the calculation is divided into a fixed number of time slices, the most natural choice of slice length being the molecular dynamics integration time step Δt. The phase space coordinates are propagated adiabatically in a time step and the phase factor

$\mathcal{W}$ is computed if the evolution is on the mean of two adiabatic surfaces. The initial positions and momenta are updated from X to $X_{\Delta t}$ at time Δt by using the classical propagator $e^{iL_{\alpha_0\alpha_0'}t}$. Then, given $\alpha_0\alpha_0'$, we choose α_1 or α_1' uniformly from the set of allowed final states. Given a choice of state, we can compute the nonadiabatic coupling matrix element d at the updated position.

At the end of each time step the probabilities [12]

$$\Pi = \theta[(P\cdot\hat{d}_{\alpha\beta})^2 + \Delta E_{\alpha\beta}M]|(P_{\Delta t}/M)\cdot d_{\alpha_0\alpha_1}(R_{\Delta t})|\Delta t \times (1 + |(P_{\Delta t}/M)\cdot d_{\alpha_0\alpha_1}(R_{\Delta t})|\Delta t)^{-1}, \tag{3.13}$$

and $1-\Pi$, respectively, are used for acceptance or rejection of a quantum transition. In this equation the Heaviside function θ accounts for transitions that are forbidden if there is insufficient energy in the bath to cause the transition. If the transition is rejected, we have

$$N_{BW}^{\alpha_0\alpha_0'}(X,\Delta t) = \mathcal{W}_{\alpha_0\alpha_0'}(\Delta t) N_{BW}^{\alpha_0\alpha_0'}(X_{\Delta t})\frac{1}{1-\Pi}. \tag{3.14}$$

If the transition is accepted, we translate the momentum to $\widetilde{P}_{\Delta t}$ by using the momentum-jump approximation of (3.10). We then get

$$N_{BW}^{\alpha_0\alpha_0'}(X,\Delta t) = \mathcal{W}_{\alpha_0\alpha_0'}(\Delta t) N_{BW}^{\alpha_0\alpha_0'}(R_{\Delta t},\widetilde{P}_{\Delta t})\frac{P_{\Delta t}\Delta t}{M} d_{\alpha_0\alpha_1}(R_{\Delta t})\frac{1}{\Pi} w_{\alpha_0\alpha_0',\alpha_1\alpha_1'}, \tag{3.15}$$

where $w_{\alpha_0\alpha_0',\alpha_1\alpha_1'}$ is an appropriate weight. This procedure may be repeated in the subsequent time slices until the Nth slice is reached. By averaging the resulting quantity over initial phase space and quantum states, we can obtain the statistical average need to compute the correlation function. This will be described in more detail in the next section.

4. Sampling Equilibrium Initial States

One difference between the quantum and classical correlation function expressions for transport coefficients is that the statistical weight, which depends on the spectral density function W, can be a complex number for quantum systems. This makes sampling from quantum initial conditions difficult. For condensed phase chemical reactions occurring at high temperatures, the sampling problem is less difficult. In this limit we may write

$$\overline{W}_A(\mathcal{X},0) = W_A(\mathcal{X}, i\hbar\beta/2) + \mathcal{O}(\beta^2). \tag{4.1}$$

From the general expressions for the rate constant in terms of projected dynamics [7], one may show that the use of this expression in the rate constant formula is exact in the long time limit [16, 17]. This approximation will be valid if there is a large time scale separation between that for the chemical interconversion rate process and other microscopic relaxation times in the system. Although this gives rise to some differences in the short time behavior of the time-dependent rate coefficient, for long times, given reactive time scale separation, the rate coefficient extracted from the plateau value of $k_{\mathrm{AB}}(t)$ should be given accurately. The spectral density function $W_A(\mathcal{X}, i\hbar\beta/2)$ is always real [8].

Depending on the system under investigation, it may be convenient to define the metastable states in terms of degrees of freedom of the quantum subsystem, classical bath, or both. In order to illustrate the formalism in a simple context, we suppose that it is possible to characterize the progress of the reaction in terms of a single coordinate R_0 of the environment $\mathcal{E}$. For the reaction $A \rightleftharpoons B$, the A and B species operators may be defined as $\widehat{N}_{AW} = \theta(-R_0)$ and $\widehat{N}_{BW} = \theta(R_0)$ where θ is the Heaviside step function and the dividing surface has been taken to lie at $R_0 = 0$.

From the quantum-classical version of (2.5), we may perform integrations over all X'coordinates to obtain the simplified expression

$$\text{(4.2)}\quad W_A^{\alpha'\alpha}\left(X, \frac{i\hbar\beta}{2}\right) = \frac{1}{(2\pi\hbar)^\nu Z_Q}\frac{i\hbar}{M_0}\int dZ\, dZ_0'\delta'(Z_0')e^{-i/\hbar P\cdot Z}$$
$$\times \langle\alpha'; R_0|\left\langle R+\frac{Z}{2}\left|e^{-\beta/2\widehat{H}}\right| -\frac{Z_0'}{2}\right\rangle\left\langle\frac{Z_0'}{2}|e^{-\beta/2\widehat{H}}|R-\frac{Z}{2}\right\rangle|\alpha; R_0\rangle.$$

To solve this equation, we assume that the imaginary time propagator may be written in separable Hamiltonians of the subsystem and environment as $\exp(-\beta\widehat{H}) \approx \exp(-\beta\widehat{H}_{sn})\exp(-\beta\widehat{H}_{b(n)})$ and we single out the barrier region around $R_0 = 0$ for special consideration. Separating the potential energy into a harmonic part and a remainder term, the following expression for $W_A^{\alpha\alpha}$ may be derived [8, 10]

$$\text{(4.3)}\qquad W_A^{\alpha\alpha}\left(X, \frac{i\hbar\beta}{2}\right) = \rho_{sn}^{\alpha\alpha}(X_0)\rho_b(X_b; R_0),$$

where

$$\text{(4.4)}\quad \rho_{sn}^{\alpha\alpha}(X_0)$$
$$= \frac{1}{Z_{sn}}\frac{\omega_0}{2\pi\sin 2u_0}\sqrt{\frac{2M_0u_0'}{\pi\beta\hbar^2}}\frac{\beta P_0}{M_0u_0'}e^{-\beta E_\alpha(R_0)-\beta/2M_0\omega_0^2R_0^2-(2M_0u_0')/(\beta\hbar^2)R_0^2-\beta/(2M_0u_0'}P_0^2),$$

and

$$\text{(4.5)}\qquad \rho_b(X_b; R_0) = \frac{1}{Z_b}\prod_j\frac{\beta\omega_j}{2\pi u_j''}e^{-\beta/(u_j'')\{1/(2M_j)P_j^2+V_{b(n)}(R_j;R_0)\}}.$$

Here M_j and ω_j are the mass and frequency for each oscillator, $u_i' \equiv u_i\cot u_i$, $u_i'' = u_i\coth u_i$, and $u_i = \beta\hbar\omega_i/2$. The partition function may be also factorized as $Z_Q = Z_{sn}Z_b$. Here we have omitted the off-diagonal contribution in W_A to the rate constant since its magnitude has been shown to be very small for the values of parameters of interest [8]. Since the functions take Gaussian forms, we can easily implement sampling from the initial quantum distributions for all degrees of freedom.

The classical distribution functions can be easily obtained by taking the high temperature limits of these expressions. Using the relations $\lim_{\beta\to 0}\sqrt{a/(\pi\beta)}e^{-(a/\beta)R_0^2} = \delta(R_0)$ and $\lim_{\beta\to 0}u_0' = \lim_{\beta\to 0}u_j'' = 1$, the classical-limit expressions are given by

$$\text{(4.6)}\qquad \rho_{sn}^{\alpha\alpha,\mathrm{cl}}(X_0) = \frac{1}{Z_{sn}^{\mathrm{cl}}}\frac{1}{2\pi\hbar}\delta(R_0)\frac{P_0}{M_0}e^{-\beta/2M_0\ P_0^2\pm\beta\Omega},$$

and

$$\text{(4.7)}\qquad \rho_b^{\mathrm{cl}}(X_b; 0) = \frac{1}{Z_b^{\mathrm{cl}}}\prod_j\frac{\beta\omega_j}{2\pi}e^{-\beta/2M_jP_j^2-\beta/2\ M_j\omega_j^2R_j^2}.$$

With help of these analytical expressions for W_A, we can calculate the rate constant using (3.2). The simulation results are presented in the next section.

5. Nonadiabatic Reaction Rate

As a specific example of the type of system discussed in general terms above, we consider a two-level system coupled to a quartic bistable oscillator which is in turn coupled to a heat bath of independent harmonic oscillators. The Hamiltonian operator, expressed in a diabatic basis $\{|\uparrow\rangle, |\downarrow\rangle\}$, is [21, 22]

$$\text{(5.1)}\quad \mathbf{H} = \begin{pmatrix} V_n(R_0) + \hbar\gamma_0 R_0 & -\hbar\Omega \\ -\hbar\Omega & V_n(R_0) - \hbar\gamma_0 R_0 \end{pmatrix} + \left(\frac{P_0^2}{2M_0} + \sum_{j=1}^{\nu-1} \frac{P_j^2}{2M_j} + \frac{M_j\omega_j^2}{2}\left(R_j - \frac{c_j}{M_j\omega_j^2}R_0\right)^2 \right)\mathbf{I}.$$

The coupling to the two-level system is given by $\hbar\gamma(R_0) = -\hbar\gamma_0 R_0$. The nonlinear quartic oscillator, $V_n(R_0) = -M_0\omega_0^2 R_0^2/2 + AR_0^4/4$ is bilinearly coupled to $\nu - 1$ independent one-dimensional harmonic oscillators. The bilinear coupling is characterized by an Ohmic spectral density, $J(\omega) = \pi \sum c_j^2/(2M_j\omega_j)\delta(\omega - \omega_j)$, where $c_j = (\xi\hbar\Delta\omega M_j)^{1/2}\omega_j$, with ξ the Kondo parameter, $\omega_j = -\omega_c \ln(1 - j\Delta\omega/\omega_c)$ and $\Delta\omega = (\omega_c)/\nu - 1(1 - e^{-\omega_{\max}/\omega_c})$ with ω_c the cut-off frequency [14].

The adiabatic states are obtained by diagonalization of the two-level system Hamiltonian in (5.1) and are given by,

$$\text{(5.2)}\quad \begin{aligned} |1; R_0\rangle &= \frac{1}{\mathcal{N}}[(1+G)|\uparrow\rangle + (1-G)|\downarrow\rangle] \\ |2; R_0\rangle &= \frac{1}{\mathcal{N}}[-(1-G)|\uparrow\rangle + (1+G)|\downarrow\rangle], \end{aligned}$$

where $\mathcal{N}(R_0) = \sqrt{2(1 + G^2(R_0))}$ with $G(R_0) = \{-\Omega + \sqrt{\Omega^2 + \gamma(R_0)^2}\}/\gamma(R_0)$. The corresponding adiabatic energies are

$$\text{(5.3)}\quad E_\alpha(R) = -\tfrac{1}{2}M_0\omega_0^2 R_0^2 + \tfrac{1}{4}AR_0^4 \mp \sqrt{\Omega^2 + \gamma(R_0)^2} + V_{b(n)}(R),$$

where the sign is minus for the ground state and plus for the excited state. For the harmonic bath $V_{b(n)}(R)$ is given by $\sum_{j=1}^{\nu-1} \frac{1}{2}M_j\omega_j^2(R_j - c_j/(M_j\omega_j^2)R_0)^2$.

When γ_0 is small and the ground and excited potential surfaces are nearly parallel, the partition function for the reactant state, $n_A^{\rm eq}Z_Q$, can be approximated by that for the mean surface and is given by

$$\text{(5.4)}\quad (n_A^{\rm eq}Z_Q)^{-1} \approx e^{\beta V_r}\sinh(\beta\hbar\omega_r/2)\prod_{j=1} 2\sinh(\beta\hbar\omega_j/2),$$

with ω_r the well frequency and V_r the bare potential at the bottom of the well. Using the high temperature form for (5.4) in the $t = 0+$ limit of the rate formula (3.2) we can determine the transition state theory rate constant $k_{\rm AB}^{\rm TST}$,

$$\text{(5.5)}\quad k_{\rm AB}^{\rm TST} \approx \frac{\omega_r}{2\pi}\frac{e^{\beta V_r}(e^{-\beta\Omega} + e^{\beta\Omega})}{2},$$

which will be used to scale the results presented in the figures. When the coupling between the two-level system and the quartic oscillator is negligible, the value of Ω vanishes and $k_{\rm AB}^{\rm TST}$ becomes the well-known value of $\omega_r e^{\beta V_r}/2\pi$. For the current symmetric oscillator, and for the parameter range under investigation, $\omega_r \approx \sqrt{2}\omega_0$.

Using these results, the transmission coefficient is given by $\kappa_{\rm AB}(t)=k_{\rm AB}(t)/k_{\rm AB}^{\rm TST}$. We label results obtained using this formula that treat the reaction coordinate and bath quantum mechanically as QRB. The limit where the reaction coordinate and bath are treated classically can be also obtained from $W_A^{\alpha\alpha}$. When this expression is used to compute the rate coefficient, we obtain the diagonal contribution to the rate obtained earlier using quantum-classical linear response theory [22]. Results obtained using this formula that treats the reaction coordinate and bath classically are labelled by CRB.

Dimensionless units with time scaled by the cut-off frequency ω_c are used in the calculations described below [22]. In particular, $t \leftarrow t\omega_c$, $R_0 \leftarrow \sqrt{M_0\omega_c/\hbar}R_0$, $\beta \leftarrow \beta\hbar\omega_c$. We set $A = 0.05$ and $\omega^\ddagger = 1$ and the bare barrier height of the $V_n(R_0)$ potential is 5 in these dimensionless units. The number of bath harmonic oscillators is $N_b = 100$ so that $\nu = 101$. We have taken $\omega_{\rm max} = 3$. The simulation time step $dt = 0.01$.

First, we compare the QRB and CRB results for two temperatures in Figure 2. For high temperatures ($\beta = 0.1$), both the QRB and CRB results are indistinguishable, except at very short times. The CRB results for the rate constant are non-zero and equal to the transition state theory value of the rate constant at $t = 0+$ as in classical theories of the reaction rate. The QRB results for the time dependent rate constant are zero at $t = 0$, typical of quantum rate processes [7]. For the lower of the two temperatures, because of quantum effects which are included thorough initial distributions, one sees that the QRB results yield a larger rate constant than the CRB simulations. The quantum rate enhancement is similar to that seen

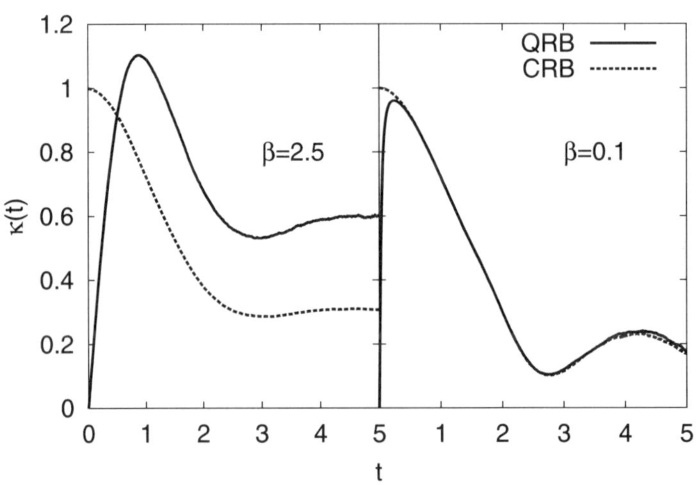

FIGURE 2. Comparison of quantum-classical nonadiabatic dynamics for cases where the equilibrium structure of the reaction coordinate and bath is treated quantum mechanically (QRB) with those where the equilibrium structure of the reaction coordinate and bath is treated classically (CRB). Parameters values: $\beta = 2.5$ (left), $\beta = 0.1$ (right), $\gamma_0 = 0.1$, $\Omega = 0.1$, and $\xi = 3$.

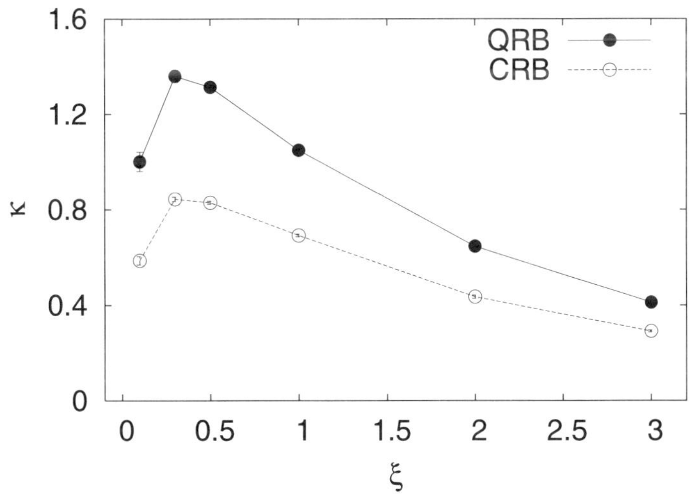

FIGURE 3. Transmission coefficient vs ξ for $\beta = 2$. The solid line denotes quantum treatments of the reaction coordinate and the bath (QRB) while the dotted line denotes classical treatments of these parts of the system (CRB).

in other studies [20, 25, 26]. Interestingly, we found that, compared to a classical description, a quantum treatment of the equilibrium structure of the reaction coordinate increases the reaction rate constant while a quantum treatment of the bath lowers the rate constant. The quantum rate enhancement is, therefore, caused by the fact that the change in the rate constant arising from quantum and classical treatments of the reaction coordinate is much greater than the corresponding change for quantum and classical treatments of the bath equilibrium structure.

In Figure 3, we plot the transmission coefficients κ_{AB}^{QRB} and κ_{AB}^{CRB} for various values of Kondo parameter ξ. Both results show similar turnover behavior for small ξ. For the model parameters used in this study, quantum effects tend to be larger for smaller values of ξ corresponding to weaker coupling to the bath modes. This trend is also consistent with the fact coupling between the reaction coordinate and the bath gives rise to decoherence and the stronger the coupling the more pronounced this effect.

6. Conclusion

The theory presented in this chapter shows how quantum chemical reaction rates of complex systems with a large number of degrees of freedom can be computed by using surface-hopping trajectories. The simulation method combines a surface-hopping dynamics derived from the quantum-classical Liouville equation with initial sampling from quantum or classical equilibrium distributions. Using this simulation scheme, the high-dimensional matrix partial differential equation that describes quantum-classical Liouville dynamics can be solved through a hybrid Monte Carlo-molecular dynamics simulation method. Future algorithmic developments in this ares must focus on the construction of algorithms that are useful for arbitrarily long times. Such developments will likely have to focus on methods for dealing with decoherence in the quantum subsystem induced by coupling to the environmental dynamics.

References

1. M.P. Allen and D.J. Tildesley, *Computer simulation of liquids*, Oxford University Press, Oxford, 1987.
2. D. Frenkel and B. Smit, *Understanding molecular simulation: from algorithms to applications*, Academic Press, San Diego, 1996.
3. G. Hanna and R. Kapral, *Quantum-classical Liouville dynamics of nonadiabatic proton transfer*, J. Chem. Phys. **122** (2005), 244505.
4. G. Hanna, H. Kim, and R. Kapral, *Quantum-classical reaction rate theory*, Quantum Dynamics of Complex Molecular Systems (D.A. Micha and I. Burghardt, eds.), Springer Ser. Chemical Phys., Springer-Verlag, Berlin, 2006, pp. 295–319.
5. R. Kapral and G. Ciccotti, *Mixed quantum-classical dynamics*, J. Chem. Phys. **110** (1999), 8919–8929.
6. ———, *A statistical mechanical theory of quantum dynamics in classical environments*, Lecture Notes in Phys., vol. 605, Springer-Verlag, Berlin, 2002, pp. 445–472.
7. R. Kapral, S. Consta, and L. McWhirter, *Chemical rate laws and rate constants*, Classical and Quantum Dynamics in Condensed Phase Simulations , Proceedings of the Internat. School of Phys. (Lerici, 1997) (B.J. Berne, G. Ciccotti, and Da.F Coker, eds.), World Scientific, Singapore, 1998, pp. 583–616.
8. H. Kim and R. Kapral, *Nonadiabatic quantum-classical reaction rates with quantum equilibrium structure*, J. Chem. Phys. **123** (2005), 194108.
9. ———, *Transport properties of quantum-classical systems*, J. Chem. Phys. **122** (2005), 214105.
10. ———, *Quantum bath effects on nonadiabatic reaction rates*, Chem. Phys. Lett. **423** (2006), 76–80.
11. R. Kubo, *Fluctuation-dissipation theorem*, Rep. Prog. Phys. **29** (1966), 255.
12. D. MacKernan, G. Ciccotti, and R. Kapral, *Surface-hopping dynamics of a spin-boson system*, J. Chem. Phys. **116** (2002), 2346–2353.
13. D. MacKernan, R. Kapral, and G. Ciccotti, *Sequential short-time propagation of quantum-classical dynamics*, J. Phys. **14** (2002), 9069–9076.
14. N. Makri, *The linear response approximation and its lowest order corrections: an influence functional approach*, J. Phys. Chem. B **103** (1999), 2823–2829.
15. N. Metropolis, A.W. Rosenbluth, M.N. Rosenbluth, A.H. Teller, and E. Teller, *Equation of state calculations by fast computing machines*, J. Chem. Phys. **21** (1953), 1087–1092
16. W.H. Miller, *Quantum-mechanical transition-state theory and a new semiclassical model for reaction-rate constants*, J. Chem. Phys. **61** (1974), 1823–1834.
17. W.H. Miller, S.D. Schwartz, and J.W. Tromp, *Quantum-mechanical rate constants for bimolecular reactions*, J. Chem. Phys. **79** (1983), 4889–4898.
18. S. Nielsen, R. Kapral, and G. Ciccotti, *Non-adiabatic dynamics in mixed quantum-classical systems*, J. Stat. Phys. **101** (2000), 225–242.
19. ———, *Statistical mechanics of quantum-classical systems*, J. Chem. Phys. **115** (2001), 5805–5815.
20. E. Rabani, G. Krilov, and B.J. Berne, *Quantum mechanical canonical rate theory: a new approach based on the reactive flux and numerical analytic continuation methods*, J. Chem. Phys. **112** (2000), 2605–2614.
21. A. Sergi and R. Kapral, *Nonadiabatic reaction rates for dissipative quantum-classical systems*, J. Chem. Phys. **119** (2003), 12776–12783.
22. ———, *Quantum-classical dynamics of nonadiabatic chemical reactions*, J. Chem. Phys. **118** (2003), 8566–8575.
23. ———, *Quantum-classical limit of quantum correlation functions*, J. Chem. Phys. **121** (2004), 7565–7576.
24. A. Sergi, D. MacKernan, G. Ciccotti, and R. Kapral, *Simulating quantum dynamics in classical environments*, Theor. Chem. Accts. **110** (2003), 49–58.

25. M. Topaler and N. Makri, *Quantum rates for a double well coupled to a dissipative bath accurate path-integral results and comparison with approximate theories*, J. Chem. Phys. **101** (1994), 7500–7519.
26. H.B. Wang, X. Sun, and W.H. Miller, *Semiclassical approximations for the calculation of thermal rate constants for chemical reactions in complex molecular systems*, J. Chem. Phys. **108** (1998), 9726–9736.
27. T. Yamamoto, *Quantum statistical mechanical theory of the rate of exchange chemical reactions in the gas phase*, J. Chem. Phys. **33** (1960), 281–289.
28. C. Zhu, A.W. Jasper, and D.G. Truhlar, *Non-Born–Oppenheimer Liouville–von Neumann dynamics. evolution of a subsystem controlled by linear and population-driven decay of mixing with decoherent and coherent switching*, J. Chem. Theory Comput. **1** (2005), 527–540.

Department of Chemistry, University of Toronto, Toronto, Ontario M5S 3H6, Canada.

E-mail address: `hkim@chem.utoronto.ca`

E-mail address: `rkapral@chem.utoronto.ca`

Centre de Recherches Mathématiques
CRM Proceedings and Lecture Notes
Volume **41**, 2007

Simulating Realistic and Nonadiabatic Chemical Dynamics: Application to Photochemistry and Electron Transfer Reactions

Denise M. Koch, Qadir K. Timerghazin, and Gilles H. Peslherbe

Abstract. Various methodologies to simulate molecular dynamics and complex chemical reaction dynamics are reviewed. In principle, the most accurate description of the time evolution of a system would be obtained by solving the time-dependent Schrödinger equation for the system of interest. This equation, however, is too difficult to solve for high-dimensional systems, such that various approximations are generally introduced to simulate reaction dynamics. In the first part of these proceedings, *first-principles* molecular dynamics simulation techniques are presented as an avenue to simulate reactions occurring on a single electronic state realistically, and application of this methodology is illustrated with simulations of the dynamics of the photochemically excited iodide-acetonitrile complex. In the second part, simulations of nonadiabatic processes, *i.e.* involving more than one electronic state, are discussed. Various techniques available to simulate nonadiabatic dynamics are reviewed, and the investigation of the effects of solvation on the electron transfer reaction occurring upon photoexcitation of sodium-iodide salt in water clusters serves to illustrate the applicability of surface hopping methods.

1. Introduction

Computer simulations play an increasingly important role in science today. Indeed, increasing the accuracy of computational models makes the results of simulations reliable enough to be directly compared with experimental measurements. In turn, simulations become an essential tool to understand and interpret experimental results. With the advent of femtosecond spectroscopy, which allows reaction dynamics to be observed in real time, computer simulations greatly contribute to our understanding of chemical reactions at the microscopic level. To investigate chemical reactions in real time, computational chemists rely on approximate molecular dynamics techniques that allow to follow the motion of atoms and molecules during

2000 *Mathematics Subject Classification.* Primary: 92E99, 92E20; Secondary: 81Q20, 81V55.

This research was supported by the Natural Sciences and Engineering Research Council (NSERC) of Canada. DMK was the recipient of a Canada Graduate Scholarship and Concordia University Graduate Fellowships, QKT was the recipient of Concordia University Graduate Fellowships, and GHP holds a Concordia University Research Chair.

the course of a reaction. In these proceedings, we review modern techniques to simulate reaction dynamics and illustrate their application with selected examples.

2. Adiabatic Dynamics

2.1. Traditional and first-principles simulation techniques. In quantum mechanics, the time evolution of a system follows the time-dependent Schrödinger equation, which is the partial differential equation [25]

$$i\hbar\frac{\partial\Psi(\mathbf{Q},\mathbf{q},t)}{\partial t}=\widehat{H}\Psi(\mathbf{Q},\mathbf{q},t), \tag{2.1}$$

where $\hbar = h/2\pi$, h being the Plank constant, $\Psi(\mathbf{Q},\mathbf{q},t)$ is the wavefunction of the entire system, $\mathbf{Q}$ are the nuclear coordinates, $\mathbf{q}$ are the electronic coordinates, t is time and $\widehat{H}$ is the Hamiltonian operator, which is the sum of the kinetic and potential energy operators for the system. The wavefunction is the cornerstone of quantum mechanics and its norm defines the probability of finding the particles in time and space. Except for very simple systems, it is not possible to solve this partial differential equation exactly in many dimensions, and computational chemists generally introduce various approximations. The *Born–Oppenheimer* (BO) approximation is the first one of a hierarchy of approximations, which allows separation of the electronic and nuclear motions. Because the mass of nuclei is orders of magnitude larger than that of the electrons, the electronic relaxation with respect to the nuclear motion is considered to be instantaneous, and the Schrödinger equation for the electrons (e) and the nuclei (n) can be solved separately. In this context, the nuclear dynamics is said to evolve *adiabatically* on the electronic potential energy surface, and the pseudo-separable Hamiltonian for the system can be expressed as

$$\widehat{H}\approx\widehat{K}_n+\widehat{H}_e, \tag{2.2}$$

where $\widehat{K}_n$ is the nuclear kinetic energy operator and $\widehat{H}_e$ the electronic Hamiltonian (including the nuclei-nuclei interactions). With a wavefunction of the form

$$\Psi=\psi_n(\mathbf{Q})\psi_e(\mathbf{q}). \tag{2.3}$$

The electronic energy, which depends parametrically on the fixed nuclear positions $\mathbf{Q}$, can be calculated from the time-independent Schrödinger equation

$$\widehat{H}_e\psi_e=V(\mathbf{Q})\psi_e. \tag{2.4}$$

and the motion of atoms can be monitored in time by solving the nuclear equation of motion

$$i\hbar\frac{\psi_n}{\partial t}=[K_n+V(\mathbf{Q})]\psi_n. \tag{2.5}$$

Electronic structure theory (or quantum chemistry) is the art of solving the time-independent electronic Schrödinger equation, Eq. (2.4), in the field of fixed nuclei. This equation, however, can only be solved exactly for systems containing a single electron, such that approximate numerical methods are required to solve this equation for chemical systems. The basis of electronic structure theories is the orbital approximation, according to which each electron is assumed to be localized in a given orbital. The wavefunction Ψ for the whole system is usually expressed as an antisymmetrized product of molecular orbitals ψ, which are constructed as a linear combination of atomic orbitals usually resulting from a basis set expansion. In order to solve the Schrödinger equation, the variational principle is invoked, and molecular orbitals are optimized through an iterative self-consistent field (SCF) procedure, known to chemists as the Hartree–Fock (HF) theory. The HF method only provides a zeroth-order wavefunction of the system because it completely ignores electron correlation, *i.e.* the orbital associated with a given electron is determined in the average field of all other electrons and details of the correlated motion of electrons are ignored. Treating the correlation between the electrons is a formidable challenge and various methods based on perturbation theory (Möller–Plesset theory) or the variational principle (configuration interaction theory) have been designed to that effect [8].

For systems containing more than a few degrees of freedom, the quantum nuclear equation of motion, Eq. (2.5), can not be solved exactly and computational chemists often employ classical mechanics instead to simulate the dynamics of the system. The motion of the atoms is then typically propagated by solving the Newtonian equation of motion

$$m\ddot{\mathbf{Q}} = -\nabla V(\mathbf{Q}), \tag{2.6}$$

where m is the mass of the atom, $\ddot{\mathbf{Q}} = \partial^2\mathbf{Q}/\partial t^2$ is the acceleration, and $\nabla V(\mathbf{Q})$ is the gradient of the potential energy V with respect to the nuclear coordinates, *i.e.* the force exerted on the atoms.

Typically, because computational chemists investigate large or complex systems, such as polyatomics, biomolecules, liquids or solids, the traditional approach to molecular dynamics has relied on model potential energy surface $V(\mathbf{Q})$ based on chemically intuitive functional forms to describe interatomic and intermolecular interactions. Such model potentials, also referred to as force fields, typically contain additive terms,

$$V(Q) = V_{\text{bond}} + V_{\text{bend}} + V_{\text{tors}} + V_{\text{coul}} + V_{\text{repdis}} + V_{\text{other}}, \tag{2.7}$$

where V_{bond} is a bonding potential between two adjacent atoms, V_{bend} is the angular bending potential between three bonded atoms, V_{tors} is the torsional interaction between four bonded atoms, V_{coul} is the Coulombic (electrostatic) interaction and V_{repdis} is the repulsion-dispersion (or Van der Waals) potential between non-bonded atoms, and V_{other} comprises all other optional many-body interactions that may be required for the chemical system of interest [8]. Each term in this sum includes one or more empirical parameters that often correspond to physical quantities that can be measured experimentally or calculated from electronic structure theory.

The classical-mechanical approach just described is commonly used by computational chemists to evaluate thermodynamic properties from molecular dynamics simulations or to follow chemical reaction dynamics with the classical trajectory approach. The former appeals to the ergodic hypothesis, which assumes that trajectories uniformly sample the phase space, and thus, the phase space average for a given property can be replaced by the time average for sufficiently long simulation times. The evaluation of time averages for static chemical properties often requires the propagation of few trajectories over a long simulation period [8]. On the other hand, to evaluate complex reaction dynamics, large ensembles of initial conditions are generated, usually by Monte Carlo sampling, in order to mimic the probabilistic nature of the quantum world [8], and many trajectories are propagated. Dynamical observables are then evaluated from the distribution of properties over the ensemble at any given time.

Many chemical processes are quantum chemical in nature because they involve the rearrangement of the electronic structure within the chemical system during the course of a reaction, as bonds break or form for instance. A more realistic approach to simulate chemical reaction dynamics is the Born–Oppenheimer first-principles or direct dynamics method [3, 33]. In this approach, instead of relying on an arbitrary model function to describe the interatomic and intermolecular interactions, an electronic structure calculation, *i.e.* the solution of (2.4), is performed at each time step to determine the energy and the gradient of the energy, while the classical equations of motion, equation 2.6, are still employed to propagate the atomic motion. Since large ensembles of trajectories are typically required to realistically describe chemical reactions, this approach is very computer intensive and is limited to systems containing a small number of atoms or to ultra-fast phenomena, but nonetheless essential to investigate reactions involving significant changes in the electronic structure. One such application of first-principles dynamics is now illustrated.

2.2. Application of first-principles dynamics: $I^-(CH_3CN)$ Photochemistry. Charge–transfer–to–solvent (CTTS) excited states, where an electron is ejected from a solute and is subsequently stabilized by the polar electric field of the surrounding solvent, has recently attracted considerable attention ([9,31], and references therein). Previously, CTTS states were considered to exist only within liquid solvents. Johnson and co-workers performed the first experiments in which bound (CTTS) excited states were observed for small clusters formed by iodide and a few polar solvent molecules, namely acetone (CH_3COCH_3), acetonitrile (CH_3CN) and water (H_2O) [10,11,32]. In bulk solution, the relaxation of CTTS excited states leads to solvated electrons which remain within the liquid. In clusters, it was found that relaxation leads to the formation of neutral iodine and a negatively charged solvent cluster [10,11]. For iodine in a cluster of n acetonitrile molecules, for example, the photochemical reaction can be summarized as

$$I^-(CH_3CN)_n \xrightarrow{h\nu} [I^-(CH_3CN)_n]^* \longrightarrow\rightarrow I^\bullet + (CH_3CN)_n^- \tag{2.8}$$

where $h\nu$ corresponds to the laser pulse excitation energy, and the asterisk refers to the electronically excited cluster before relaxation.

The relaxation mechanism of the CTTS excited state of iodide in water clusters has been studied extensively by means of computational chemistry by inferring the mechanism from quantum-chemically determined potential energy profiles ([4, 34], and references therein). However, due to the floppiness of the solvent clusters, the potential energy surface for the photochemical reaction is very complex. Hence, it has recently been shown that a realistic description of the highly complex relaxation dynamics for these clusters can only be obtained by performing full-scale first-principles molecular dynamics simulations [35]. Since first-principles molecular dynamics simulations, as described earlier, are computationally intensive, a small prototype system, $I^-(CH_3CN)$, was initially employed to investigate the dynamics of the relaxation mechanism, and to benchmark various quantum chemical procedures that could eventually be applied to larger systems [36].

Upon photoexcitation of the $I^-(CH_3CN)$ complex, an electron from the valence $5p$ orbital of iodide is transferred to a diffuse orbital around the solvent, which is supported by the large dipole moment of the acetonitrile molecule, as shown in Fig. 1. Snapshots of the relaxation dynamics following photoexcitation along a

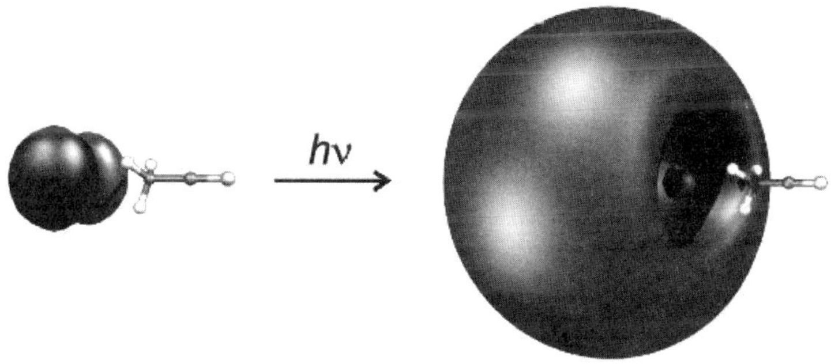

FIGURE 1. Schematic representation of the transfer of an electron from the 5p orbital of iodide to a solvent-supported diffuse orbital.

representative trajectory are shown in Fig. 2. The slow ejection of the iodine atom and the formation of the stable negatively charged acetonitrile molecule is clearly apparent from these snapshots. This study clearly demonstrates the importance of performing first-principles dynamics simulations to observe the dynamics of the stabilization of an electron post-photoexcitation, and to gain a full understanding of the relaxation mechanism. Since the relaxation mechanism is dynamical in nature, the full mechanism could not be directly inferred from experiment nor from static electronic structure calculations. Furthermore, the complex electronic structure of the CTTS excited states requires the use of state-of-the-art high-level electronic structure calculations, and can certainly not be modeled by empirical potentials. This application thus illustrates the power of first-principles dynamics in allowing simulations of real-time photochemical problems.

3. Nonadiabatic Dynamics

3.1. Nonadiabatic dynamics simulation techniques. As mentioned in the previous section, traditional and first-principles molecular dynamics simulation techniques are based on the Born–Oppenheimer (BO) approximation, which allows for the separation of nuclear and electronic degrees of freedom. For many chemical reactions, however, a breakdown of the decoupling of the motion of the nuclei and the electrons may occur in processes involving a change of electronic state during the course of a reaction. Examples of such *nonadiabatic* processes include photochemically-induced reactions such as the electron transfer reaction discussed later. The dynamics of these reactions involves multiple electronic surfaces, *i.e.* the electronic state on which the nuclear motion proceeds may change during the reaction, normally because of crossings of electronic states, as illustrated in Fig. 3 for a one-dimensional problem. For instance, the potential energy curves of two adiabatic electronic states obtained as a linear combination of two *diabatic* states with given electronic character (covalent and ionic, for example) give rise to an avoided crossing where the diabatic curves cross, and nonadiabatic transitions between adiabatic electronic states may be possible in this avoided crossing region.

In order to propagate the dynamics of chemical reactions on multiple electronic states, the adiabatic potential energy surfaces must be well defined, and the probability of undergoing a transition between adiabatic states must be known. The

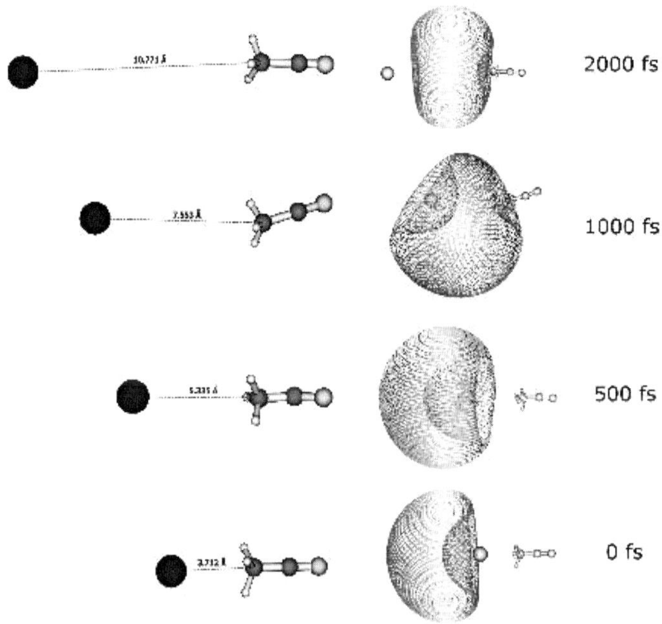

FIGURE 2. Snapshots of the relaxation dynamics of $I^-(CH_3CN)$ following photoexcitation along a typical trajectory.

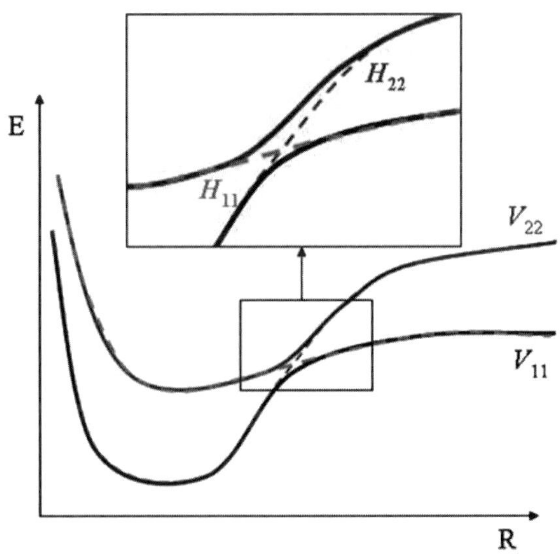

FIGURE 3. Schematic representation of a one-dimensional avoided crossing between two adiabatic electronic states E_i (red and blue curves), formed by the crossing of two diabatic states V_i (orange and green curves), along the internuclear coordinate R.

time-dependent wavefunction for two electronically-coupled (adiabatic or diabatic) states represented by ψ_1 and ψ_2 can generally be expressed as

$$\Psi(t) = [c_1(t)\psi_1 + c_2(t)\psi_2]\exp\left[-\frac{i}{2}\int^t (H_{11} + H_{22})\,dt'\right] \tag{3.1}$$

where $H_{ij} = \langle\psi_i|\widehat{H}|\psi_j\rangle$ [26]. The probability amplitudes c_i can be obtained by substituting this wavefunction in the time-dependent Schrödinger equation, Eq. (2.1), such that

$$\begin{aligned} i\frac{dc_1}{dt} &= \frac{1}{2}\Delta V c_1 + H_{12}c_2, \\ i\frac{dc_2}{dt} &= H_{12}c_1 - \frac{1}{2}\Delta V c_2 \end{aligned} \tag{3.2}$$

where $\Delta V(R) = H_{11} - H_{22}$ and H_{12} is the coupling-matrix element. In this case, the probability to undergo a transition from state 1 to state 2 is given as $P = |c_2(+\infty)|$. Equation (3.2), however, cannot be solved analytically, and many semiclassical and analytical models have been developed to approximate the transition probability [19], one of the most extensively used being the Landau–Zener model [23, 41]. For a one-dimensional problem, an analytic expression can be derived from (3.1) and (3.2), assuming that the two diabatic state energies change linearly in the curve-crossing region (as shown in the inset of Fig. 3), that the diabatic coupling term H_{12} is constant throughout time, and that the nuclei move at constant relative velocity. The resulting Landau–Zener probability P of undergoing a transition between two electronic states that cross at finite R is

$$P = \exp\left[\frac{-2\pi H_{12}^2}{\hbar\mathbf{v}|\mathbf{F}_1 - \mathbf{F}_2|}\right], \quad \mathbf{F}_i = \frac{-dH_i}{dR}, \tag{3.3}$$

where $\mathbf{v}$ is the relative velocity at which the nuclei are traveling along the internuclear coordinate R, and the $\mathbf{F}_i$ are the forces acting on the nuclei at the curve crossing point. The Landau–Zener model suggests that the nonadiabatic transition probability between states increases when (1) the avoided curve crossing occurs at large internuclear distances where the electronic coupling is usually smaller, (2) the nuclei travel at high velocities, and (3) the shape of the potential energy surfaces change suddenly in the region of the avoided crossing. Chemical reactions rarely proceed along a single coordinate, however, and many methods have been and are currently being developed for multi-dimensional problems, and to investigate the dynamics of nonadiabatic reactions [19].

The most direct approach to nonadiabatic dynamics is a full quantum mechanical treatment of the system of interest. This approach relies on directly solving the time-dependent Schrödinger equation, Eq. (2.1) [12, 19]. However, due to the complexity of solving this partial differential equation as mentioned earlier, a fully quantum mechanical treatment of nonadiabatic dynamics has been limited to systems containing at most 4 atoms. One of the most popular quantum mechanical methods to obtain accurate transition probabilities relies on time-independent quantum mechanical scattering theory [42], and the time evolution of the system is obtained by numerically solving the time-dependent Schrödinger equation by finite element methods [29], for example.

Since the full quantum mechanical treatment of nonadiabatic reaction dynamics is limited to very small systems, various mixed, or hybrid, quantum-classical techniques have also been developed, where the electrons involved in the reaction are treated quantum mechanically and the nuclei classically [19]. In general, the interactions between the nuclei are described by model potentials (cf. (2.7)), and at each time step, the electronic structure is evaluated for fixed nuclear positions and is used to determine on which electronic state the classical nuclei evolve. The main problem with decoupling the classical nuclei and the quantum electrons is in defining the quantum backreaction, that is the effect of the quantum degrees of freedom on the classical ones. One technique to treat the quantum backreaction relies on the De Broglie–Bohm or hydrodynamic interpretation of quantum mechanics [1, 2, 6, 7, 30], and is referred to as Bohmian dynamics. In Bohmian dynamics, the total wavefunction for the system is expressed in a polar coordinate form, and inserted into the time-dependent Schrödinger equation. The resulting equations couple the classical and quantum modes through a quantum force which acts on the classical degrees of freedom. Another equally promising and very recently developed hybrid approach is based on the Liouville formulation of quantum mechanics, which combines the commutator structure of the quantum subsystem with phase-space distribution functions of the classical subspace ([16] and references therein). Finally, another method to fully couple the quantum and classical degrees of freedom was developed by Pechukas a long time ago [27, 28], and applied only more recently by Xiao and Coker [39]. The Pechukas approach relies on the path integral formulation of quantum mechanics [14], which represents the probabilistic nature of the quantum world by considering all possible paths joining two points in phase space. Pechukas semiclassical theory is based on a reduced propagator that governs the propagation of the bath or the nuclear motion over different adiabatic states. This approach to nonadiabatic reaction dynamics has the main advantage of being much less computationally intensive than the previous quantum-classical techniques.

A number of simpler semiclassical trajectory approaches to simulate nonadiabatic dynamics have been developed that involve the complete separation of the propagation of nuclear degrees of freedom and that of the electronic wavefunction, and account for the quantum backreaction in an *ad hoc* fashion [19]. Although these methods are approximate, they offer the advantage of being computationally inexpensive, and can therefore be applied to large systems. These approaches can be separated into two general categories: the time-dependent self-consistent field (TDSCF) and trajectory surface hopping methods. A specific implementation of the latter method will be described in detail shortly. In the TDSCF approach, the classical nuclei evolve on an effective potential energy surface determined by the quantum subsystem, which is an average of the electronic states for the system. For this reason, these methods are often referred to as Mean Field or Ehrenfest techniques [19]. Propagating the nuclear motion on the mean potential surface, however, is only appropriate in the avoided crossing region where the coupling between adiabatic states is strong, but is unphysical in other regions of the potential energy surface.

Trajectory surface hopping methods allow for a more accurate representation of the electronic structure of the chemical system in all regions of the potential energy surfaces, because the nuclei evolve on individual adiabatic states, which are also determined by the quantum subsystem, rather than on an average of states as for the TDSCF method. They can also take full advantage of the first-principles molecular dynamics methods discussed above for the propagation of the nuclear degrees of freedom. A stochastic (probabilistic) algorithm is typically used to determine whether a switch (or hop) between states occurs. These methods offer the important advantage that the transitions are nonlocalized, since hops between adiabatic states can occur anywhere along the electronic surfaces, and they are known to describe well fast nonadiabatic processes.

One of the most elegant of these methods is the Molecular Dynamics with Quantum Transitions (MDQT) method developed by Tully and co-workers [15,38], which will be applied below to explore the photochemistry of sodium iodide (NaI) in water clusters. In the MDQT method, the time-dependent wavefunction Ψ of the system is expanded in terms of the electronic adiabatic states ψ as

$$\Psi(\mathbf{Q},\mathbf{q},t) = \sum_j C_j(t)\psi_j(\mathbf{Q},\mathbf{q}). \tag{3.4}$$

The wavefunction $\Psi(\mathbf{Q},\mathbf{q},t)$ is propagated together with the classical nuclear motion, and solution of the time-dependent Schrödinger equation results in the *classical path* equations of motion

$$i\hbar\dot{C}_j = \sum_j C_j(V_{kj} - i\hbar\dot{\mathbf{Q}} \bullet \mathbf{d}_{kj}) \tag{3.5}$$

where

$$V_{kj} = \langle\psi_k|\widehat{H}_e|\psi_j\rangle \tag{3.6}$$

are the Hamiltonian matrix elements over the adiabatic electronic states, and

$$\mathbf{d}_{kj} = \langle\psi_k|\nabla_{\mathbf{Q}}\psi_j\rangle \tag{3.7}$$

is the nonadiabatic coupling vector between the electronic states. The propagation of the quantum equations of motion, Eq. (3.5), reduces to a two-point boundary value problem (between times t and $t+\Delta t$), which can be solved by a second-order finite-difference equation numerical method [29].

The nonadiabatic transition (hopping) probability is defined as

$$g_{kl} = \frac{f_{kl}\Delta t}{\rho_{kk}}, \tag{3.8}$$

where Δt is the integration time step for the classical nuclear equations of motion, the density matrix ρ elements are obtained from the time-dependent wavefunction expansion coefficients C_j as

$$\rho_{kj} = C_k(t)C_j^*(t), \tag{3.9}$$

and the f_{kl} ingredients are related to the time evolution of the density matrix by

$$f_{kl} = 2\hbar^{-1}Im(\rho_{kl}^* V_{kl}) - 2\,\mathrm{Re}(\rho_{kl}^*\dot{\mathbf{Q}} \bullet \mathbf{d}_{kl}), \quad \dot{\rho}_{kk} = \sum_{l\neq k} f_{kl}. \tag{3.10}$$

A stochastic fewest-switches algorithm is used to determine whether a hop between states should occur [38]. This algorithm minimizes the number of switches between states, in order to ensure the correct statistical distribution of state populations, by comparing the transition probability with a randomly generated number ξ between 0 and 1 (a switch between states occurs if $g_{kl} \leq \xi$). To conserve the energy of the system when a transition between states occurs, the energy difference between the states is redistributed among the classical degrees of freedom along the nonadiabatic coupling vector, (3.7), as

$$\dot{\mathbf{Q}}_i = \dot{\mathbf{Q}}_i - \gamma_{kk'} \frac{\mathbf{d}_{kk'}}{m_i}, \tag{3.11}$$

where m_i is the mass of atom i, and

$$\gamma_{kk'} = \frac{b_{kk'} \pm \sqrt{b_{kk'}^2 - 4a_{kk'}[V_{kk}(\mathbf{Q}) - V_{k'k'}(\mathbf{Q})]}}{2a_{kk'}}, \quad a_{kk'} = \frac{1/2 \sum_i |\mathbf{d}_{kk'}|^2}{m_i} \quad \text{and} \quad b_{kk'} = \sum_i \dot{\mathbf{Q}}_i \bullet \mathbf{d}_{kk'}. \tag{3.12}$$

Following this equation, no transition can occur if $b_{kk'}^2 - 4a_{kk'}[V_{kk}(\mathbf{Q}) - V_{k'k'}(\mathbf{Q})] < 0$, which, in this case, gives rise to a so-called *frustrated* transition, as will be explained in more detail shortly.

Quantum decoherence is a reflection of how the classical bath around a quantum particle affects its behavior. An open quantum system, *i.e.* a system embedded within a bath of more massive particles for example, tends toward a mixture of states and the final state of the system is not a purely-defined quantum state. The system must, however, formally evolve on one adiabatic state, and not a mixture of states, a phenomenon that translates as the zeroing of the off-diagonal elements of the density matrix. In the MDQT method, each trajectory is propagated completely coherently, *i.e.* the values of the wavefunction coefficients are retained throughout the propagation. Many methods exist to account for the quantum decoherence of the system, such as the Decay of Mixing approach, where terms are added in order to dampen the off-diagonal density matrix elements along trajectories [19]. These methods require some *a priori* knowledge of the timescale on which quantum decoherence arises, which, despite much effort in the field, remains difficult to evaluate. The most straightforward strategy within the MDQT method is to propagate swarms of (or multiple) trajectories for each initial classical condition. The many possible outcomes of a given reaction (for a single classical initial condition) are therefore explored, ensuring a thorough mixture of final states over the ensemble average.

Despite its robustness, the MDQT method has a few pitfalls. One major problem concerns frustrated transitions inherent in the fewest-switches algorithm described earlier, which may occur when the momentum along the nonadiabatic coupling vector is insufficient to allow for a hop between states. To remediate this problem, a fewest-switches algorithm with time uncertainty has been implemented [18], in which the fewest-switches algorithm is reapplied to the system at one time step forward or backward in time, and the system is allowed to hop at either times according to the nonadiabatic transition probability. If a transition still doesn't occur, and the signs of the nuclear force and the nuclear momentum along the nonadiabatic coupling vector $\mathbf{d}_{kj}$ are the same, the frustrated transition is ignored; othermise the nuclear momentum is redistributed along the nonadiabatic coupling vector without hopping between states [15]. Another drawback of the MDQT method is that classically-forbidden regions of phase space are not accessible, and tunneling phenomena may be ignored. Finally, large ensembles of trajectories must be propagated in order to ensure a proper description of quantum decoherence, such that the MDQT method may become increasingly computational intensive with system size. However, the computational cost of MDQT simulations

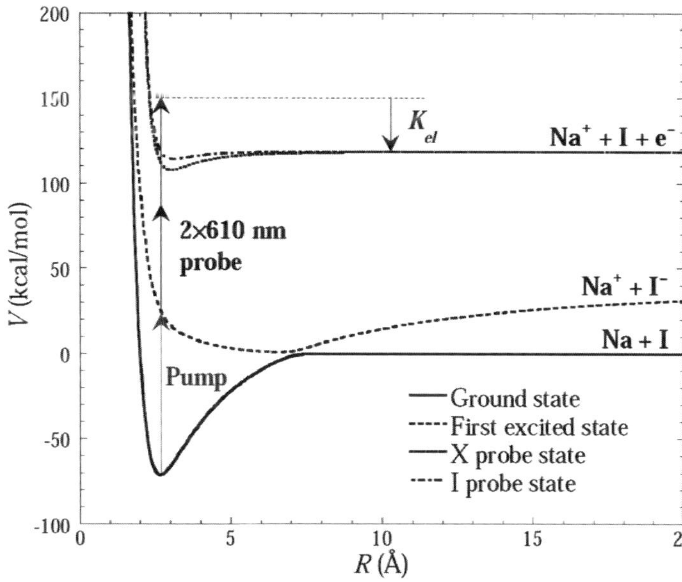

FIGURE 4. NaI potential energy curves for the ground and first excited adiabatic states, and probe state. Also illustrated is the two-photon (2×610nm) probe scheme.

does not grow exponentially with system size as traditional quantum mechanical methods do. Despite these drawbacks, the MDQT method remains a very elegant and efficient means to investigate ultrafast nonadiabatic reaction dynamics of large systems, such as that described below, while retaining conceptual simplicity, in that trajectories are propagated on a single electronic potential energy surface at a time.

3.2. Application of the MDQT method: NaI photochemistry in water clusters. Sodium iodide (NaI) has long served as a prototype system for the investigation of electron transfer reactions because it is characterized by a curve crossing between ionic and covalent diabatic curves at a small NaI internuclear distance of 7 Å, where the electronic coupling is large enough to govern in part the adiabatic motion. For these reasons, NaI was chosen as a paradigm by Zewail and co-workers during the development of femtosecond spectroscopy, for which Zewail was awarded the 1998 Nobel Prize in Chemistry [40]. In femtosecond experiments of NaI, the excited-state population evolution, subsequent to photoexcitation from the ground state, could be monitored in time by promoting the freshly excited NaI to a probe state, whose population can in turn be monitored by laser-induced fluorescence of the sodium D line [40]. Alternatively, the excited-state population can be monitored by promoting NaI to an ionized probe state, as depicted in Fig. 4, and collecting sodium ions by mass spectrometry [21]. These detection schemes give rise to oscillatory signals — because of the transient trapping of the NaI population in the excited-state well resulting from the crossing of the ionic and covalent states — with decreasing amplitude, because of the leakage of excited-state population to form dissociated atoms via nonadiabatic transitions. Many previous theoretical studies of gas-phase NaI have been performed to aid in the interpretation of the

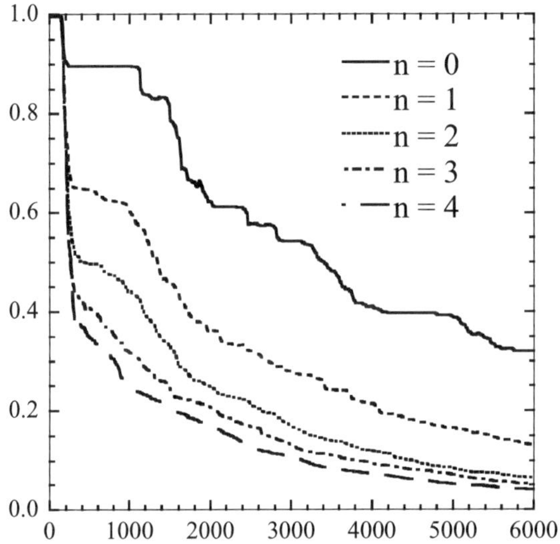

FIGURE 5. Excited-state population as a function of simulation time obtained from MDQT simulations of $NaI(H_2O)_n$ clusters of size 0 (solid), 1 (dash), 2 (dot), 3 (dash-dot) or 4 (thin solid).

experimental probe signals ([5, 13], and references therein). Our interests lie in solvation effects on the NaI photodissociation dynamics since polar solvents, such as water, are likely to significantly affect the NaI ionic state, and consequently its photodissociation dynamics. In fact, Marcus showed that solvation plays a key role in electron transfer reactions, and the electron transfer reaction may proceed differently in a liquid solvent than in the gas phase [24]. The objective of this theoretical study is to investigate the effects of solvation on the NaI electron transfer process by gradually increasing the number of solvent molecules n surrounding NaI [22].

The MDQT method discussed above is employed to simulate NaI photodissociation dynamics in water clusters. A semiempirical valence bond (VB) approach is used to describe the electronic structure of NaI, in which only two valence electrons are treated explicitly, the 3s electron of Na and the unpaired 5p electron of I. The NaI ionic state is characterized by both valence electrons in the iodine orbital, whereas the valence electrons are distributed in the Na and I orbitals in the covalent state. Within VB theory, the adiabatic wavefunction for the system is expressed as a linear combination of the two states, $\psi = \sum_i C_{ij}\Phi_i$, where $\Phi_1 = N_1|\overline{X}X|$ for the ionic state and $\Phi_2 = N_2(|\overline{M}X| - |M\overline{X}|)$ for the covalent state, with N_i normalization constants, X the 5p orbital of I and M the 3s orbital of sodium. Furthermore, the solvent-solvent and the solute-solvent interactions are treated with model potentials, where the water molecule is represented by the TIP4P model and Optimized Parameters for Liquid Simulations (OPLS) are employed for the solute-solvent interactions [20]. The parameters in the semiempirical VB approach and in the solute-solvent interaction models were adjusted to reproduce properties

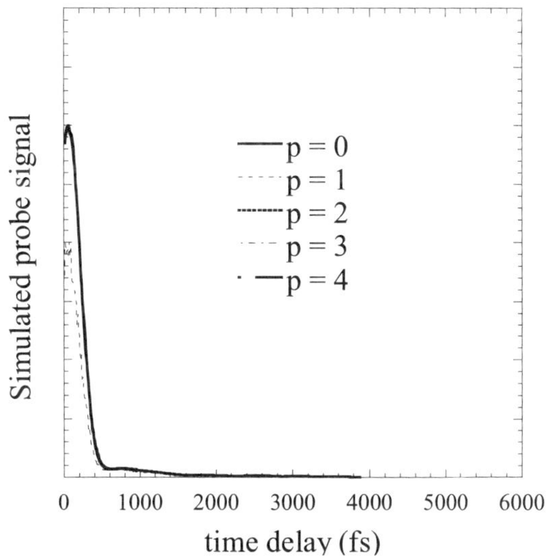

FIGURE 6. Simulated $Na^+(H_2O)_p$ probe signals, where p designates the product cluster size after water evaporation, for an initial cluster size of 4 with the two-photon (2×610nm) probe scheme.

predicted by high-level quantum chemistry, and the MDQT approach was benchmarked by simulating the femtosecond probe signals for gas-phase NaI, which were found to be in very good agreement with experiment [22].

Figure 5 displays the $NaI(H_2O)_n$ $[n = 0 - 4]$ excited-state population decay as a function of simulation time, averaged over $2000 \times n$ trajectories. The graph clearly demonstrates that the excited-state population decay increases significantly with cluster size when NaI reaches the curve-crossing region for the first time (at around 200 fs). The enhanced decay is due to the greater stabilization of the NaI ionic branch of the excited state, relative to the covalent branch, due to the presence of surrounding water molecules. Consequently, the curve-crossing region will shift towards larger NaI internuclear separations due to solvation (*cf.* Fig. 4). Since the curve crossing occurs at larger NaI internuclear separations in the presence of water molecules and the electronic coupling decreases exponentially with increasing internuclear separation, the nonadiabatic transition probability increases and the excited-state population decay is enhanced accordingly. The long tail of the cluster excited-state population, on the other hand, parallels that of bare NaI for all cluster sizes, because of water evaporation from the cluster following photoexcitation and energy redistribution within the cluster [22]. The simulated probe signals employing the 2×610 nm scheme, illustrated in Fig. 4 and shown in Fig. 6, were found to be in good agreement with experimental results [21]: the probe signals decay exponentially such that the NaI excited-state dynamics, *i.e.* the NaI excited-state vibrational motion and population decay, are no longer observable beyond the first NaI excited-state vibrational period, as shown in Fig. 6. Simulation results demonstrate that the disappearance of the probe signals is caused by the enhanced stabilization of the NaI ionic state in the presence of water molecules, and to the resulting enhancement in the nonadiabatic transition probability [22].

4. Conclusion

Simulations of chemical reaction dynamics remain a great challenge for computational chemists, mainly because the time-dependent Schrödinger equation, which describes the time evolution of quantum systems, is usually a high-dimensional partial differential equation that can hardly be solved in practice. Computational chemists therefore rely on a hierarchy of approximations to describe reaction dynamics, and much work is still required in this field to obtain more accurate and computationally affordable techniques. In these proceedings, various techniques to simulate realistic chemical reaction dynamics have been reviewed. Traditional and first-principles Born–Oppenheimer molecular dynamics techniques were first presented, and the use of first-principles molecular dynamics simulations was illustrated with the investigation of the relaxation dynamics on the charge-transfer-to-solvent excited state of the $I^-(CH_3CN)$ complex. In the second part, various approaches to nonadiabatic dynamics, where multiple electronic states are involved, were discussed. One of the most efficient and chemically intuitive techniques, the Molecular Dynamics with Quantum Transitions (MDQT) approach was described in more detail, and investigation of the electron transfer reaction in $NaI(H_2O)_n$ clusters following photoexcitation served as an example to illustrate its applicability.

References

1. D. Bohm, *A suggested interpretation of the quantum theory in terms of "hidden" variables. I*, Phys. Rev. **85** (1952), 166–179
2. ———, *A suggested interpretation of the quantum theory in terms of "hidden" variables. II*, Phys. Rev. **85**, 180-193.
3. K. Bolton, W.L. Hase, and G.H. Peslherbe, *Direct dynamics simulations of reactive systems*, Multidimensional Molecular Dynamics Methods (D.L. Thompson, ed.), World Scientific Publishing Co., Hackensack, 1998.
4. S.E. Bradforth and P. Jungwirth, *Excited States of Iodide Anions in water: a comparison of the electronic structure in clusters and in bulk solution*, J. Phys. Chem. A **106** (2002), 1286–1298.
5. M. Braun, C. Meier, and V. Engel, *The reflection of predissociation dynamics in pump/probe photoelectron distributions*, J. Chem. Phys. **105** (1996), 530–534.
6. L. deBroglie, *La structure atomique de la matière et du rayonnement et la mécanique ondulatoire*, C. R. Acad. Sci. Paris **183** (1926), 447–448.
7. ———, *TITLE ????*, C. R. Acad. Sci. Paris **185** (1927), 580–???.
8. C.J. Cramer, *Essentials of computational chemistry: theories and models*, John Wiley & Sons, Ltd, West Sussex, 2002.
9. R.A. Crowell, R. Lian, I.A. Shkrob, D.M. Bartels, X. Chen, and S.E. Bradforth, *Ultrafast dynamics for electron photodetachment from aqueous hydroxide*, J. Chem. Phys. **120** (2004), 11712–11725.
10. C.E. Dessent, H., C.G. Bailey, and M.A. Johnson, *Observation of the dipole-bound excited state of the iodide-scetone ion-molecule complex*, J. Chem. Phys. **102** (1995), 6335–6338.
11. C.E. Dessent, H., C.G. Bailey, and M.A. Johnson, *Dipole bound excited states of the* $I^-(CH_3CN)$ *and* $I^-(CH_3CN)_2$ *ion-molecule complexes: evidence for asymmetric solvation*, J. Chem. Phys. **103** (1995), 2006.
12. N.L. Doltsinis and D. Marx, *First principles molecular dynamics involving excited states and nonadiabatic transitions*, J. Theor. Chem. **1** (2002), 319–349.

13. V. Engel and H. Metiu, *The study of* NaI *predissociation with pump-probe femtosecond laser pulses: the use of an ionizing probe pulse to obtain more detailed dynamic information*, Chem. Phys. Lett. **155** (1985), 77–82.
14. R.P. Feynman and A. Hibbs, *Quantum mechanics and path integrals*, McGraw Hill, New York, 1965.
15. S. Hammes-Schiffer and J.C. Tully, *Proton transfer in solution: molecular dynamics with quantum transitions*, J. Chem. Phys. **101** (1994), 4657–4667.
16. G. Hanna and R. Kapral, *Nonadiabatic dynamics of condensed phase rate processes*, Acc. Chem. Res. **39** (2006), 21–27.
17. J.O. Hirschfelder and W.J. Meath, *Intermolecular forces*, Adv. Chem. Phys. **12** (1967), 3–106.
18. A.W. Jasper, S.N. Stechmann, and D.G. Truhlar, *Fewest-switches with time uncertainty: a modified trajectory surface-hopping algorithm with better accuracy for classically forbidden electronic transitions*, J. Chem. Phys. **116** (2002), 5424–5431.
19. A.W. Jasper, C. Zhu, S. Nangia, and D.G. Truhlar, *Introductory lecture: Nonadiabatic effects in chemical dynamics*, Faraday Discussions **127** (2004), 1–22.
20. W.L. Jorgensen, J. Chandrasekhar, J.D. Madura, R.W. Impey, and M.L. Klein, *Comparison of simple potential functions for simulating liquid water*, J. Chem. Phys. **79** (1983), 926–935.
21. C. Jouvet, S. Martrenchard, D. Solgadi, C. Dedonder-Lardeux, M. Mons, G. Grégoire, I. Dimicoli, F. Piuzzi, J.P. Visticot, J.M. Mestdagh, P. D'Oliveira, P. Meynadier, and M. Perdrix, *Experimental femtosecond photoionization of* NaI, J. Phys. Chem. A **101** (1997), 2552–2554.
22. D.M. Koch, Q.T. Timerghazin, G.H. Peslherbe, B.M. Ladanyi, and J.T. Hynes, *Nonadiabatic trajectory studies of* $NaI(H_2O)_n$ *photodissociation dynamics*, J. Phys. Chem. A **110** (2006), 1438–1454.
23. L.D. Landau, *TITLE ???*, Physik Z. Sowjetunion **2**, 46–51.
24. R.A. Marcus, *On the theory of oxidation-reduction reactions involving electron transfer*, J. Chem. Phys. **24** (1956), 966–978.
25. D.A. McQuarrie, *Quantum chemistry*, University Science Books, Mill Valey, 1983.
26. H. Nakamura, *Nonadiabatic dynamics: concepts, basic theories, and applications*, World Scientific, Singapore, 2002.
27. ———, *Time-dependent semiclassical scattering theory.* I. *Potential scattering*, Phys. Rev. **181** (1969), 166–174.
28. P. Pechukas, *Time-dependent semiclassical scattering theory.* II. *Atomic collisions*, Phys. Rev. **181** (1969), 174–185.
29. W.H. Press, S.A. Teukolsky, W.T. Vetterling, and B.P. Flannery, *Numerical recipes: the art of scientific computing*, Cambridge University Press, Cambridge, 1992.
30. O.V. Prezhdo and C. Brooksby, *Quantum backreaction through the Bohmian particle*, Phys. Rev. Lett. **86** (2001), 3215–3219.
31. W.H. Robertson and M.A. Johnson, *Molecular aspects of Halide ion hydration: the cluster approach*, Annu. Rev. Phys. Chem. **54** (2003), 173–213.
32. D. Serxner, C.E.H. Dessent, and M.A. Johnson, *Percursor of the* I^-aq CTTS *band in* $I^-(H_2O)_n$ *clusters*, J. Chem. Phys. **105** (1996), 7231–7234.
33. J.J.P. Stewart, L.P. Davis, and L.W. Burggraf, *Semi-empirical calculations of molecular trajectories: method and applications to some simple molecular systems*, J. Comput. Chem. **8** (1987), 1117–1123.
34. Q.K. Timerghazin and G.H. Peslherbe, *Theoretical investigation of charge transfer to solvent in Photoexcited iodide-acetonitrile clusters*, Chem. Phys. Lett. **354** (2002), 31–37.
35. ———, *Further insight into the relaxation dynamics of photoexcited* $I^-(H_2O)_n$ *clusters*, J. Am. Chem. Soc. **125** (2003), 9904–9905.
36. ———, *Photoexcitation and charge-transfer-to-solvent relaxation dynamics on the* $I^-(CH_3CN)$ *complex*, J. Phys. Chem. A (2006) (to appear).
37. J.C. Tully, *Dynamics of molecular collisions. Part* B, (W.H. Miller, ed.), Plenum, New York. 1976, pp. 1–17.
38. J.C. Tully, *Molecular dynamics with electronic transitions*, J. Chem. Phys. **93** (1990), 1061–1072.
39. L. Xiao and D.F. Coker, *Methods for molecular dynamics with nonadiabatic transitions*, J. Chem. Phys. **102** (1995), 496–511.
40. A.H. Zewail, *Femtochemistry*, World Scientific Publishing, Singapore, 1994.

41. C. Zener, Non-adiabatic crossing of energy levels, Proc. Roy. Soc. Lond. A **137** (1932), 696–702.
42. J.Z.H. Zhang, *Theory and applications of quantum molecular dynamics*, World Scientific Publishing, Singapore, 1999.

Centre for Research in Molecular Modeling (CERMM) and Department of Chemistry and Biochemistry, Concordia University, Richard J. Renaud Science Complex, 7141 Sherbrooke St. West, Montréal, Québec, H4B 1R6,Canada.

E-mail address, D.M. Koch: denise@arsenic.concordia.ca

E-mail address, Q.K. Timerghazin: Qadir.Timerghazin@CERMM.concordia.ca

E-mail address, G.H. Peslherbe: ghp@alcor.concordia.ca

Centre de Recherches Mathématiques
CRM Proceedings and Lecture Notes
Volume **41**, 2007

A Maxwell–Schrödinger Model for Non-perturbative Laser-molecule Interaction and Some Methods of Numerical Computation

Emmanuel Lorin, Szczepan Chelkowski, and André Bandrauk

ABSTRACT. We present in this paper a numerical Maxwell–Schrödinger model to simulate intense ultrashort laser pulses interacting with a H_2^+-gas in the nonlinear non-perturbative regime. After presentation of the model and a short mathematical study, we propose a numerical approach for its computation. In particular we focus on the polarization computation allowing the coupling between the Maxwell and Schrödinger equations, and on the boundary conditions problem for the time dependent Schrödinger equations, TDSE. Some comparisons with existing methods are also addressed.

1. Introduction

From the theoretical point of view to the very practical point of view (controlled fusion by inertial confinement [13], quantum dynamic imaging [6], attosecond pulse generation [1], filamentation [27]), there exist numerous applications of ultrashort and intense laser pulses. Indeed current laser technology allows to create ultrashort pulses with intensities exceeding molecular and atomic internal electric fields. Typically, for the hydrogen atom the period of electron circulation is 24.6×10^{-18}s (attosecond) and the intensity of its Coulomb field is 3.5×10^{16}W/cm^2. Current laser intensities can reach around 10^{20}W/cm^2 with pulse durations of ~ 150 attoseconds. The main goal of this work is to study high order non-perturbative nonlinear phenomena: ATI (Above Threshold Ionization), HOHG (High Order Harmonic Generation) [12], obtained with very intense lasers interacting with molecules and to study their dynamics. HOHG is one of these nonlinear phenomena that appears when the electric field is greater than the Coulomb potential field and leads to ionization. It is, moreover, the current source of coherent "attosecond" pulses [1, 11]. In a first step the electron is excited by an intense electric field and then leaves through tunnelling the ion vicinity and enters in the ionized continuum with an initial velocity equal to zero [12]. Then the free electron is accelerated by the strong electric field and gains energy. In a last step the electron is driven back to the vicinity of the parent ion and recombines with it, leading to the creation of

2000 *Mathematics Subject Classification.* Primary: 65Z05, 81V10; Secondary: 65Y05, 81V55.
This is the final form of the paper.

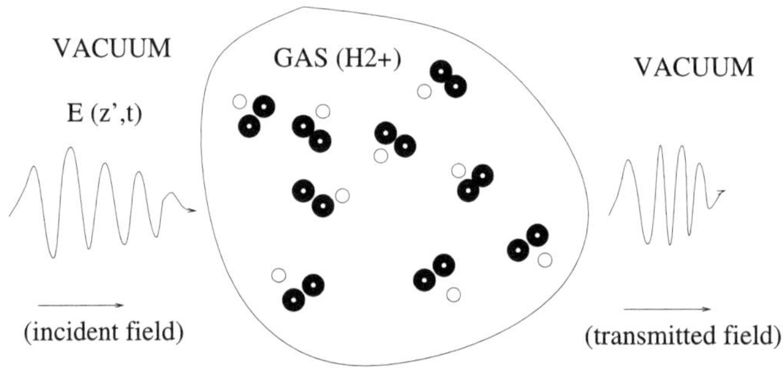

FIGURE 1. Laser-molecule interactions in a H_2^+ gas.

high order harmonics (see [24] for description of HOHG or [17] for their numerical aspects).

In this paper we will introduce a precise numerical model to describe this dynamics, including ionization, dissociation of the molecule and including propagation effects in order to take into account phase matching [12]. We propose here to study ultrashort pulses interacting with one electron H_2^+-molecules. To model the nonlinear response, the most natural way is to consider a coupling between the Maxwell equations describing the behavior of the electric and magnetic fields and the time dependent Schrödinger equation (TDSE) describing the electron motions and ionization. In order to increase the precision of the model, we propose to go beyond the Born–Oppenheimer approximation taking into account the proton motions. The Maxwell and the Schrödinger equations are then coupled with the polarization of molecules that describes the relative position of all the particles constituting the molecule. The model is written first in its whole generality and then some approximations are proposed in order to solve it numerically. Note that the model we propose although close to the atomic models presented in [28, 29] (see also [15, 16]) includes both electron and nuclear motion. The present work is the first to treat molecules completely.

In Section 2. we will present our model and we will briefly evoke some other existing ones. Then we will focus on the existence of solutions for the TDSE we consider. Finally in a last part we will give a simple presentation of the numerical method we have used to solve the coupled system. In particular we will detail the polarization computation (and how to reduce it in CPU time) and on the numerical boundary conditions for the TDSE, crucial in this framework. In [25] an exhaustive presentation of both the model and numerical approach will be proposed in a real 3-D configuration.

2. Mathematical Modeling

There exist many models describing the laser-matter interaction problems described above. The simplest ones are the *nonlinear Maxwell models*, and consist in calculating the polarization as an expansion of the susceptibility (linear, quadratic, cubic, and so on):

$$P(z',t) = \chi^{(1)} \cdot E + \chi^{(2)} \cdot E^2 + \chi^{(3)} \cdot E^3 + \cdots, \tag{2.1}$$

where $\chi^{(i)}$ are the nonlinear polarizabilities (susceptibilities) [12]. Some slowly varying envelope (SVE) versions leading to nonlinear TDSE, also exist. These models allow for example to simulate the behavior of low intensity lasers with low order harmonic generation. As the transmitted fields possess high order harmonics at high intensities these models are in fact not valid here as they do not include ionization. The Maxwell–Bloch equations constitute a more precise model where the polarization is obtained from the TDSE *via* the Bloch equation coupled with the Maxwell equations. The polarization is then deduced from the Bloch equation (ODE for the density matrix).

Unfortunately with such a model it is *a priori* not possible to include the continuous part of the Hamiltonian (ionization) and therefore it is not adapted to compute high order harmonics with very intense laser pulses (typically of the order of the Coulomb potential intensity). Such a strong approximation induces then a "loss of informations" hard to evaluate. Details can be found in [26].

The Maxwell–Schrödinger model we consider is then much more precise than the usual models briefly presented above in the sense that it considers much more physical phenomena in the non-perturbative regime where the continuous spectrum must be included. Furthermore, we propose to include nuclear motion thus allowing to go beyond the Born–Oppenheimer approximation and take into account dissociative ionization [7].

2.1. Presentation of the Maxwell–Schrödinger model. The 3-D TDSE describing the H_2^+ molecule response excited by a laser field is given by [7]:

$$i\partial_t\psi(\mathbf{r},R,\mathbf{r}',t) = \Big[-\frac{1}{2}\triangle_{\mathbf{r}} + V_c(\mathbf{r},R) - \frac{i}{c}\mathbf{A}(\mathbf{r}',t)\cdot\boldsymbol{\nabla}_{\mathbf{r}} + V_i(R) - \frac{1}{m_p}\partial^2_{RR} + \frac{\|\mathbf{A}(z',t)\|^2}{2c^2}\Big]\psi(\mathbf{r},R,\mathbf{r}',t), \tag{2.2}$$

where ψ represents the wave function, m_p denotes the H^+ mass. The Coulomb and nuclear potentials are given by:

$$V_c(\mathbf{r},R) = -\frac{1}{\sqrt{x^2+(y-R/2)^2+z^2}} - \frac{1}{\sqrt{x^2+(y+R/2)^2+z^2}}, \quad V_i(R) = \frac{1}{R}. \tag{2.3}$$

We use atomic units, where $e = h/2\pi = m_e = 1$ and $m_p = 1837$ and $c = 137$. The electron position in the ion center of mass coordinates is denoted by $\mathbf{r} = (x,y,z)^T$. R represents the proton relative position. We assume in this paper that the ion motion R, is 1-D. The term $\mathbf{A}\cdot\boldsymbol{\nabla} + \|\mathbf{A}\|/2c^2$ denotes the field interaction, $\mathbf{A}$ being the vector potential, and $\mathbf{r}'$ denotes the field space variable. In the following we will suppose that the electric field propagation is 1-D (in the direction z') and polarized in the direction y, so that the TDSE becomes (see Fig. 2):

$$i\partial_t\psi(\mathbf{r},R,z',t) = \Big[-\frac{1}{2}\triangle_{\mathbf{r}} + V_c(\mathbf{r},R) - \frac{i}{c}A_{y'}(z',t)\partial_y + V_i(R) - \frac{1}{m_p}\partial^2_{RR} + \frac{|A_{y'}(z',t)|^2}{2c^2}\Big]\psi(\mathbf{r},R,z',t). \tag{2.4}$$

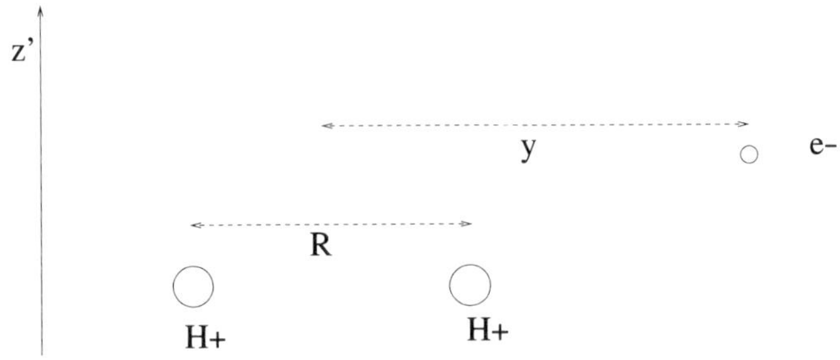

FIGURE 2. Coordinates.

Note that in our coordinates, $\mathbf{e}_y \cdot \mathbf{e}_{z'} = 0$, $\mathbf{e}_z = \mathbf{e}_{z'}$, $\mathbf{e}_y = \mathbf{e}_{y'}$ and $A_{y'}$ is the y'-component of $\mathbf{A}$. Now, if the laser wavelength λ, is large enough we can assume that the electric and magnetic fields are constant in space at the molecular scale ℓ, $(\lambda/\ell \ll 1)$, allowing us to reduce the numerical and mathematical complexity of the problem. Indeed, in this situation the TDSE can be written in its dipole form[1]:

$$(2.5)\quad i\partial_t\psi_{z'}(\mathbf{r},R,t) = \Big[-\frac{1}{2}\triangle_{\mathbf{r}} + V_c(\mathbf{r},R) - \frac{iA_{y',z'}(t)}{c}\partial_y + V_i(R) - \frac{1}{m_p}\partial^2_{RR} + \frac{|A_{y',z'}(t)|^2}{2c^2}\Big]\psi_{z'}(\mathbf{r},R,t),$$

z' is now a parameter denoting the molecule position in the "Maxwell domain". For each z' in the gas, one has to solve a 4D-TDSE (3-D for the electron and 1-D for the proton). Let us remark that the dipole approximation leads to $[\triangle_{\mathbf{r}}, A_{y',z'}(t)\partial_y] = [\partial^2_{RR}, A_{y',z'}(t)\partial_y] = 0$; very useful feature to have a free error numerical splitting method based on this commutation.

The above TDSE is given is the so-called *velocity gauge* where the electron-field interaction is through the electron momentum $p_y = i\partial/\partial_y$. Another possible formulation is in the so-called *length gauge*, see [7]. Setting

$$(2.6)\qquad \psi \mapsto \psi\exp\left(\frac{i}{2c^2}\int_0^t |A_{y',z'}|^2(s)\,ds + \frac{i}{c}yA_{y',z'}(t)\right),$$

we get a new formulation of the TDSE:

$$(2.7)\quad i\partial_t\psi_{z'}(\mathbf{r},R,t) = \Big[-\frac{1}{2}\triangle_{\mathbf{r}} + V_c(\mathbf{r},R) + yE_{y',z'}(t) + V_i(R) - \frac{1}{m_p}\partial^2_{RR}\Big]\psi_{z'}(\mathbf{r},R,t).$$

with $cE_{y',z'}(t) = -\partial A_{y',z'}(t)/\partial t$ denoting the electric field. The interaction with the electric field is now through the electron dipole y.

Ideally the electric field dynamics modeling is given by the microscopic Maxwell equations where the molecular density $n = n(\mathbf{r}', t)$, coupled with the TDSE and will be done in a future work. A full description of the microscopic Maxwell equations can be found in [20].

An approximate approach is proposed in [22], where the authors study the microscopic Maxwell equations but coupled with classical dynamics equations to

[1]The notation $A_{y',z'}$ denotes that z' is a parameter of the y'-component of $\mathbf{A}$.

describe the particle motions. We will here consider the macroscopic Maxwell equations that correspond to a spatial average of the microscopic ones. These equations are typically valid in a domain equal or exceeding a size of 10^{-18}cm^3 with a sufficiently high molecular density (see again [20]). The macroscopic 3-D-Maxwell equations written in atomic units are:

$$\begin{cases} \partial_t \mathbf{B} = -c\nabla \times \mathbf{E}, \\ \partial_t \mathbf{E} = c\nabla \times \mathbf{B} - 4\pi\partial_t \mathbf{P}, \\ \nabla \cdot \mathbf{B} = 0, \\ \nabla \cdot (\mathbf{E} + 4\pi\mathbf{P}) = 0. \end{cases} \tag{2.8}$$

We consider here a 1-D propagation (direction $\mathbf{e}_{z'}$ orthogonal to $\mathbf{e}_y$) with c denoting the light velocity in the vacuum and we neglect the transversal effects (justified in a gas due to large inter-molecular distances). We then have $\mathbf{E} = E_{y'}\mathbf{e}_{y'}$, $\mathbf{P} = P_{y'}\mathbf{e}_{y'}$ and $\mathbf{B} = B_x\mathbf{e}_{x'}$. Denoting $E = E_{y'}$, $P = P_{y'}$ and $B = B_{x'}$, we have the simple equations (the fields are not independent of x' and y'):

$$\begin{cases} \partial_t B(z',t) = c\partial_t E(z',t), \\ \partial_t E(z',t) = c\partial_t B(z',t) - 4\pi\partial_t P(z',t). \end{cases}$$

Under the dipolar approximation the polarization P for a molecule located at z' is effectively parallel to $\mathbf{e}_{y'}$ and is given by:

$$P(z',t) = -n(z')\int \psi_{z'}(\mathbf{r},R,t)y\psi^*_{z'}(\mathbf{r},R,t)\,dR\,d\mathbf{r}, \tag{2.9}$$

and the dipolar acceleration is

$$d(z',t) = -n(z')\int \psi_{z'}(\mathbf{r},R,t)\left(\frac{\partial V}{\partial y} + E(t)\right)\psi^*_{z'}(\mathbf{r},R,t)\,dR\,d\mathbf{r}, \tag{2.10}$$

where $n(z')$ denotes the molecular density given for example for all x' and y', by:

$$n(z') = \begin{cases} 0, & \text{if } \ell < |z'| < L, \\ n_0 \sin^2\left(\dfrac{\pi(|z'|-\ell)}{2(a-\ell)}\right) & \text{if } a < |z'| < \ell, \\ n_0 & \text{if } |z'| < a, \end{cases} \tag{2.11}$$

where $n_0 \in \mathbb{R}^*_+$ (mol/cm^3). Such a density choice will allow us in particular to numerically reduce the reflections of the incoming electric field at the "frontiers" of the gas.

We assume now, that the Maxwell computational domain is given by $[-L,L]$, $L > 0$. We introduce some real constants ℓ, a such that $L > \ell > a > 0$, where the molecules are located in $[-\ell,\ell]$, and the vacuum region is $[-L,-\ell] \cup [\ell,L]$. Initially for the laser-pulse of frequency ω, the electric field is defined as follows:

$$\begin{cases} E(z',0) = E_0(z')f_\omega(z'), & z' \in [-L,-\ell],\ \omega \in \mathbb{R}^*_+, \\ E(z',0) = 0, & \text{elsewhere}. \end{cases} \tag{2.12}$$

Where $\int E(t)\,dt = 0$.

The function f, usually a sinusoidal function, and E_0, the envelope, is a Gaussian function $E_0(z') = e^{-c_z(z'-z_c)^2}$, where z_c is the molecule center of mass and c_z a positive constant depending on the pulse radius. In fact this function should depend on x' and y' ($E_0(x',y',z') = e^{-c_x(x'-x_c)^2 - c_y(y'-y_c)^2 - c_z(z'-z_c)^2}$) but in most

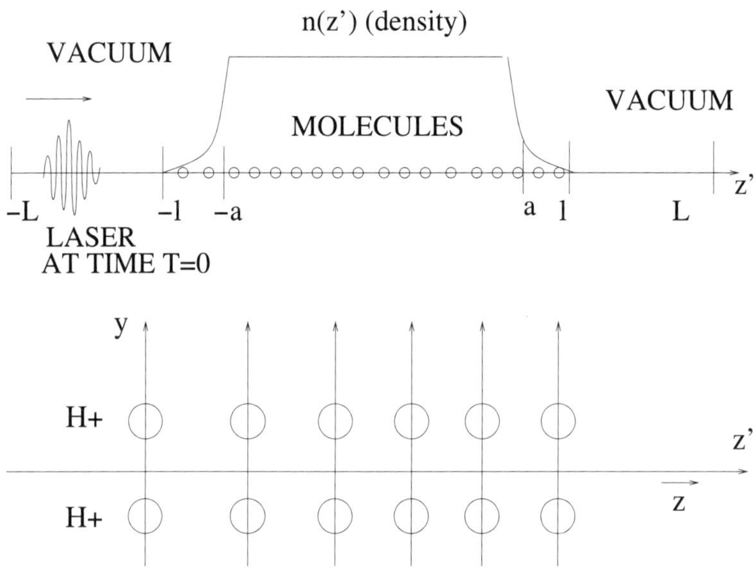

FIGURE 3. Physical configuration

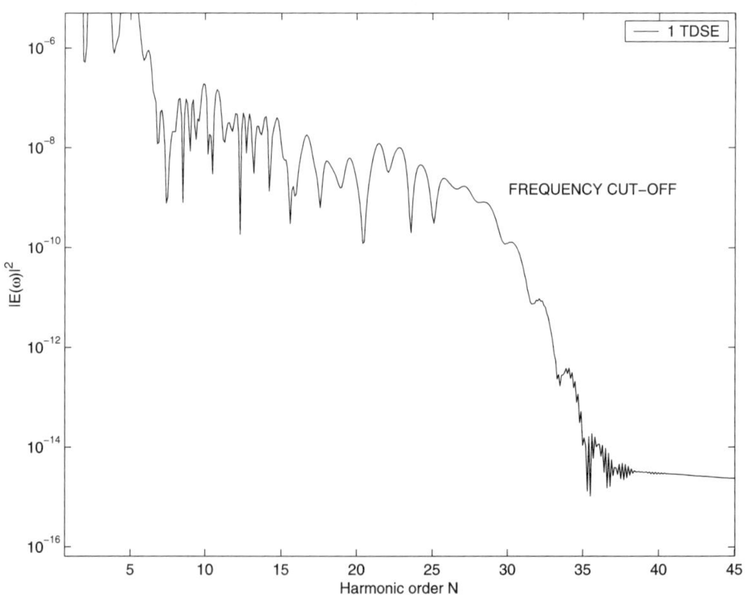

FIGURE 4. $|\widehat{E}(\omega)|^2$, for a single molecule, and a 5-cycle laser pulse at $I = 10^{14}\text{W/cm}^2$ and $\lambda = 800\text{nm}$ ($\omega_0 = 0.057$). In abscissa: harmonic order N, with $\omega = N\omega_0$, the harmonic frequency, and ω_0 the incident electric field frequency $R_0 = 3.2\text{a.u.}$

usual cases the beam is sufficiently large so that along the propagation axis z', we can again neglect the x' and y' dependencies.

The harmonic spectra of the transmitted field, denoted by E_T, possesses in theory [24], a frequency cut-off denoted by ω_c (see Fig. 4). This cut-off corresponds to the maximum return energy acquired by an electron in the field of $E(t)$, $N_m\omega = 3.17U_p + I_p$ [14]. It is then possible to filter this field around ω_c and to create a shorter and "intense" pulse denoted by E_F, thus leading to attosecond pulses through the uncertainty relation $\Delta\omega\Delta t \sim 1$ [1, 16]:

$$E_F(t) = \frac{1}{2\pi}\int_{\omega_c-\Delta\omega}^{\omega_c+\Delta\omega} \widehat{E}_T(\omega)e^{i\omega t}\,dt. \tag{2.13}$$

See for instance, [31] or [5] or [11] for the control aspects of such new pulses.

Note that, this problem constitutes typically a multiscale physical problem as the characteristic length and time of the TDSE are much smaller ($\|\mathbf{r}\| \sim 10^{-10}$m) than the Maxwell equations characteristic ones ($\lambda \sim 10^{-6}\text{m} = 1\mu$m).

2.2. Existence and regularity for the TDSE model. In this section we are interested in the existence and regularity of solutions for the TDSE (2.7). To do this, we will use the Fourier transform set of H^1:

$$H_1 = \left\{u \in L^2(\mathbb{R}^4),\ \int_{\mathbb{R}^4}(1+\|(\mathbf{r},R)\|^2)|u(\mathbf{r},R)|^2 d\mathbf{r}\,dR < \infty\right\} \tag{2.14}$$

From Baudouin, Kavian and Puel [9, 10], we can easily deduce that for $\mathcal{L} \in L^\infty(0,T;C_b^2(\mathbb{R}^4))$ and if $u_0 \in H^1 \cap H_1$, then there exists a unique solution u in $C^0(0,T;H^1\cap H_1)$ such that

$$\left(i\partial_t + \frac{\triangle_{\mathbf{r}}}{2} + \frac{\partial^2_{RR}}{m_p} + \mathcal{L}(\mathbf{r},R,t)\right)u(\mathbf{r},R,t) = 0, \quad u(\mathbf{r},R,0) = u_0(\mathbf{r},R). \tag{2.15}$$

Furthermore, for $K > 0$ such that

$$\|V\|_{L^\infty(0,T;C_b^2(\mathbb{R}^4))} \le K \tag{2.16}$$

then there exists $C_{T,K}$ such that

$$\|u\|_{C^0(0,T;H^1\cap H_1)} \le C_{T,K}\|u_0\|_{H^1\cap H_1}. \tag{2.17}$$

Considering now (2.7), let us suppose that for all $T > 0$, $E \in L^\infty(0,T)$ and $\partial_t E \in L^1(0,T)$. The laser field interaction $yE(t)$ we consider is *a priori* non-zero in the whole space as it is defined for all $y \in \mathbb{R}$. Physically it is obviously not true, so that we will consider a function χ defined as follows: $\chi\colon (\mathbf{r},R) \mapsto \chi(\mathbf{r},R)$ in $C_b^2(\mathbb{R}^4)$ and $\chi(\mathbf{r},R) = y$ on a compact set Ω_1 of $\mathbb{R}^4$ and $\chi(\mathbf{r},R)$ is zero outside a set Ω_2 containing strictly Ω_1. Such a function can easily be constructed by convolution of y and a plateau function. Then we can prove:

Theorem 2.1. *Let us consider the following* TDSE:

$$\begin{aligned} &i\partial_t\psi(\mathbf{r},R,t) \\ &\quad = \left[-\frac{1}{2}\triangle_{\mathbf{r}} + V_c(\mathbf{r},R) - \frac{1}{m_p}\partial^2_{RR} + V_i(R) - E(t)\chi(\mathbf{r},R)\right]\psi(\mathbf{r},R,t). \end{aligned} \tag{2.18}$$

$\forall T > 0$, suppose that $E \in L^\infty(0,T)$ and $\partial_t E \in L^1(0,T)$. Then there exists $C_T > 0$ such that, for all $\psi_0 \in H^1\cap H_1$, there exists a solution unique $\psi \in L^\infty(0,T;H^1\cap H_1)$ and

$$\|\psi(t)\|_{L^\infty(0,T;H^1\cap H_1)} \le C_T\|\psi_0\|_{H^1\cap H_1}. \tag{2.19}$$

PROOF. The proof is based an energy estimate in $H_1 \cap H^1$ *via* a Gronwall inequality and a compactness argument to extract a convergent subsequence. As it can be derived with some elementary modifications from the proof in Ref. [10, Theorem 2.1] and is very close to existing results, for instance [19], we do not present the proof. □

The existence of solutions for the coupled Maxwell–Schrödinger system has yet to be proved; in particular the regularity of the polarization P, equation (2.9) (linking the two systems) is useful in our numerical approach. We can expect that the polarization regularity on z' comes from the incoming electric field spatial regularity. Some interesting results about this coupling can be found for instance in [18], where the existence of global smooth and weak Maxwell–Bloch solutions is presented.

3. Some Issues on the Numerical Approach

We propose in this section to give a non exhaustive description of our numerical method to approximate a 1-D version of the Maxwell–Schrödinger system presented above. In this study the electronic variable becomes y, instead of $\mathbf{r}$ in 3-D. The extension to higher dimensions can be found in [25]. We will focus here on the polarization P computation that allows the coupling between the Maxwell and TDSE and on the boundary conditions for the TDSE.

The TDSE is approximated by a finite difference Crank–Nicolson scheme in time, and the Laplace operator is approximated using a 3-point stencil. Such a scheme, allows to preserve the ℓ^2-norm and is a second order scheme and is unconditionally stable. The Yee scheme is used to solve the Maxwell equations [32] that consists in a finite difference scheme where the electric E, and magnetic B, fields are computed on two spatial and temporal staggered grids. Under a Courant–Friedrichs–Lewy (CFL) condition, this is a stable and order two scheme (see for details [25]). At this point it is important to recall that to be valid the macroscopic Maxwell equations have to be applied on a sufficiently large domain (as it is obtained by a spatial average on microscopic Maxwell equations). Typically if we denote by $\Delta z'_M$ the Maxwell space step should be such that $\Delta z'_M \times n_0^{1/3}$ is much larger than 1 but also should be smaller than $\lambda_{\min}$, with $\lambda_{\min}$ corresponding to the highest harmonics created during the HOHG process. In each Maxwell cell, a large number of molecules is included, but in practice we solve only one TDSE to determine a local polarization. The problem has then some multiple scales as $\Delta z'_M$ is much larger than Δy_S where Δy_S is the Schrödinger grid space step. Also, Δt_M is much larger than Δt_S where Δt_S, Δt_M are respectively the Schrödinger and Maxwell solver time steps. In the numerical scheme approximating the Maxwell equations, the time step is taken, for stability reasons, equal to $D\Delta z'/c$, with D a positive constant. We then set

$$\overline{N} = E\left[\frac{\Delta t_M}{\Delta t_S}\right]. \tag{3.1}$$

Now denoting by $\mathcal{S}$ the Schrödinger operator and by $\mathcal{B}$ the Maxwell operator, the global coupled numerical scheme can be written as the following splitting scheme (here written at order 1 but computed at order 2, to preserve the Maxwell and

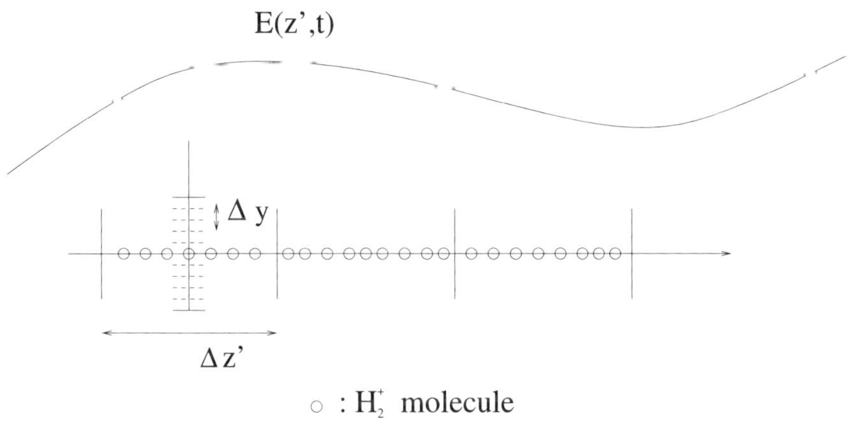

FIGURE 5. Maxwell and Schrödinger scales.

Schrödinger solvers order), from time t^n to time t^{n+1}:

$$(3.2)\quad \begin{cases} \mathbf{E}^{n+1} \leftarrow e^{\Delta t_M \mathcal{B}^n} \mathbf{E}^n, \ \forall z' \in [-L, L], \\ \qquad\qquad \text{solving the Maxwell equations,} \\ \psi^{n+1} \leftarrow e^{[\Delta t_M - (\overline{N}-1)\Delta t_S]\mathcal{S}^n} e^{(\overline{N}-1)\Delta t_S \mathcal{S}^n} \psi^n, \\ \qquad\qquad \text{solving the TDSE in each ``Maxwell cell''}, \\ \mathbf{P}^{n+1} \leftarrow \psi^{n+1}, \\ \qquad\qquad \text{by definition of the polarization(2.9).} \end{cases}$$

Globally, this then constitutes an order 2 scheme, stable under a CFL condition given by the Yee scheme (see [32]).

3.1. Polarization computation. The polarization computation is central in this work as it allows to couple the Maxwell equations and the TDSE. As said above the polarization is deduced from the TDSE by (2.9).

We note that for a 100 cycle laser pulse with a wavelength of 800nm interacting with molecules, we have to solve numerically at each iteration and in 1-D 8000 (80×100) TDSE's, for a Maxwell cell size equal to $10nm$ (one TDSE per Maxwell cell). As noticed above, even in 1-D, the TDSE we have to solve, because of the ion motions, are 2-D equations which is costly in CPU time. We therefore propose a technique based on a simple Taylor expansion that allows to reduce the numerical cost, assuming the polarization smooth in space. As discussed above, this assumption seems *a priori* valid when the electric field is smooth enough over dimensions (wavelength, λ) larger than molecular size (y). This satisfies also the dipole approximation, equation (2.5).

The idea is the following: we make a partition of the domain $[-\ell, \ell] = \cup_{i=1,\dots,I-1}[z_i', z_{i+1}'[$ with $z_1' = -l$ and $z_I' = l$ and I is small integer. We suppose that each sub-domain $[z_i', z_{i+1}'[$ contains a sufficiently large number of Maxwell cells. For each interval $[z_i', z_{i+1}'[$, we choose a reference cell $z_{i,0}'$ (located for example in the middle of the interval). For this cell we compute the corresponding TDSE which can be written in this framework:

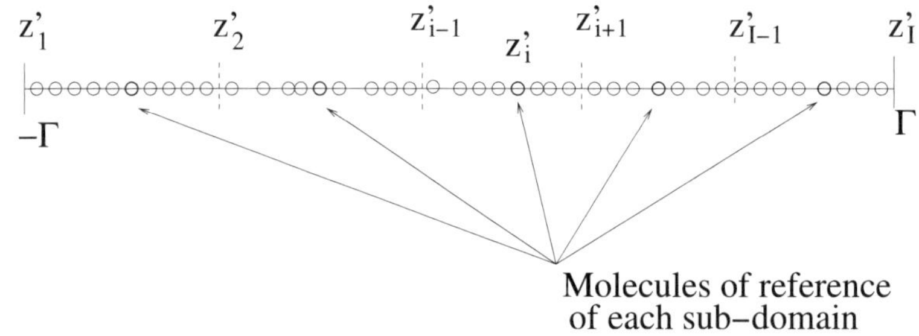

FIGURE 6. Decomposition of the domain.

$$i\partial_t\psi_{z'_{i,0}}(y,R,t)=\Big(-\frac{1}{2}\partial^2_{yy}+V(y,R)-\frac{1}{m_p}\partial^2_{RR}+\frac{1}{R}+yE_{z'_{i,0}}(t)\Big)\psi_{z'_{i,0}}(y,R,t). \tag{3.3}$$

and the corresponding polarization:

$$P(z'_{i,0},t)=-n(z'_{i,0})\int\psi_{z'_{i,0}}(y,R,t)y\psi^*_{z'_{i,0}}(y,R,t)\,dy\,dR. \tag{3.4}$$

An equivalent expression is obtained by using the acceleration form [14]. The present dipolar formulation is more convenient for performing a Taylor expansion as follows. Now for every other cell located in $z'\in[z'_i,z'_{i+1}[$ we have:

$$P(z',t)=P(z'_{i,0},t)+(z'-z'_{i,0})\partial_{z'}P(z'_{i,0},t)+\mathcal{O}\big((z'-z'_{i,0})^2\big). \tag{3.5}$$

Then we deduce the value of the polarization P, for every cell of the domain $[z'_i,z'_{i+1}[$. However to do this, it is necessary to compute $\partial_{z'}P(z'_{i,0},t)$. With this purpose, by differentiation in z' of (2.9), we get

$$\begin{aligned}\partial_{z'}P(z'_{i,0},t)=-n(z'_{i,0})&\int\partial_{z'}\psi_{z'_{i,0}}(y,R,t)y\psi^*_{z'_{i,0}}(y,R,t)\,dy\,dR\\ &-n(z'_{i,0})\int\psi_{z'_{i,0}}(y,R,t)y\partial_{z'}\psi^*_{z'_{i,0}}(y,R,t)\,dy\,dR\\ &-(\partial_{z'}n)(z'_{i,0})\int\psi_{z'_{i,0}}(y,R,t)y\psi^*_{z'_{i,0}}(y,R,t)\,dy\,dR.\end{aligned}$$

To compute $\partial_{z'}P(z'_{i,0},t)$ we have to compute $\partial_{z'}\psi_{z'_{i,0}}(y,R,t)$ then to solve the following equation obtained by derivation in z' of the TDSE:

$$\begin{aligned}&i\partial_t\big(\partial_{z'}\psi_{z'_{i,0}}(y,R,t)\big)\\ &=\Big(-\frac{1}{2}\partial^2_{yy}+V(y,R)-\frac{1}{m_p}\partial^2_{RR}+\frac{1}{R}+yE_{z'_{i,0}}(t)\Big)\partial_{z'}\psi_{z'_{i,0}}(y,R,t)\\ &\qquad+iy\big(\partial_{z'}E_{z'_{i,0}}(t)\big)\psi_{z'_{i,0}}(y,R,t).\end{aligned} \tag{3.6}$$

Numerically the computation of the previous equation is very close to the computation of the TDSE.

$$\begin{aligned}\big(\partial_{z'}\psi_{z'_{i,0}}\big)^{n+1}&=\frac{(\partial_{z'}\psi_{z'_{i,0}})^{n+1}}{2}+\frac{(\partial_{z'}\psi_{z'_{i,0}})^{n}}{2}\\ &-\frac{i}{2}H^{n+1}(\partial_{z'}\psi_{z'_{i,0}})^{n+1}-\frac{i}{2}H^n\partial_{z'}\psi^n_{z'_{i,0}}+iy(\partial_{z'}E_{z'_{i,0}})^{n+1}\psi^{n+1}_{z'_{i,0}}.\end{aligned} \tag{3.7}$$

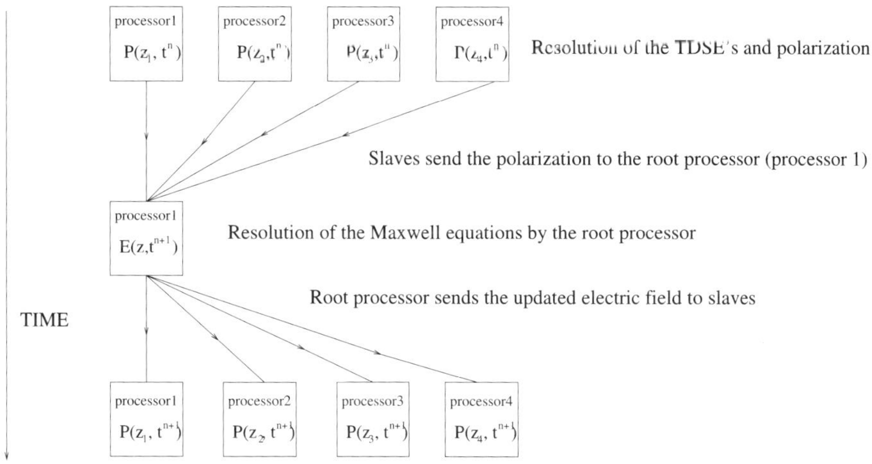

FIGURE 7. Parallelization for 4 TDSE's (molecules).

In (3.7) the quantity $\psi_{z'_{i,0}}^{n+1}$ is previously computed by the numerical scheme approximating the TDSE and with H^n naturally given by:

$$H^n = -\frac{1}{2}\partial_{yy}^2 + V(y, R) - \frac{1}{m_p}\partial_{RR}^2 + \frac{1}{R} + yE_{z'_{i,0}}^n, \quad n \geq 0. \tag{3.8}$$

Finally, for each sub-domain $[z'_i, z'_{i+1}[$ we compute $\psi_{z'_{i,0}}(y, R, t)$ and $\partial_{z'}\psi_{z'_{i,0}}(y, R, t)$ to deduce linearly from $P(z'_{i,0}, t)$ the polarization for each cell located in z' of this sub-domain. The error due to this process is naturally here of order one. It is possible to increase to the second order using a higher order Taylor expansion. In this case, it would be necessary to compute $\partial_{z'z'}^2\psi_{z'_{i,0}}(y, R, t)$ obtained by double differentiation in z' of the TDSE. We can then deduce easily the following result:

Proposition 3.1. *Assuming the polarization and the molecular density smooth enough, the approximation proposed in each $z' \in [z'_i, z'_{i+1}[$, for $z' \neq z'_{i,0}$ is of order 1 or 2 depending on the Taylor expansion order.*

In practice we can expect in some cases, a strong reduction of the number of TDSE's to solve. Typically the sub-domains size will depend on the transmitted electric field E_T harmonics, then on the intensity and frequency of the incoming laser. The higher the harmonics will be, the smaller the size of the sub-domains will be chosen.

Taking into account the previous method to reduce the computational cost of the polarization, that is the number of TDSE's to solve, we have the following temporal scheme from t^n to t^{n+1}, denoting by $I - 1$ the number of sub-domains, with I much smaller than the number of cells in the Maxwell grid:

$$
(3.9)\quad \begin{cases}
\mathbf{E}^{n+1} \leftarrow \mathbf{E}^{n}, \quad \forall z' \in [-L, L], \\
\psi^{n+1}_{z'_{i,0}} \leftarrow \psi^{n}, \quad \text{for } z'_{i,0} \in [z'_i, z'_{i+1}[, \quad \forall i \in [1, I], \\
\mathbf{P}^{n+1}_{z'_{i,0}} \leftarrow \psi^{n+1}_{z'_{i,0}} \quad \text{for } z'_{i,0} \in [z'_i, z'_{i+1}[, \quad \forall i \in [1, I], \\
\partial_{z'}\psi^{n+1}_{z'_{i,0}} \leftarrow \psi^{n+1}_{z'_{i,0}} \quad \text{and} \quad \text{eq. (3.6)}, \quad \text{for} \quad z'_{i,0} \in [z'_i, z'_{i+1}[, \quad \forall i \in [1, I], \\
\partial_{z'}\mathbf{P}^{n+1}_{z'_{i,0}} \leftarrow \partial_{z'}\psi^{n+1}_{z'_{i,0}} \quad \text{for } z'_{i,0} \in [z'_i, z'_{i+1}[, \quad \forall i \in [1, I], \\
\mathbf{P}^{n+1}_{z'} \leftarrow \mathbf{P}^{n+1}_{z'_{i,0}}, \quad (\partial_{z'} n)(z'_{i,0}) \quad \text{and} \quad \partial_{z'} \boldsymbol{P}^{n+1}_{z'_{i,0}}, \quad \forall z' \in [z'_i, z'_{i+1}[, \\
\qquad \text{and} \quad \forall i \in [1, I].
\end{cases}
$$

Remark on the parallelization of the polarization computation. Many approaches are possible to parallelize the Maxwell–Schrödinger equations; in particular in [25], we will present some possible issues. One of the most effective ones is as follows. To simplify the presentation let us suppose that the Maxwell-grid possesses N cells and that we solve N TDSE's (one per Maxwell-cell) with a code running on N processors. At each temporal Maxwell iteration, each processor solves one single TDSE, and computes the corresponding polarization P. Then it sends it to the root processor. The root processor "having" now the polarization in the whole physical domain, it solves the Maxwell equations (note that here our configuration the CPU cost of the Maxwell equations computation is negligible compared to the TDSE's computation cost). Then the root processor sends to the slaves the

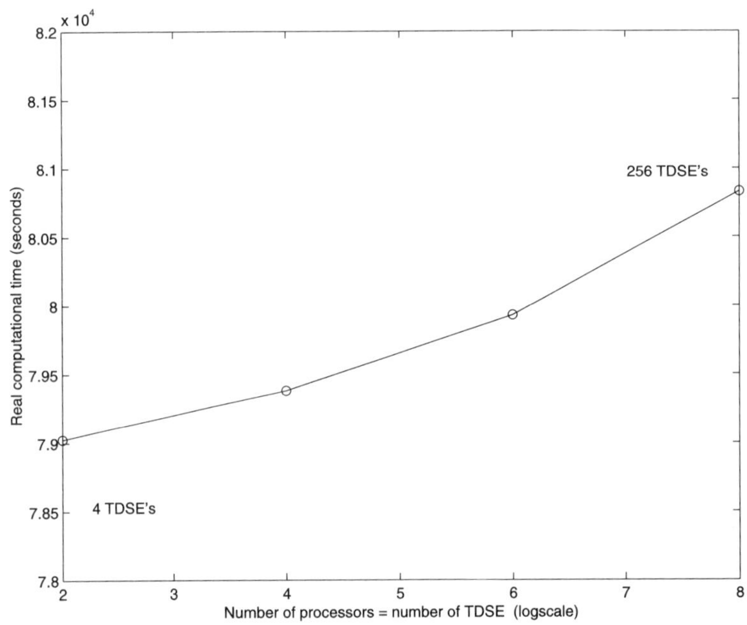

FIGURE 8. Computation of the 3-D Maxwell–Schrödinger equations with one TDSE (molecule) per processor. In ordinate the real time (in *seconds*) necessary for the numerical computation. Representation on log-scale.

updated electric field $E(t)$ (this process is summed-up in Fig. 7). Such a simple parallelization scheme allows us to obtain a very efficient speed-up. Note also that the data are distributed between the computer nodes allowing to consider a large number of molecules with sufficiently large spatial grids.

The Fig. 8 represents for a given mesh: in abscissa the number of TDSE's (4, 16, 64, 256 TDSE's) to solve also equal to the number of processors (one equation per processor) and in ordinate the real time for the 3-D code to solve the corresponding Maxwell–Schrödinger equations. The ideal configuration would be y=constant, corresponding to a speed-up equal to the number of processors. The real time of computation of our Maxwell–Schrödinger code with 256 TDSE's running on 256 processors is only 2.5% slower than the code running on 4 processors with 4 TDSE's.

The coupling of our proposed reduced computation of the polarization with this parallelization allows us to expect efficient simulations with a very large number of molecules. Note also that the parallelization is almost totally independent of the way we solve numerically the TDSE and Maxwell equations.

3.2. Boundary conditions for the TDSE. Let us now discuss the boundary conditions problem for the TDSE. The initial wavefunction (initial state) $\psi(\cdot, 0)$ support is located in the Schrödinger computational domain center. Then the laser field E interacts with the molecules and delocalizes the wavefunctions whose supports can become very large due to dissociative ionization into continuum states of both electrons and nuclei (beyond Born–Oppenheimer). Numerically it means that it is necessary to discretize the TDSE in a very large domain. To overcome this well-known problem in numerical scattering theory [6], we have to find an adapted method. The usual idea to circumvent this difficulty is to reduce the computational domain and to impose some particular numerical boundary conditions on the reduced domain. It is well-known that imposing Dirichlet or Neumann boundary conditions leads to important numerical oscillations and reflections on the boundary on the domain interacting with "physical" waves inside the domain. Absorbing boundary conditions are also used in order to absorb the numerical spurious reflections. Even if this kind of method allows effectively to reduce the spurious reflections, these are often empirical (see for instance [14] in this framework), as some "parameters" have to be adapted for each numerical situation. Moreover the spurious reflections can be made to vanish but a part of the wavefunction can also be partially or totally absorbed by the absorber. Outside the reduced domain the potential is supposed to be zero and the TDSE with laser interaction can then be solved "exactly" using for instance the Volkov state propagator [14]. As absorbing boundary conditions globally do not preserve the ℓ^2-norm of the wavefunction (as they absorb it) it can hardly be preserved. Ideally we would like to impose particular boundary conditions such that the solution of the whole space problem restricted to the compact domain (containing the solution support) is equal to the total solution on the compact domain (that is without spurious reflections). Then outside the reduced domain the wavefunction-solution is yet correct and can be updated using for instance the Volkov state propagator. To be more precise let us consider the following simplified model *without laser field*:

$$(3.10)\qquad \begin{cases} (i\partial_t u + \partial^2_{yy} u + V(y)\cdot u)(y,t) = 0, \\ u(y,0) = u_0(y), \quad u_0 \in H^1(\mathbb{R}). \end{cases}$$

We suppose that the supports of u_0 and V are strictly included in a compact set $\overline{\Omega}$. One then considers the domain $\Omega \times [0,T]$ with $\overline{\Omega} \subset \Omega$ and one denotes by Γ the boundary of Ω. One then looks for v solution of

$$\begin{cases} (i\partial_t v + \partial^2_{yy} v + V(y)\cdot v)(y,t) = 0, \quad y \in \Omega, \\ \mathcal{B}(y, \partial_y, \partial_t) v(y,t) = 0, \quad y \in \Gamma, \\ v(y,0) = u_0(y), \quad y \in \Gamma, \quad u_0 \in H^1(\mathbb{R}) \end{cases} \tag{3.11}$$

such that

$$u_{|\Omega \times [0,T]} = v. \tag{3.12}$$

The main problem consists then in finding an adequate (pseudo-)differential boundary operator $\mathcal{B}$ on Γ such that (3.12) occurs. As is well-known these conditions, called Dirichlet–Neumann, are nonlocal in time (and in space in multidimension).

Denoting by $\mathbf{n}$ the outward normal of Γ and $\partial_{\mathbf{n}}$ is the trace operator on Γ we obtain:

$$\begin{cases} (i\partial_t v + \partial^2_{yy} v + V(y)\cdot v)(y,t) = 0, \quad y \in \Omega, \\ v(y,t) = -e^{i\pi/4} \displaystyle\int_0^t \frac{\partial_y v(y,\tau)}{\sqrt{\pi(t-\tau)}} d\tau, \quad y \in \Gamma. \end{cases} \tag{3.13}$$

This approach has been very well described in particular in [2], and some numerical issues can be found in [3,4]. We also refer to [8] for the first presented discretization of nonlocal transparent boundary conditions for the TDSE. As unfortunately these conditions are nonlocal in time, many studies have been devoted to find effective numerical approximations. At each iteration we can for instance decompose the convolution product as the sum of a local part and historical part as proposed in [21]. Thus at time t^{n+1}:

$$v(y_\Gamma, t^{n+1}) = -e^{i\pi/4} \int_0^{t^{n+1}} \frac{\partial_y v(y_\Gamma,\tau)}{\sqrt{\pi(t^{n+1}-\tau)}} d\tau \tag{3.14}$$

is decomposed into

$$v(y_\Gamma, t^{n+1}) = -e^{i\pi/4} \int_0^{t^{n}} \frac{\partial_y v(y_\Gamma,\tau)}{\sqrt{\pi(t^{n+1}-\tau)}} d\tau - e^{i\pi/4} \int_{t^n}^{t^{n+1}} \frac{\partial_y v(y_\Gamma,\tau)}{\sqrt{\pi(t^{n+1}-\tau)}} d\tau. \tag{3.15}$$

It is then possible to treat accurately each integral (see again [21]).

Coupling with the laser. Now we present some ideas related to the boundary conditions for the TDSE coupled with a laser interaction $yE(t)$. As the laser interaction does not depend on the inter-nuclear distance R, then the TDSE we consider here simply writes:

$$i\partial_t v(y,t) = \big(- \partial^2_{yy} + V(y) + yE(t) \big) v(y,t). \tag{3.16}$$

It is important to note that the term $V(y)+yE(t)$ does not have a compact support, so that it is no more possible to solve "easily" the free potential TDSE outside a bounded domain as done above. Indeed, outside a given domain containing the support of the Coulomb potential V, the potential free equation becomes:

$$i\partial_t v(y,t) = \big(- \partial^2_{yy} + yE(t) \big) v(y,t). \tag{3.17}$$

Using as above a Laplace transform in time would lead to a convolution product between the Laplace transforms of E and v. We then propose not to solve exactly this equation but to give an approximate condition based on a splitting operator. Given the solution on the boundaries at time t^n we look for it at time t^{n+1} splitting the equation (3.16):

$$\begin{cases} i\partial_t v(y,t) = yE(t)v(y,t), & y \in \Gamma,\ t \in [t^n, t^{n+1/2}[, \\ i\partial_t v(y,t) = \big(-\partial_{yy}^2 + V(y)\big)v(y,t), & y \in \Gamma,\ t \in [t^{n+1/2}, t^{n+1}]. \end{cases}$$

The first equation provides the following solution:

$$v(y_\Gamma, t^{n+1/2}) = e^{-iy_\Gamma \int_{t^n}^{t^{n+1}} E(s)\,ds} v(y_\Gamma, t^n). \tag{3.18}$$

To approximate the second one we consider the solution at time t^{n+1} when the laser is *null*:

$$v(y_\Gamma, t^{n+1}) = -e^{i\pi/4} \int_0^{t^{n+1}} \frac{\partial_y v(y_\Gamma,\tau)}{\sqrt{\pi(t^{n+1}-\tau)}}\,d\tau. \tag{3.19}$$

We then decompose this integral in two parts corresponding to the local and historical parts.

$$v(y_\Gamma, t^{n+1}) = -e^{i\pi/4} \int_0^{t^n} \frac{\partial_y v(y_\Gamma,\tau)}{\sqrt{\pi(t^{n+1}-\tau)}}\,d\tau - e^{i\pi/4} \int_{t^n}^{t^{n+1}} \frac{\partial_y v(y_\Gamma,\tau)}{\sqrt{\pi(t^{n+1}-\tau)}}\,d\tau. \tag{3.20}$$

The historical part depends on v at time t^n and is then known at time t^{n+1}. For the local part we use $v(y_\Gamma, t^{n+1/2})$ in order to compute $v(y_\Gamma, t^{n+1})$. More precisely and following approximations proposed by Greengard in [21], the historical part is approximated by:

$$-e^{i\pi/4} \int_0^{t^n} \frac{\partial_y v(y_\Gamma,\tau)}{\sqrt{\pi(t^{n+1}-\tau)}} \sim -e^{i\pi/4} \sum_{j=1}^{M} w_j c_j(n), \tag{3.21}$$

where $\big(c_j(n)\big)_j$ and $(w_j)_j$ are some sequences described in [21].

The local part is then approximated by Gauss-Legendre quadrature:

$$\begin{aligned} -\,e^{i\pi/4} \int_{t^n}^{t^{n+1}} \frac{\partial_y v(y_\Gamma,\tau)}{\sqrt{\pi(t^{n+1}-\tau)}}\,d\tau \sim -e^{i\pi/4}\sqrt{\frac{\Delta t^{n+1}}{\pi}}&(\alpha v_{y,\Gamma}^{n+1/2} \\ &+ (2-\alpha)v_{y,\Gamma}^n), \quad 0 \le \alpha \le 2. \end{aligned} \tag{3.22}$$

With such an approach we then obtain the following boundary conditions. If J denotes the last cell index: $J\Delta y = y_\Gamma \in \Gamma$.

$$\begin{cases} v_J^{n+1} = -e^{i\pi/4}\left(\sum_{j=1}^{N_c} w_j c_j(n) + \sqrt{\frac{\Delta t^{n+1}}{\pi}}\left(\alpha \frac{v_J^{n+1/2} - v_{J-1}^{n+1/2}}{\Delta y} e^{-iy_\Gamma \int_{t^n}^{t^{n+1}} E(s)\,ds} + (2-\alpha)\frac{v_J^n - v_{J-1}^n}{\Delta y}\right)\right), \\ c_j(n+1) = e^{s_j^2 \Delta t^n} c_j(n) + \frac{\Delta t^n}{2}\left(e^{-s_j^2\Delta t^{n+1}} \frac{v_J^n - v_{J-1}^n}{\Delta y} + e^{-s_j^2(\Delta t^n + \Delta t^{n+1})} \frac{v_J^{n+1} - v_{J-1}^{n+1}}{\Delta y}\right). \end{cases}$$

Where,

$$v_J^{n+1/2} = e^{-iy_\Gamma \int_{t^n}^{t^{n+1}} E(s)\,ds} v_J^n, \quad v_{J-1}^{n+1/2} = e^{-iy_\Gamma \int_{t^n}^{t^{n+1}} E(s)\,ds} v_{J-1}^n. \tag{3.23}$$

Note that by induction, the historical part depends also on the laser field E. A symmetric approach is proposed in $-y_\Gamma$. These 2 discrete equations in $-J$ and J close the Crank–Nicolson system. As a consequence the Hermitian structure of the sparse matrix is lost so that a GMRES method (see [30]) is used to solve the linear system. Our strong assumption consists then in searching for a numerical boundary condition that behaves like $\int_0^t \partial_y v(y,\tau)/\sqrt{\pi(t-\tau)}\,d\tau$. Note also that it is straightforward to apply the technique presented above in the velocity gauge.

To illustrate this technique we propose a simple benchmark on the TDSE under the Born–Oppenheimer approximation (fixed R). We assume here that the Coulomb potential is equal to zero.

$$\begin{cases} i\partial_t u(y,t) + \partial_{yy} u(y,t) + yE(t)u(y,t) = 0, & y \in [-10,10], \quad t \geq 0, \\ u(y,0) = u_0(y) = e^{8iy}e^{-y^2}. \end{cases} \tag{3.24}$$

The benchmark we propose is as follows. The global domain is $[-10;10]$ and the space grid τ contains 800 cells with a space step equal to 0.05. The reduced domain $\Omega = [-5,5]$ contains the initial data support. We impose the Dirichlet–Neumann boundary conditions coupled the laser as described above, at $x_{-\Gamma} = -5$ and $x_\Gamma = 5$. Outside Ω we solve the TDSE taking as boundary conditions the solution computed $x_{\mp\Gamma}$. We compare our numerical solution with the solution obtained using Dirichlet boundary conditions at $x_{-\Gamma} = -5$ and $x_\Gamma = 5$ and with a reference solution. We consider a two-cycle laser pulse, with an intensity equal to 0.05 and a duration of 0.76 (note that we use *a.u.*).

The Dirichlet–Neumann numerical (Fig. 9) solution is then far less reflected (even if a small reflection exists) than the Dirichlet solution. Here, note that the grid is coarse and small, so that the influence of spurious reflections can be obviously diminished using a larger grid and smaller space steps. Another approach to make decrease the reflections consists in making a better approximation of the historical part in the method presented above. In order to do this, it is necessary to approximate this historical integral using the solution at the boundaries at all previous times (t^n, t^{n-1}, t^{n-2}, *etc*). This then would lead to the approach of Antoine and Besse in [3] where the authors are enable to reduce drastically the spurious reflections. The price to pay is however to store the solution at the boundaries at all times, which would be problematic in higher dimensions.

We note that the benchmark we have considered above corresponds to a total ionization (the wavefunction leaves totally the reduced domain because of the wavenumber equal to 8). This is not of interest in our model at this time. Indeed we want the electrons to stay inside the domain in order to take into account the plasma effects due to free electrons.

Finally note that another approach based on the exact solution of the TDSE without Coulomb potential but with an electric field (see Volkov wavefunction) will be developed in a forthcoming paper. Typically this would involve the computation

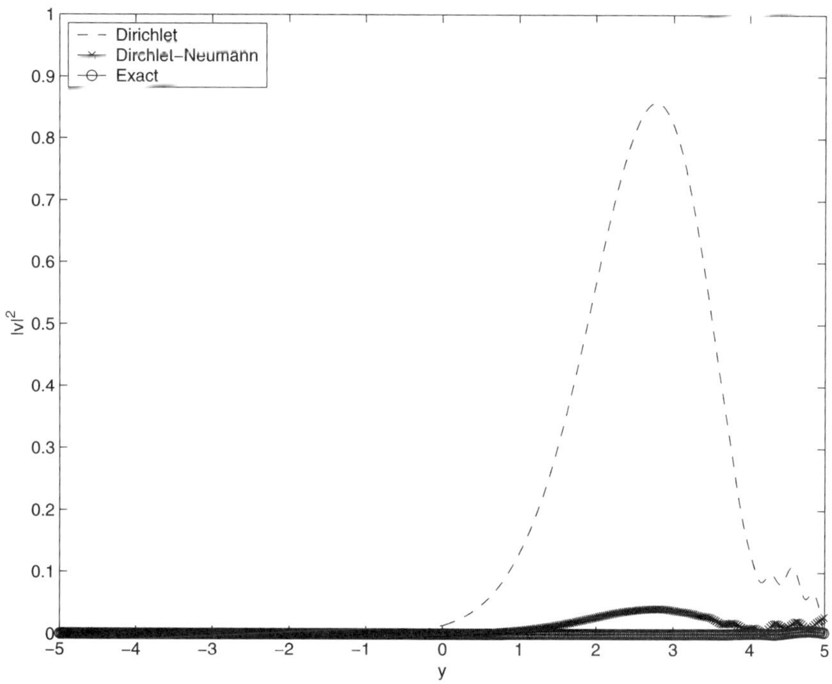

FIGURE 9. Comparison between a reference solution and the numerical solutions obtained with Dirichlet and Dirichlet–Neumann coupled with the laser.

at the boundary y_Γ of the quantity:

$$
\begin{aligned}
(3.25)\quad & v(y_\Gamma, t^{n+1}) \\
&= \frac{1-i}{2\sqrt{\pi t^{n+1}}} \exp\left(-iy_\Gamma \int_0^{t^{n+1}} E(s)\,ds - \frac{i}{2}\int_0^{t^{n+1}}\left(\int_0^s E(\tau)d\tau\right)^2 ds\right) \\
&\quad \times \int_{\mathbb{R}} v_0(y')\exp\left(\frac{i(y_\Gamma - \int_0^{tn+1}\left(\int_0^s E(\tau)\,d\tau\right)\,ds - y')^2}{2t^{n+1}}\right) dy,
\end{aligned}
$$

corresponding to the exact solution of the TDSE without potential. Note that in the convolution product in the previous formula, we only integrate over the initial wavefunction support.

3.3. Some preliminary numerical tests. We present here two numerical results obtained when solving the Maxwell–Schrödinger equations in 1-D and 3-D. The first result Fig. 10 represents for a 1-D computation, the harmonic orders contained in the transmitted electric field $\widehat{E}(\omega)$ of intensity $|\widehat{E}(\omega)|^2$ ($\omega = N\omega_0$, with ω_0 the incident laser frequency), for a few-cycle (2–3 cycles) laser propagation in a H_2^+ medium of length equal from 14nm (corresponding to 1 TDSE), to 28.7μm (2048 TDSE's) at $R = 2$. Note that we obtain a good linear scaling of the low order harmonic intensities, when we increase the domain length, as expected for such a coherent process [15,16]. Note also that the seventh harmonics is particularly excited as remarked in numerous computations (see [23] for instance). In this case, as the density is very small (10 Torr, that is $\sim 3.54 \times 10^{17}$mol/cm^3) the attopulse shapes are not modified (only linearly amplified) by the medium. In opposite for a much larger molecular density ($\sim 10^{20}$mol/cm^3, with the same data than above and $R = 2$) we observe a modification on Fig. 11 of the linear scaling even for low order harmonic intensities (see for instance the seventh harmonics) This phenomenon related to the fact that the wavelength of the highest harmonics approach the inter-molecule distance [25].

The second numerical result Fig. 12 is the transmitted electric field harmonics (squared) for a 3-D computation, with a 5-cycle pulse propagation in a H_2^+ media of length respectively equal to 15nm (corresponding to 4 TDSE's), 60nm (16 TDSE's), 240nm (64 TDSE's). The pressure is here equal to 640 Torr. The code was run during approximately 22 hours respectively on 4, 16 and 64 Xeon (Intel) processors (see www.ccs.usherbrooke.ca/mammouth). Again we remark that the computation gives us a linear scaling for the low order harmonics with respect to the propagation length. These two results show also the high order harmonic generation by the electric field acting on the H_2^+ molecules and the high intensity of some of these harmonics as a result of nonlinear non-perturbative laser-molecule interaction. This observation confirms the necessity for the use of our Maxwell–Schrödinger model instead of perturbative Maxwell–Schrödinger models to study intense ultrashort lasers behaviours in nonlinear media where ionization occurs. Indeed for these

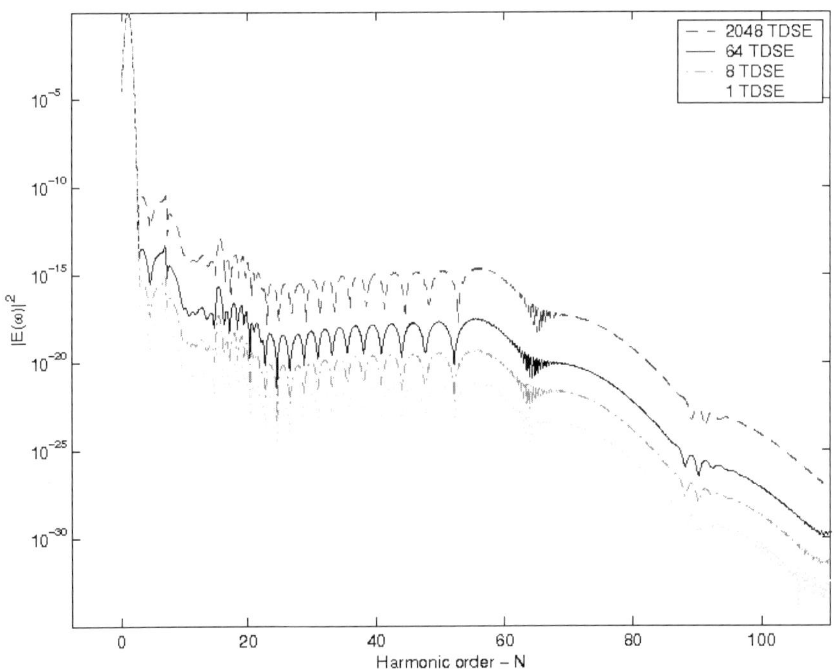

FIGURE 10. $|\widehat{E}(\omega)|^2$, for 14nm (1 molecule), 112nm (8 molecules), 896nm (64 molecules) and 28.7μm (2048 molecules) and $n_0 \sim 3.54 \times 10^{17}$mol/cm^3 in 1-D. In abscissa: harmonic order N, with $\omega = N\omega_0$ and ω_0 incident electric field frequency.

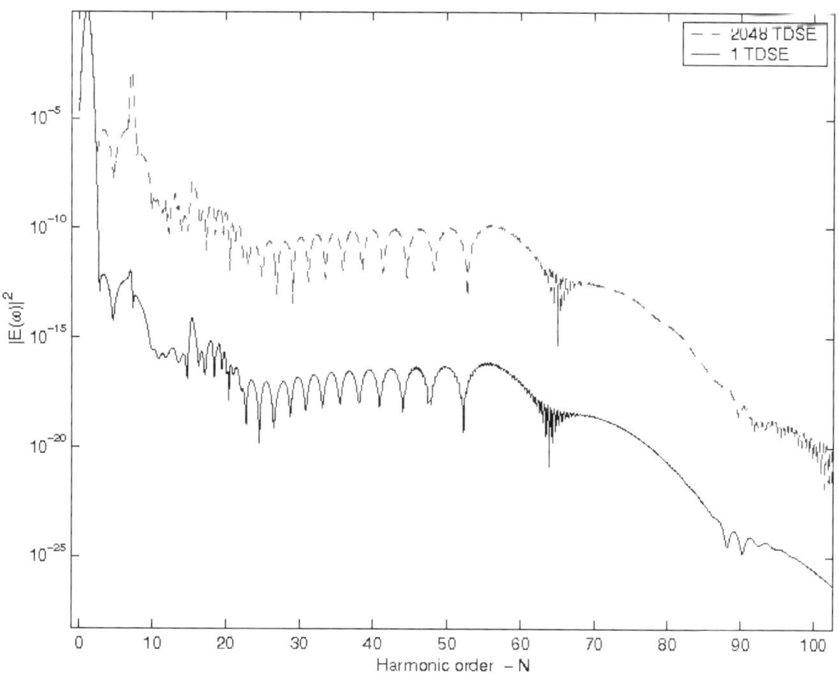

FIGURE 11. $|\widehat{E}(\omega)|^2$, for $2nm$ (1 molecule) and 4μm (2048 molecules) and $n_0 \sim 10^{20}$mol/cm^3 in 1-D. In abscissa: harmonic order N, with $\omega = N\omega_0$, harmonic frequency, and ω_0 incident electric field frequency

last models only small order harmonics are taken into account (one equation for each harmonic). Our numerical results show that higher harmonics can have an important intensity. Our model can then be for instance, an issue for filamentation [27]. Indeed, this multidimensional phenomenon is usually treated using nonlinear TDSE's (one for each harmonic) with a plasma term modeled with a nonlinear function of the electric field. However, in practice only small harmonics are taken into account and the plasma term is often poorly modeled. Finally, note that by filtering around high intense harmonics it is also possible to create shorter pulses (much higher frequency) than the incident one (2.13). See for instance that in the 3-D case, the 70th harmonics is relatively intense allowing to create by filtering an "intense" pulse of frequency 70 times larger than the incident laser one (2.13) [1]. Details of theses simulations and others will be presented in [25].

4. Conclusion

In this short paper, we have presented a Maxwell–Schrödinger model for laser-molecule interaction in the nonlinear nonperturbative regime and some aspects of its numerical computation. In particular, we have proposed some avenues to reduce the algorithmic complexity of the numerical problem due to the multiple scales (polarization computation for instance) but also due to the boundary conditions involving continuum (dissociative ionization) states both electronic and nuclear.

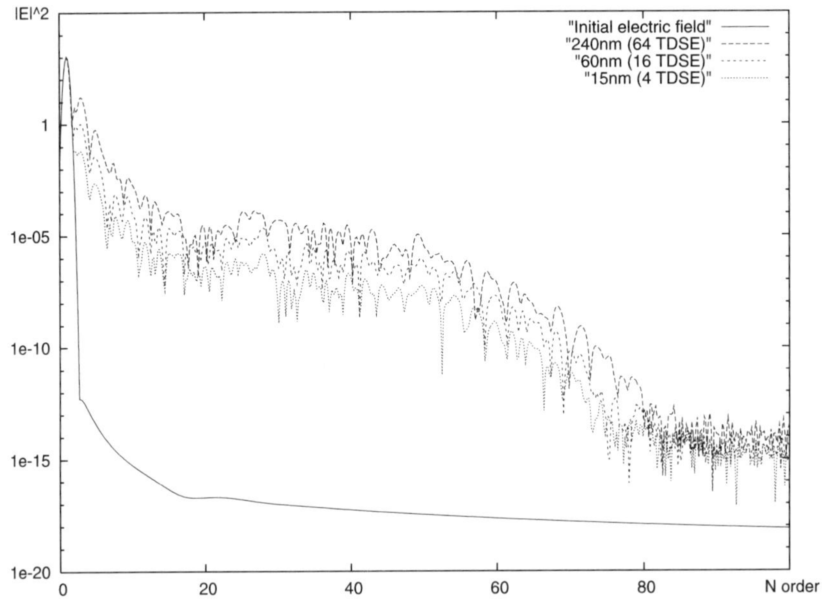

FIGURE 12. $|\widehat{E}(\omega)|^2$, for 15nm (4 molecules), 60nm (16 molecules), 240nm (64 molecules) in 3-D. In abscissa: harmonic order N, with $\omega = N\omega_0$, harmonic frequency, and ω_0 incident electric field frequency.

In this framework, we have presented a 1-D approach to manage the "dispersive" effects for the TDSE which is obviously applicable on cubic box faces in 3-D. However to overcome the difficulties due to the cubic domain singularities, a better approach consists in applying some really multi-dimensional Dirichlet–Neumann boundary conditions. Then some new numerical difficulties arise due to the fact that such conditions are non-local in space (and in time): the condition on each boundary point of the domain depends on all the points of the domain at all previous times. In [4] the authors proposed in 2-D for laser free problems to use an approximation of the Dirichlet–Neumann nonlocal pseudodifferential operator by some local differential operators (involving in particular the Beltrami operator on the boundary). They are then able to obtain order two local approximations of the "exact" nonlocal operator. In [25] we will show how we propose to apply a derivation of their technique in the framework of laser-molecule interactions. We will also focus on transversal effects, that is spatial-temporal pulse shape effects, of the laser field interacting with a gas not taken into account in this note. Another point to investigate comes from the following remark. For a 10 Torr pressure, and supposing an ideal molecular equidistribution, the space between two molecules is given by 12nm. In HOHG by a 800nm-laser, the 67th harmonic is approximately equal to $12nm$. This specific configuration, that corresponds exactly to one electric field cycle between two molecules presents challenges for the inclusion of microscopic details into Maxwell–Schrödinger equations.

References

1. P. Agostini and L.F. DiMauro, *The physics of attosecond light pulses*, Rep. Progr. Phys. **67** (2004), no. 813, 813–855.
2. X. Antoine, H. Barucq, and L. Vernhet, *High-frequency asymptotic analysis of a dissipative transmission problem resulting in generalized impedance boundary conditions*, Asymptot. Anal. **26** (2001), no. 3-4, 257–283; MR1844544 (2002e:35036)
3. X. Antoine and C. Besse, *Unconditionally stable discretization schemes of non-reflecting boundary conditions for the one-dimensional Schrödinger equation*, J. Comput. Phys. **188** (2003), no. 1, 157–175; MR1985023 (2004e:65104)
4. X. Antoine, C. Besse, and V. Mouysset, *Numerical schemes for the simulation of the two-dimensional Schrödinger equation using non-reflecting boundary conditions*, Math. Comp. **73** (2004), no. 248, 1779–1799 (electronic). MR2059736 (2005c:65067)
5. A.D. Bandrauk, S. Chelkowski, and N.H. Shon, *Nonlinear photon processes in molecules at high intensities-route to* XUV*-attosecond pulse generation*, J. Mol. Struct. **735–736** (2005), 203–209.
6. A.D. Bandrauk and S. Chelkowski, *Dynamic imaging of nuclear wavefunctions with ultrashort* UV *laser pulses*, Phys. Rev. Lett. **87** (2001), no. 27, 113–120.
7. A.D. Bandrauk and H.-Z. Lu, *Numerical methods for molecular time-dependent Schrödinger equations—bridging the perturbative to nonperturbative regime*, Handbook of Numerical Analysis, vol. X, North-Holland, Amsterdam, 2003, pp. 803–832.
8. V.A. Baskakov and A.V. Popov, *Implementation of transparent boundaries for numerical solution of the Schrödinger equation*, Wave Motion **14** (1991), no. 2, 123–128. MR1125294 (92g:78001)
9. L. Baudouin, O. Kavian, and J.-P. Puel, *Régularité dans une équation de Schrödinger avec potentiel singulier à distance finie et à l'infini*, C. R. Math. Acad. Sci. Paris **337** (2003), no. 11, 705–710. MR2030406 (2004k:35308)
10. ———, *Regularity for a Schrödinger equation with singular potentials and application to bilinear optimal control*, J. Differential Equations **216** (2005), no. 1, 188–222. MR2158922
11. A. Ben Haj-Yedder, C. Le Bris, O. Atabek, A.D. Bandrauk, and S. Chelkowski, *Optimal control of attosecond pulse synthesis from high order harmonic generation*, Phys. Rev. A **69** (2004), 041802-1–041802-4.
12. T. Brabec and F. Krausz, *Intense few-cycle laser fields: frontier of nonlinear optics*, Rev. Modern Phys. **72** (2000), no. 545–591.
13. S. Chelkowski, A. D. Bandrauk, and P.B. Corkum, *Muonic molecules in superintense laser fields*, Phys. Rev. Lett. **93** (2004), 083602-1–083602-4.
14. S. Chelkowski, C. Foisy, and A.D. Bandrauk, *Electron-nuclear dynamics of multiphoton* H_2^+ *dissociative ionization in intense laser fields*, Phys. Rev. A **57** (1998), no. 2, 1176–1185.
15. I. Christov, *Generation and propagation of attosecond* x*-ray pulses in gaseous media*, Phys. Rev. A **57** (1998), no. 4, 2285–2288.
16. ———, *Enhanced generation of attosecond pulses in dispersion-controlled hollow-core fiber*, Phys. Rev. A **60** (1999), no. 4, 3244–3250.
17. M. Defranceschi and C. Le Bris (eds.), *Mathematical models and methods for ab initio quantum chemistry*, Lecture Notes in Chem., vol. 74, Springer-Verlag, Berlin, 2000, pp. 95–119. MR1857459 (2003e:82002)
18. E. Dumas, *Global existence for Maxwell–Bloch systems*, J. Differential Equations **219** (2005), no. 2, 484–509. MR2183269
19. R.J. Iório, Jr. and D. Marchesin, *On the Schrödinger equation with time-dependent electric fields*, Proc. Roy. Soc. Edinburgh Sect. A **96** (1984), no. 1-2, 117–134. MR741652 (85k:35179)
20. J.D. Jackson, *Classical electrodynamics*, 2nd ed., John Wiley & Sons Inc., New York, 1975. MR0436782 (55 #9721)
21. S. Jiang and L. Greengard, *Fast evaluation of nonreflecting boundary conditions for the Schrödinger equation in one dimension*, Comput. Math. Appl. **47** (2004), no. 6-7, 955–966. MR2060329 (2004m:35217)
22. M. Jungreuthmayer, C.M. Geissler, J. Zanghelini, and T. Brabec, *Microscopic analysis of large-cluster explosion in intense laser fields*, Phys. Rev. Lett. **92** (2004), no. 13, 133401-1–133401-4.

23. G.L. Kamta and A.D. Bandrauk, *Three-dimensional time-profile analysis of high-order harmonic generation in molecules: nuclear interferences in* H_2^+, Phys. Rev. A **71** (2005), no. 5, 53407-1–19.
24. M. Lewenstein, Ph. Balcou, M. Yu, A. Huillier, and P.B. Corkum, *Theory of high-harmonic generation by low frequency laser fields*, Phys. Rev. A **49** (1994), no. 3, 2117–2132.
25. E. Lorin, S. Chelkowski, and A.D. Bandrauk, *Numerical Maxwell–Schrödinger model for intense laser-molecule interaction and propagation*, Preprint CRM (to appear).
26. A. Maimistov E. Kazantseva, and J.-G. Caputo, *Reduced Maxwell–Duffing description*, Phys. Rev. E **71** (2005), 056622-1–056622-12.
27. G. Mechain, A. Couairon, M. Franco, B. Prade, and A. Mysyrowicz, *Organizing multiple femtosecond filaments in air*, Phys. Rev. Lett. **93** (2004), no. 3, 035003.
28. S.C. Rae and K. Burnett, *Harmonic generation and phase matching in the tunnelling limit*, J. Phys. B **26** (1993), 1509–1518.
29. ______, *Generation and propagation of high-order harmonics in a rapidly ionizing medium*, Phys. Rev. A **50** (1994), 3438–3446.
30. Y. Saad and W. Schultz, GMRES: *a generalized minimal algorithm for solving nonsymmetric linear systems*, SIAM J. Sci. Static. Comput. **7** (1986), no. 3, 856–869.
31. N.H. Shon, A. Suda, and K. Midorikawa, *Generation and propagation of attosecond pulses in He gas with sub-10-fs driver pulses*, Phys. Rev. A **60** (1999), no. 3, 2587–2590.
32. K.S. Yee, *Numerical solution of initial boundary value problems involving Maxwell's equations in isotropic media*, IEEE Trans. Antennas and Propagation **AP-16** (1966), 302–307.

Centre de Recherches Mathématiques, Université de Montréal, C.P. 6128, succ. Centre-ville, Montréal, Québec, H3C 3J7, Canada.
E-mail address: `lorin@CRM.UMontreal.CA`

Laboratoire de Chimie Théorique, Université de Sherbrooke, Sherbrooke, Québec, J1K 2R1, Canada.
E-mail address: `Szczepan.Chelkowski@USherbrooke.ca`

Département de Chimie, Université de Sherbrooke, Sherbrooke, Québec, J1K 2R1, Canada.
E-mail address: `Andre.Bandrauk@USherbrooke.ca`

Centre de Recherches Mathématiques
CRM Proceedings and Lecture Notes
Volume **41**, 2007

Parareal in Time Algorithm for Kinetic Systems Based on Model Reduction

Yvon Maday

Abstract. The parareal in time algorithm allows for large speedups for the simulation of general time dependent problems from the combined use of (i) a decomposition of the interval of propagation in time into temporal slabs and of (ii) discrete propagation methods over each slab in parallel on different processors. This is done by choosing suitably the seed values at the beginning of each temporal slab through a (sequential) coarse predictor. The efficiency of the algorithm depends on the choice of the coarse solver used in the predictor step. We report here some ways of choosing this coarse solver on an application to a simple kinetic stiff systems. The choice is guided from the availability of a reduced model that governs the main feature of the evolution at equilibrium. This choice of a simplified dynamics maintains the convergence performance of the parareal in time algorithm but allows for a reduction in the global CPU time.

1. Introduction

In 2001, Lions, Maday and Turinici have proposed in a CRAS note [17] a new algorithm for the simulation of time dependent problems that allows to take benefit of the parallel architectures of the large scale computing resources that are now available in order to decrease the solution time to obtain the numerical results. This method was named parareal-in-time algorithm based on the prospect of being able to perform **real** time simulations with the help of **para**llel processinCentre de Recerca Matematica, Barcelonag. An improved version of the same method was proposed later in [1, 4, 20] much better tuned to the treatment of nonlinear problems. The method appears now as a predictor corrector scheme, where the predictor is based on a coarse grain (in time) and inexpensive simulation, solved in a serial manner and the expensive corrector can be spread on different processors and treated independently at the same time on different processors. This is done by updating suitably the seed values at the beginning of each slab through a coarse iterative method.

The method finds its roots in the domain decomposition method from which, (i) the decomposition of tasks based on the splitting of a geometric entity and (ii)

2000 *Mathematics Subject Classification.* Primary: 65L06, 65Y05; Secondary: 68W10, 80A30.

This is the final form of the paper.

the notion of coarse solver, predominant for the scalability of the algorithms, have been adapted to the time domain (see e.g. [27]).

In a recent paper, Gander and Vandewalle [7] have pointed out how the parareal algorithm relates to some earlier algorithms that have appeared in the literature on parallel time-integration methods. The parareal algorithm can be interpreted as a variant of the multiple shooting method (see e.g. [24]), with a Jacobian matrix that is computed approximately by finite differences on a coarse mesh. The superlinear convergence of the parareal in time algorithm on bounded intervals noted by [7] can thus find an heuristical explanation from this interpretation as a Newton-type method for solving a nonlinear system.The algorithm can also be rewritten as a two-level multigrid method. It is characterized by a possibly "aggressive coarsening" in time that is not restricted to a small integer for the method to be effective by the use of a rather unusual smoother combined with most simple restriction and prolongation operators. As a multigrid method for general time-dependent problems, parareal can be interpreted as a type of parabolic multigrid method, a class of solvers for parabolic partial differential equations that includes the space-time multigrid, time-parallel multigrid and multigrid waveform relaxation methods ([9, 13, 14, 16, 25, 28, 29]). Similar to those methods, parareal shares the linear convergence rate on long time-intervals. In contrast to those methods, and as a result of having its roots in the domain decomposition community more than in the multigrid or multiple shooting ones, parareal in time algorithms are applicable to a much wider range of problems than parabolic partial differential equations.

Recent advances include stability analysis [3, 26], coupling of the algorithm in time with other iterative methods such as domain decomposition methods [22] and control problem [19, 20]. Most of the applications treated up to now have dealt with time dependent partial differential equations. This short paper is a contribution to the use of the method for solving complex differential equations. We are considering a simple model of chemical reactions such as the ones encountered in the frame of air pollution, combustion processes, biochemistry or metabolism issues.

The current models involve a large number of chemical species and reactions that usually have much different time scales, resulting in what is known as a stiff problem. The numerical solution of such systems requires indeed very small time steps leading to enormous computational times. For many problems however, there is, most of the time, a simple behavior that is hidden ; when the simple dynamics in the good unknown is discovered, then the computational cost is reduced by a large factor. The method we are aware of, presented in [5], aims at discovering this simple dynamics is a systematic and constructive reduction method for kinetic systems where the chemistry is monomolecular with respect to a set of limitant species and whose concentrations have to be evaluated. The literature in this direction is quite wide, we are far from being specialists of the field so we prefer to refer to the reference list quoted in [5]. The goal there is to construct, without any change of basis, a smaller and no longer stiff differential system involving slow species and to express the other fast species in terms of the slow ones. As recalled in Section 3, the partitioning of the species is completely algorithmic and explicit. The dimension of the system is reduced at the same time the stiffness diminishes. Fast species are recovered, at the end from the evolution of the slow ones in a very cheap way. This reduction procedure is valid after some times, once the species have equilibrated. The simulation of this period to get equilibrium is stiff based on the original model but short. Of course each time there is a burst of sources that perturb the system, the reduced model is not valid any more and the full system should be used some times after the burst happens.

The use or the combination of model reduction and parareal algorithm is illustrated on this example, chosen as being an archetype of other situations where the reduction procedure may have other names such as adiabatic reduction in quantum chemistry, effective equations or reduced basis methods in complex dynamics. The global added computational time is (of course) larger for the parareal in time algorithm than a single fine solution. It is indeed at least equal to the number of parareal iterations multiplied by the total cost of a fine propagation. The net win is on the *restitution time* since the coarse propagator is the only one that is used sequentially and — assuming the corresponding cost is negligible — the restitution time is slightly larger than the number of iteration times the solution to solve on a time slab. In all simulations we did, the number of iterations is much smaller than the number of time slab.

2. Parareal Methodology

Consider the following time dependent problem

$$\frac{\partial y}{\partial t} + \mathcal{A}y = 0, \quad y(0) = y^0$$

where, for the sake of simplicity, we assume that $\mathcal{A}$ does not depend on time.

We introduce the propagator $\mathcal{E}$ such that, $\mathcal{E}_\tau(\mu) = y(\tau)$ is the solution, at time τ of the problem

$$\frac{\partial y}{\partial t} + \mathcal{A}y = 0, \quad y(0) = \mu.$$

Due to time invariance, the classical semi group property on the propagator takes the simple form

$$\forall \tau' < \tau, \quad \mathcal{E}_\tau = \mathcal{E}_{\tau-\tau'} \circ \mathcal{E}_{\tau'}.$$

For instance, let $0 = T_0 < T_1 < \cdots < T_n < \cdots < T_N = T$ be special times at which we are interested to consider snapshots of the solution $y(T_n)$, then we have

$$y(T_{n+1}) = \mathcal{E}_{T_{n+1}}(y^0) = \mathcal{E}_{T_{n+1}-T_n}\big(y(T_n)\big).$$

In most cases $\mathcal{E}$ is not achievable but we have to refer to approximations based on a time discretization and the use of an Euler-type or a more involved scheme.

For instance we can introduce the fine and precise discretization scheme

$$\frac{y^{m+1} - y^m}{\delta t} + \mathcal{A}y^{m+1} = 0 \quad \text{for an implicit scheme,}$$

based on a small time step δt. For any time $T = M\delta t$ the approximated propagator $\mathcal{F}_T$ is then defined from the iterative solution of M problems as above.

Similarly to the continuous solution, the approximations $Y_n = \mathcal{F}_{T_n}(y^0)$ of $y(T_n)$ satisfies

$$Y_{n+1} = \mathcal{F}_{T_{n+1}-T_n}(Y_n).$$

Assuming, for the sake of simplicity that $T_{n+1} - T_n$ is constant ($= \Delta T \gg \delta t$), then this reads

$$Y_{n+1} = \mathcal{F}_{\Delta T}(Y_n)$$

where it appears clearly that the approximated solution process is fundamentally sequential, which, a priori, prevents it from being amenable to any parallelization.

In what follows we propose an algorithm that allows to define a sequence $\{\mathbf{Y}^k\}_k, \mathbf{Y}^k \equiv (Y_1^k, Y_2^k, \ldots, Y_N^k)$, such that, for any n $Y_n^k \to Y_n$ as k goes to infinity. The definition of this sequence requires the datum of another approximated propagator, denoted as $\mathcal{G}$. This second propagator is assumed to be cheaper than $\mathcal{F}$ at the price of some inaccuracy. One can think immediately about a fine propagator $\mathcal{F}$ based on an Euler scheme with a very small time step δt and $\mathcal{G}$ based on a similar Euler scheme with the larger time step $\Delta\tau, \delta t \ll \Delta\tau \leq \Delta T$. But other possibilities are also offered, as e.g. for the application to evolution partial differential equations, in addition to the solution scheme in time, a discretization in space has to be performed. For instance, the spacial discretization can be based on a finite element method and the fine solver $\mathcal{F}$ can be based on a fine time step and a fine spacial discretization while the coarse solver $\mathcal{G}$ can be based on a coarse time step and a coarse spacial discretization, we refer to [6, 18]. Another possibility on which we shall elaborate in this paper consists in providing $\mathcal{F}$ with all the physics of the phenomenon but $\mathcal{G}$ will be based on a simplified physics. Before this, let us proceed in the definition of the sequence

Let us initiate $\mathbf{Y}^0$ by $Y_n^0 = \mathcal{G}_{n\Delta T}\big(y(0)\big)$, for $n = 1, \ldots, N$. Assume now $\mathbf{Y}^k$ is known, the iterative process proceeds by setting $Y_0^k = y(0)$, and,

$$Y_{n+1}^{k+1} = \mathcal{G}_{\Delta T}(Y_n^{k+1}) + \mathcal{F}_{\Delta T}(Y_n^k) - \mathcal{G}_{\Delta T}(Y_n^k). \tag{2.1}$$

The algorithm converges, under the following natural hypothesis: denoting by $\delta\mathcal{F}$ (resp. $\delta\mathcal{G}$) the difference $\delta\mathcal{F} = \mathcal{E} - \mathcal{F}$ (resp. $\delta\mathcal{G} = \mathcal{E} - \mathcal{G}$), assume that

$$\forall\tau, \quad |\delta\mathcal{F}_\tau(x)| \leq C\tau\eta(1 + |x|), \quad |\delta\mathcal{G}_\tau(x)| \leq C\tau\varepsilon(1 + |x|), \tag{2.2}$$

where η and ε are small quantities, typically, for the Euler scheme these are $\eta = \delta t$ and $\varepsilon = \Delta T$. In addition let us assume that

$$\forall\tau, \quad |\delta\mathcal{F}_\tau(x) - \delta\mathcal{F}_\tau(y)| \leq C\tau\eta|x - y|, \quad |\delta\mathcal{G}_\tau(x)| \leq C\tau\varepsilon|x - y|, \tag{2.3}$$

where C is a generic constant, independent of the quantities on the right hand side of the previous inequalities.

The convergence is stated in the following

Theorem 2.1. *Assume that the discrete propagators $\mathcal{F}$ and $\mathcal{G}$ satisfy* (2.2) *and* (2.3)*, then the error between the exact solution and the solution provided by the parareal scheme* (2.1) *satisfies*

$$|\lambda_n^k - y(T^n)| \leq C(\varepsilon^k + \eta)(1 + |y_0|). \tag{2.4}$$

Remark 2.5. For Euler schemes based on a time step δt for $\mathcal{F}$ and ΔT for $\mathcal{G}$, hypothesis (2.2) and (2.3) hold with $\eta = \delta t$ and ΔT for $\mathcal{G}$ respectively.

We now comment on the parallelization of the above algorithm. The initialization step

$$Y_{n+1}^0 = \mathcal{G}_{\Delta T}(Y_n^0)$$

is sequential but involves the cheap coarse propagator. Then assuming that every $(Y_n^k)_n$ is known at step k, we first solve over each $]T_n, T_{n+1}[$ the totally independent propagation problems that provide $\mathcal{F}_{\Delta T}(Y_n^k)$; we can either propagate again (and now in parallel) the coarse solver to get $\mathcal{G}_{\Delta T}(Y_n^k)$, or just remember that these quantities were computed during the previous step, any way this is again parallel work. The corrector $\mathcal{F}_{\Delta T}(Y_n^k) - \mathcal{G}_{\Delta T}(Y_n^k)$ can then be built, for any n and the step can proceed. We finally solve the predictor based on the coarse solver $\mathcal{G}_{\Delta T}(Y_n^{k+1})$

— sequentially — and immediately correct it by adding the previously computed corrections until convergence. Note that in this process, the fine solutions are only used on small propagation intervals and can be performed in parallel.

It is one of the aims of this paper to elaborate on the way the coarse propagator should be defined. Since it is the only contribution that is propagated sequentially over the full time range $[0, T]$, the cheaper, the better.

3. Pararealization of the Simulation of the Chemical Reaction Model

Many complex problems in e.g. physics, chemistry and natural sciences involve dynamical systems that depend on different scales that range on a wide spectrum. The full understanding of the behavior requires a simulation of all scales and lead to stiff problems resulting from the very large difference between the time the propagation should occur (typically a few units of the scale of the *slow* evolution) and the time step size used for the approximation scheme (typically a few times smaller than the *fastest* evolution).

In this paper we shall deal with a simple model for chemical reactions as an archetype of systems that display such a separation of scales for which the theory to derive reduced evolutions is well understood. There are many contributions in this field and the choice we have made (resulting from local discussions in the department corridor) is just an illustration of what can be made in this direction. We believe that the conclusions are general.

3.1. Reduction of linear kinetic systems with multiple scales. Let us present here the example given by the reduction procedure for stiff molecular kinetic reactions proposed in [5]. The problem of interest consists in an isothermal mixture of N species $\mathcal{A}_1, \ldots, \mathcal{A}_N$ with concentrations $y_1, \ldots, y_N$. These generally numerous species interact together through reactions symbolized as r:

$$\mathcal{A}_{i_r} + \mathcal{N}_{i'_1} + \mathcal{N}_{i'_2} + \cdots + \mathcal{N}_{i'_q} \to \mathcal{A}_{j_r} + \mathcal{N}_{j'_1} + \mathcal{N}_{j'_2} + \cdots + \mathcal{N}_{j'_q}$$

where the species $\mathcal{N}_\ell$ are in excess and are thus not measured. The rate ω_r of the reaction r appears in the balance equation

$$\frac{dy_i}{dt} = -\sum_{r|i_r=i} \omega_r + \sum_{r|j_r=i} \omega_r, \tag{3.1}$$

where the first sum (with a negative sign) is taken over all reactions where $\mathcal{A}_i$ is a reactant while the second is taken over those reactions where $\mathcal{A}_i$ is a product. We assume that the reaction rate satisfies $\omega_r = k_r y_{i_r}$ where k_r are given constants. This results in a matrix system

$$\frac{d\mathbf{y}}{dt} = J\mathbf{y} \tag{3.2}$$

where J is a kinetic matrix since it satisfies:

$$\begin{aligned}
&\text{(i) } \forall i, && 1 \le i \le N,\ J_{ii} \le 0,\\
&\text{(ii) } \forall i,j, && 1 \le i,j \le N,\ i \ne j J_{ij} \ge 0,\\
&\text{(iii) } \forall j, && 1 \le j \le N,\ \sum_{i=1}^{N} J_{ij} = 0
\end{aligned} \tag{3.3}$$

hence, in particular, all its eigenvalues have negative real parts and zero is an eigenvalue with multiplicity equal to the number of invariants in (3.1).

The reduction process then proceeds as follows. We rank the eigenvalues in increasing order of real part

$$\mathrm{Re}(\lambda_1) \le \mathrm{Re}(\lambda_2) \le \cdots \le \mathrm{Re}(\lambda_p) = \mathrm{Re}(\lambda_{p+1}) = \mathrm{Re}(\lambda_N) = 0.$$

We choose a non-zero left eigenvector $\mathbf{b}$ of the matrix $J \equiv J_0$ for the eigenvalue λ_1. Let i_1 be the smallest integer such that $|\mathbf{b}_{i_1}| = \max|\mathbf{b}_j|$ we define the $(N-1) \times (N-1)$ matrix J_1 by

$$\forall i,j, \quad 1 \le i,\ j \le N,\ i,j \ne i_1, \qquad J_{1,ij} = J_{0,ij} - \frac{\mathbf{b}_j}{\mathbf{b}_{i_1}} J_{0,i_1 j}.$$

We now remark that the eigenvalues of J_1 are $\lambda_2, \dots, \lambda_N$ so that we can proceed exactly as before if all eigenvalues are real and treat the conjugate complex eigenvalues in the row in a slightly modified manner.

After removing the highest eigenvalues, $1 \le i \le I$, we are left with a non stiff problem of smaller size (system $N - I \times N - I$) to be solved:

$$\frac{dy^R}{dt} = J^R y^R. \tag{3.4}$$

Once it is solved approximately, we can reconstruct recursively the concentrations that have not been simulated (and are at equilibrium) through the following algebraic relations

$$y^R_{i_k} = -\sum_{j=i_k+1}^{N} \frac{\mathbf{b}^{k-1}_j}{\mathbf{b}^{k-1}_{i_k}} y^R_j \tag{3.5}$$

from the one involved in (3.4).

The method proposed in [5] proceeds as follows. We

- first propagate the full system (3.2) over a small time range $]0, t_0[$ with $t_0 \simeq \mathcal{O}(\mathrm{Re}(\lambda_I)^{-1} \ln[\mathrm{Re}(\lambda_I)])$ with an accurate scheme and a time step tuned to deal with the stiffness of the problem
- then, solve the reduced system (3.4) from t_0 until the final time T.

This process is interesting since the complex computations are done only on a short time range, the time the fast species equilibrate, then the reduced system is sufficient to get the full evaluation. Note that the non propagated species can be recovered from (3.5) at any time it is required and not at every $m\delta t$.

This methodology stands on firm ground as is explained in [5, Theorem 2.3]: assuming that at time t_0 the species are such that $|y^R_j(t_0) - y_j(t_0)| \le \varepsilon$ for any $j, N - I + 1 \le j \le N$, then, the propagated species satisfy $|y^R_j(t) - y_j(t)| \le \varepsilon + [\mathrm{Re}(\lambda_j)]^{-1}$ for any simulated species $(j, I+1 \le j \le N)$ or reconstructed ones $(j, 1 \le j \le I)$ and any $t, t_0 \le t \le T$.

3.2. Combining the reduction process with parareal in time evolution. The reduction method presented in the previous subsection allows us to solve problem (3.2) efficiently by reducing the complexity of the system to be solved. The method can be extended to the solution of a system as

$$\frac{d\mathbf{y}}{dt} = J\mathbf{y} + S \tag{3.6}$$

where S is some given source term, in case the variations of S are at the scale of the slow species. It is of interest to be able to solve (3.6) when S represents releases of pollution or medicine that are represented as Dirac masses or more regular functions with evolution scale that is faster than the rapid species. In this case, the system has to re equilibrate after the emission of fresh reactant. In order to get an accurate simulation, we are obliged to shift back to the full model each time an emission occurs and this has to be done during a length time t_0. Of course if there are (e.g.) 10 time such emissions, the shift has to be done 10 times from the reduced simulation to the full one and back over to the reduced model.

In case a parallel architecture is available, we propose to define the coarse solver from the reduced model (3.4) (without the initial fine propagation period of the procedure above) and solve the exact problem on short time intervals in parallel. Let $\Delta T \geq t_0$ and choose for $\mathcal{F}$ the solution procedure (stiff problem) based on (3.6). For $\mathcal{G}$ we solve approximately (3.5) with an adjunction of a reduced source S^R that is obtained from the same procedure as the reduction of the matrix J.

4. Numerical Results

The first system we consider is associated with the 3×3 matrix

$$J = \begin{pmatrix} -310 & 100 & 0 \\ 300 & -100 & 0 \\ 10 & 0 & 0 \end{pmatrix} \tag{4.1}$$

with eigenvalues $\lambda_1 \simeq -407$, $\lambda_2 \simeq -2.5$ and $\lambda_3 = 0$. The reduction of one order leads to eliminate λ_1, and the simulation with $J^R = J_1$ is simplified. Figure 1 on the left represents the fine evolution of two of the concentrations of the second and third species when no source term is added while Fig. 1 on the right represents the evolution of these species when a source term corresponding to 10 releases at times 0.4, 0.8, 1.2,... is added. In these simulations, the time step is 10^{-4}.

We now use our parareal strategy combined with the use of the reduced system for the coarse operator to get the evolution. The small time step remains the same $\delta t = 10^{-4}$ and the coarse time step is $\Delta\mathcal{T} = \Delta T = 10^{-2}$. Figure 2 represents the evolution of the discrete solution (+) compared with the exact one after the first and second iteration

An almost exact solution is obtained after iteration 3 as can be seen on Fig. 3

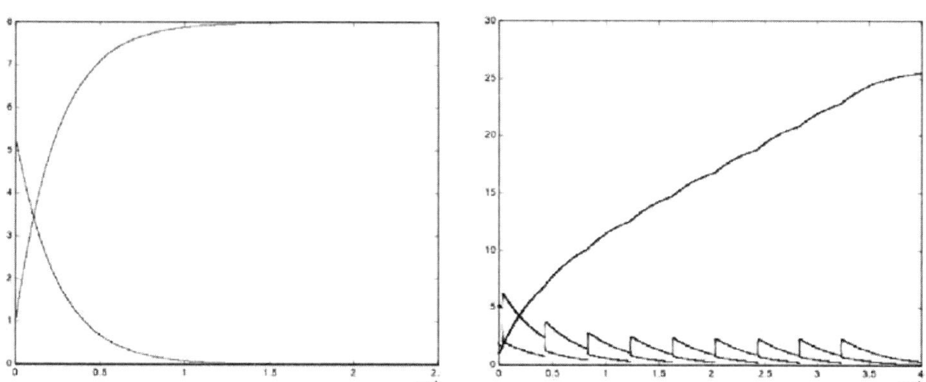

FIGURE 1. Fine approximation of the source free problem (left) and the problem with the sources on with 10 emissions (right).

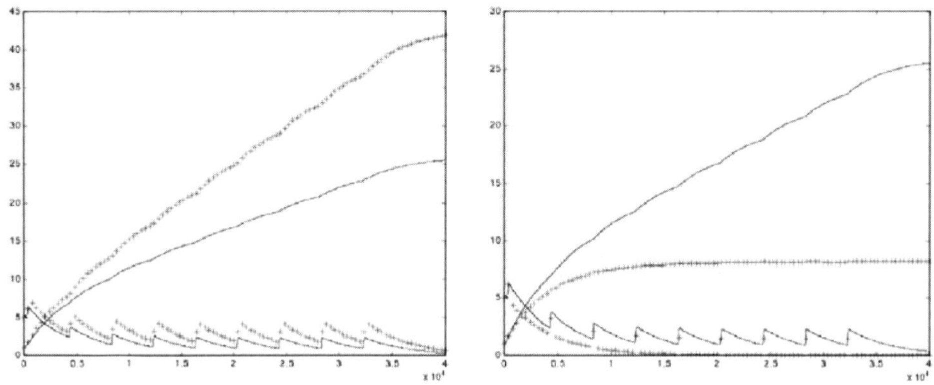

FIGURE 2. Result given by the parareal algorithm after iteration 1 (left) and iteration 2 (right). The exact solution is in solid line and the parareal iterates are represented with (+).

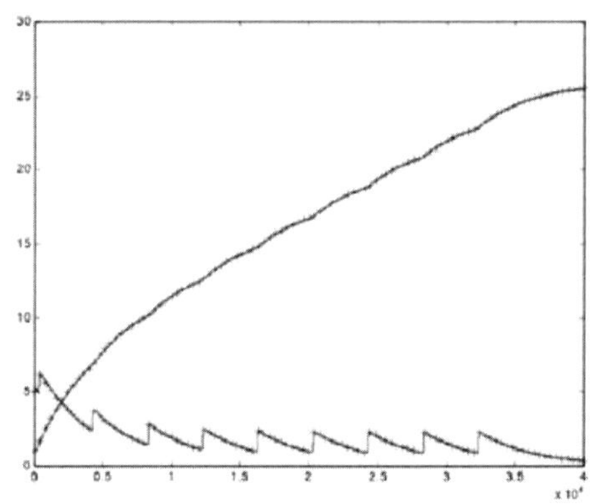

FIGURE 3. Result given by the parareal algorithm after iteration 3. The two curves coincide.

The second system we consider arises in biochemistry and models the dynamics of thyroid hormones (see [23]). The reaction involves 8 species (among them 2 are at disposal) and 14 exchanges. The matrix on the 6 variables that are not at disposal is the following

$$J = \begin{pmatrix} -5.1 & .01 & 0 & 0 & .06 & 0 \\ 0 & -2.516 & 0 & 0 & 0 & .0008 \\ 0 & 0 & -1.3 & .001 & .0003 & 0 \\ 0 & 0 & 0 & -1.091 & 0 & .00008 \\ 5 & 0 & 1 & 0 & -.0603 & 0 \\ 0 & 2.5 & 0 & 1 & 0 & -.00088 \end{pmatrix}.$$

The eigenvalues are $\lambda_1 \simeq -5.2$, $\lambda_2 \simeq -2.5$, $\lambda_3 \simeq -1.3$, $\lambda_4 \simeq -1$, $\lambda_5 = -1.210^{-3}$ and $\lambda_6 = 0$. We are led to a reduction process on the 4 fastest variables that provides a reduced system of size 2×2, the evolution of which is quite easy. Again the parareal algorithm is combined with the reduced system as a coarse propagator and the coarse time steps is here 100 times larger than the fine time step ($\delta t = 5.10^{-3}$,

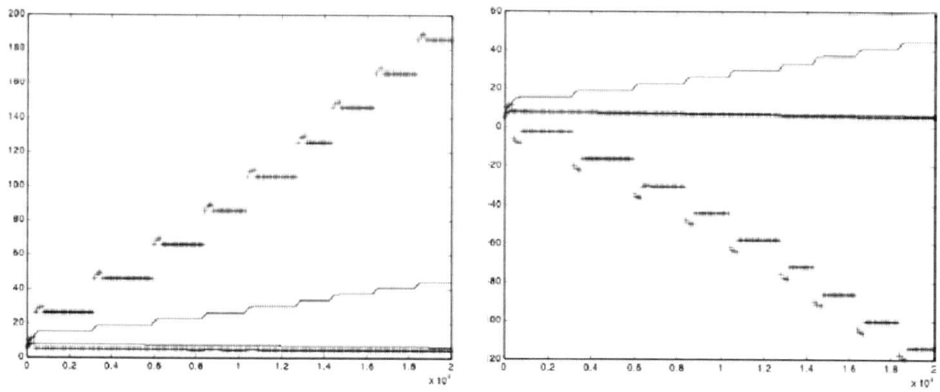

FIGURE 4. A more complex system, after 1 and 2 parareal iterations. The exact solution is in solid line and the parareal iterates are represented with (+).

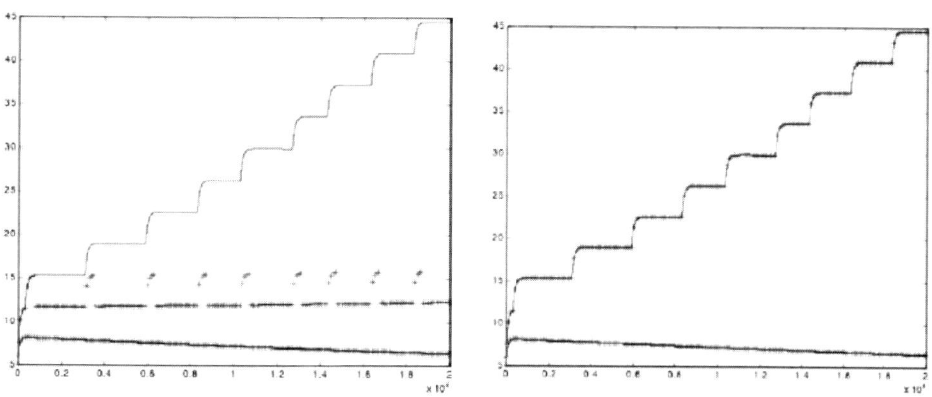

FIGURE 5. A more complex system, after 3 and 5 parareal iterations.

$\Delta\tau = \Delta T = 0.5$) and have not been optimized. The results again illustrate the good behavior of the parareal scheme in the case the coarse solver is constructed from the reduced model. As is shown on Figs. 4 and 5, only 5 iterations are required in this more difficult context and this figure is still rather independent of the number of "pollution releases" leading to a potential for large speedups.

It is interesting to notice that the second iterate is very far from the fine simulation, indeed farther than the first iterate but "*something*" is being done during this step since the next step and very few steps after, the solution is much better.

A final remark is related to the robustness of the approach described here. The very simple case of the framework (linear monomolecular) allows us to get very good and reliable reduced systems and efficient reconstruction procedures. In more complex situations it is quite certain that the reduction process may not be so accurate. In order to evaluate the robustness of the use of reduced system for the definition of the coarse propagator, we have moved slightly the entries of the reduced system J^R and the reconstruction procedure (3.5). The 2×2 system

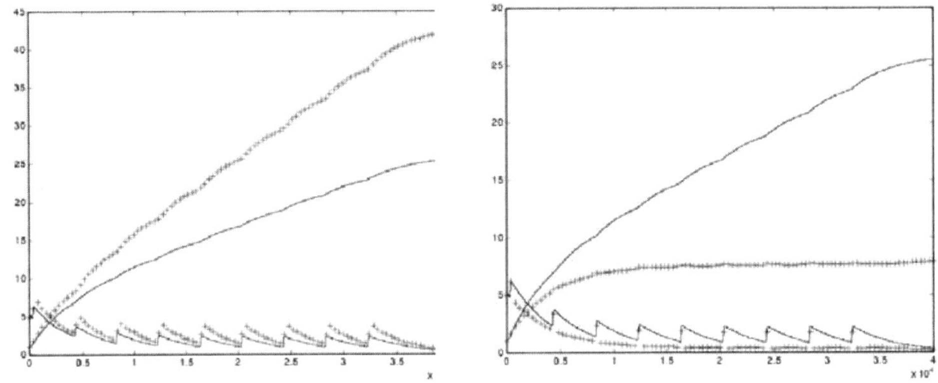

FIGURE 6. The result of the parareal scheme with modified reduced model after iteration 1 (left) and 2(right).

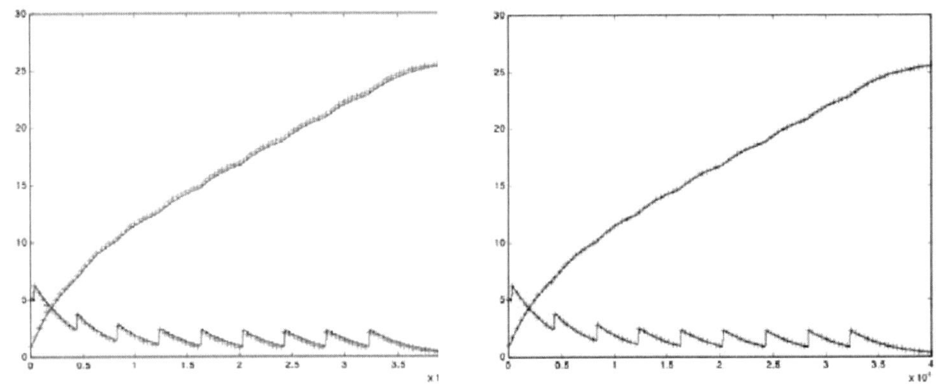

FIGURE 7. The result of the parareal scheme with modified reduced model after iteration 3 (left) and 4(right).

associated with (4.1) being multiplied by a uniform factor equal to 1.2 (this has an effect on the coarse dynamics), the third entry of the eigenvector corresponding to the eigenvalue that is eliminated being multiplied by 1.1 (this has an effect on the reconstruction of the fastest concentration) the convergence is nevertheless merely sensitive to these changes as is displayed on Figs. 6 and 7

References

1. L. Baffico, S. Bernard, Y. Maday, and G. Turinici, G. Zérah, *Parallel in time molecular dynamics simulations*, Phys. Rev. E (3) **66** (2002), 057701–057704.
2. G. Bal, *Parallelization in time of (stochastic) ordinary differential equations* (2002) (to appear).
3. ———, *On the convergence and the stability of the parareal algorithm to solve partial differential equations*, Domain Decomposition Methods in Science and Engineering, Lect. Notes Comput. Sci. Eng., vol. 40, Springer, Berlin, 2005, pp. 425–432.
4. G. Bal and Y. Maday, *A "parareal" time discretization for non-linear* PDE*'s with application to the pricing of an American put*, Recent Developments in Domain Decomposition Methods

(L.F. Pavarino and A. Toselli, eds.), Lecture Notes in Comput. Sci. and Engrg., vol. 23, Springer-Verlag, Berlin, 2002, pp. 189–202.

5. A. Blouza, F. Coquel, and F. Hamel, *Reduction of linear kinetic systems with multiple scales*, Combust. Theory Model. **4** (2000), no. 3, 339–362.
6. P. Fischer, F. Hecht, and Y. Maday, *A parareal in time approximation of the Navier Stokes equations*, Domain Decomposition Methods in Science and Engineering (R. Kornhuber, R. Hoppe , J. Périaux , O. Pironneau , O. Widlund, and J. Xu, eds.), Lecture Notes in Comput. Sci. and Engrg., vol. 40, 2004, pp. 433–440.
7. M. Gander and S. Vandewalle, *Analysis of the parareal time-parallel time-integration method*, 8th European Multigrid Conf., Scheveningen (The Hague), The Netherlands, 2005.
8. I. Garrido, G. Fladmark, and M. Espedal, *Parallelization in time for reservoir simulation* (2003) (to appear).
9. W. Hackbusch, *Parabolic multi-grid methods*, Computing Methods in Applied Sciences and Engineering (R. Glowinski and J.-L. Lions, eds.), vol. VI, North-Holland, 1984, pp. 189–197.
10. E. Hairer, C. Lubich, and G. Wanner, *Geometric numerical integration*, Structure-preserving Algorithms for Ordinary Differential Equations, Springer Ser. Comput. Math., vol. 31, Springer-Verlag, Berlin, 2002.
11. E. Hairer, S.P Nørsett, and G. Wanner, *Solving ordinary equations* I. *Nonstiff problems*, 2nd ed., Springer Ser. Comput. Math., vol. 8, Springer-Verlag, Berlin, 1993.
12. E. Hairer and G. Wanner, *Solving ordinary equations* II. *Stiff and differential-algebraic problems*, 2nd ed., Springer Ser. Comput. Math., vol. 14, Springer-Verlag, Berlin.
13. G. Horton, *The time-parallel multigrid method*, Comm. Appl. Num. Methods **8** (1992), 585–595.
14. G. Horton and S. VandewallecA space-time multigrid method for parabolic partial differential equations, SIAM J. Sci. Comput. **16** (1995), 848–864.
15. G. Horton, S. Vandewalle, and P. Worley, *An algorithm with polylog parallel complexity for solving parabolic partial differential equations*, SIAM J. Sci. Comput. **16** (1995), 531–541.
16. C. Lubich and A. Ostermann, *Multi-grid dynamic iteration for parabolic equations*, BIT archive **27** (1987), no. 2, 216–234.
17. J.-L. Lions , Y. Maday, and G. Turinici, *Résolution d'*EDP *par un schéma en temps pararéel*, C. R. Acad Sci. Paris Sér. I Math. **332** (2001), 661–668.
18. Y. Maday , E. M. Rønquist, and G. Staff, *The parareal-in-time algorithm: basics, stability analysis and more* (in preparation).
19. Y. Maday , J. Salomon, and G. Turinici, *Monotonic time-discretized schemes in quantum control*, Numer. Math. **103** (2006), no. 2, 323–338.
20. Y. Maday and G. Turinici, *A parareal in time procedure for the control of partial differential equations*, C. R. Acad Sci. Paris Sér. I Math. **335** (2002), 387–392.
21. ———, *A parallel in time approach for quantum control: the parareal algorithm*, Decision and Control (Nevada, 2002), Proc. of the 41st IEEE Conf., vol 1, IEEE, (2003), 62–66.
22. Y. Maday and G. Turinici, *The parareal in time iterative solver: a further direction to parallel implementation*, Domain Decomposition Methods in Science and Engineering (R. Kornhuber, R. Hoppe, J. Périaux , O. Pironneau , O. Widlund, and J. Xu, eds.) Lecture Notes Comput. Sci. and Eng., vol. 40, Springer, Berlin, 2005, pp. 441–448.
23. P.H. Mark and J.J. Di Stefano, *Optimal control policies for the prescription of thyroid hormones*, Math. Biosci. **42** (1978), 159–186.
24. J.H. Saltz and V.K. Naik, *Towards developing robust algorithms for solving partial differential equations on mimd machines*, Parallel Comput. **6** (1988), 19–44.
25. J. Simoens and S. Vanderwalle, *Waveform relaxation with fast direct methods as preconditioner*, SIAM J. Sci. Comput. **21** (2000), 1755–1773.
26. G. Staff and E.M. Rønquist, *Stability of the parareal algorithm*, Domain Decomposition Methods in Science and Engineering (R. Kornhuber, R. Hoppe, J. Périaux, O. Pironneau, O. Widlund, and J. Xu, eds.), Lecture Notes Comput. Sci. and Eng., vol. 40, 2004, pp. 449–456.
27. A. Toselli,O. Widlund, *Domain decomposition methods—algorithms and theory*, Springer Ser. Comput. Math., vol. 34, Springer-Verlag, Springer-Verlag, Berlin, 2005
28. S. Vanderwalle, *Parallel multigrid waveform relaxation for parabolic problems*, Stuttgart, 1993.

29. S. Vandewalle and R. Piessens, *Efficient parallel algorithms for solving initial-boundary value and time-periodic parabolic partial differential equations*, SIAM J. Sci. Stat. Comput. **13** (1992), 1330–1346.

Université Pierre et Marie Curie-Paris 6, UMR 7598 Laboratoire Jacques-Louis Lions, B.C. 187, Paris, F-75005 France; Division of Applied Mathematics, Brown University, Providence, RI, USA.

E-mail address: maday@ann.jussieu.fr